T0321454

Amphibian Morphogenesis

Bioscience

Amphibian Morphogenesis, by Harold Fox, 1983

AMPHIBIAN MORPHOGENESIS

HAROLD FOX

University College, London

HUMANA PRESS • Clifton, New Jersey

Library of Congress Cataloging in Publication Data

Fox, Harold.
  Amphibian morphogenesis.

  (Bioscience)
  Bibliography: p.
  Includes indexes.
    1. Amphibians—Metamorphosis. 2. Amphibians—Larvae.
  3. Morphogenesis. I. Title. II. Series: Bioscience
  (Clifton, N.J.)
  QL669.2.F69 1984   597.6'04332   83-26526
  ISBN 0–89603–043–1

  ©1984 The HUMANA Press Inc.
    Crescent Manor
    PO Box 2148
    Clifton, NJ 07015

Printed in the United States of America.

This book is dedicated to

OLGA

Preface

This book came about as a result of a review I had written earlier on features of cellular changes occurring during anuran metamorphosis. Only a limited treatment of this subject was possible in such a circumscribed work and only specific examples of organic change were dealt with. Thus the sins of omission weighed heavily, for so much information could not be included to provide a more comprehensive and authenticated account of the elaborate, complex, and far-reaching changes that an aquatic larva undergoes to become a terrestrial froglet.

A good deal of my working life has been spent investigating amphibians, especially their larval developmental morphology during metamorphosis, first at the level of light microscopy and in later years by electronmicroscopy. Initially I was particularly concerned with morphological homologies of a variety of larval structures, such as the cranial and pharyngeal skeleton and the nerves and musculature, in order to learn more about amphibian phylogeny, for during my pre- and early postgraduate years G. R. Beer and D. M. S. Watson inspired an undying interest in and respect for vertebrate comparative anatomy. However, it now seems to be that amphibian phylogenetic relationships are best dealt with by the paleontologists, so ably demonstrated by D. M. S. Watson and A. S. Romer and the contemporary enthusiasts in this field like A. L. Panchen, R. L. Carroll, E. Jarvik, and K. S. Thompson among a host of others, particularly in the USA.

Section I of this work introduces the reader to the Amphibians, albeit briefly. It should not be forgotten, however, that the specific tissue cells described in the subsequent sections are part of a living whole animal; they are not independent blood cells, cell membranes, or ribosomes: people working at the subcellular level in their enthusiasm often forget this obvious fact. Any consideration of metamorphosis in the Amphibia should also recognize that this major tetrapod group of animals has very likely arisen from aquatic crossopterygian fish-like ancestors hundreds of millions of years ago, and that the amphibian life cycle in its development, in the most general terms, demonstrates its phylogeny. Ontogeny does indeed recapitulate phylogeny, though not as Haeckel supposed through the adult ancestral stages. The anuran larva, as we know it, is an invention in evolution and the elaborate metamorphic morphological changes are fitted into the life cycle. The program terminating in the dramatic events at climax is a specialized development of adaptive changes resulting in a froglet—a terrestrial, air-breathing, jumping animal fitted to its habitat. The urodeles develop less dramatically; the adults are more similar anatomically and physiologically to their ancestral progenitors.

But in order to understand how a frog becomes what it is, the ontogeny of the larva to completion of metamorphosis needs to be considered. Section II deals with staging of amphibians from the fertilized egg to the end of metamorphosis, and reviews the relevant literature available for those amphibian species that have been described. Amphibian developmental stages are generally based on external form, particularly of the hindlimbs during the later period of metamorphosis. Stages are invaluable for workers wishing to record the exact level of development of their experimental animals and the method eliminates the vagaries apparent when larval age is used, for differences in feeding or temperature or other environmental factors influence the rate of development of individuals.

Section III deals with the origin, development, and modification of a variety of organs and tissues throughout embryonic and larval life. Not all of the organs are described, though most of the important ones are considered to a greater or lesser degree. However, the subject of immunology, including the thymus, is best left to the immunologists; and the reproductive system is only barely mentioned, in part because together with breeding behavior it could legitimately claim to be treated separately, but also by the fact that the gonads and ducts seem to be relatively independent of the thyroid hormones, though not completely so, at least in relation to the maturation of the ovarian oocytes. Section III clearly shows that the froglet is for all intents and purposes a newly created animal, fashioned by the modification of existing organs and the *de novo* development of new structures originating near or amid the vestiges of larval organs.

The significant fact about amphibian metamorphic change is its almost total dependence on thyroid hormones. Indeed the discovery by J. F. Gudernatsch, in 1912, that feeding thyroid glandular tissue to anuran tadpoles hastens metamorphic change, provided an initial impetus for the future work in the field of thyroid endocrinology. The study of the widespread organic changes in amphibian larvae, effected by treatment with thyroid hormones, has proved to be a fruitful field of research in the analysis of cellular differentiation and its causality during ontogeny. The problem can be posed quite simply: How does a zygote develop into a complex highly integrated froglet with its multitude of different kinds of cells? The same statement *mutatis mutandis* can be made for any animal group.

As a student of G. R. de Beer at University College, London, around 1950, and trained as an embryologist—admittedly an old-fashioned one—cellular differentiation, and thus cellular diversity, originating during the ontogeny of vertebrates was of major interest. At that time the earlier classical work on the organizer and induction, by Hans Spemann and Hilde Mangold, followed later by major contributions from J. Holtfreter, C. H. Waddington, and S. Horstadius among others, dominated our thinking. G. Ross Harrison's experiments with amphibian larval cells in tissue culture and C. M. Child's gradient phenomena during an animal's development, seemed to explain so much in terms of causality. In retrospect many of our earlier concepts now appear to be rather vague and unsatisfactory. Then we were so impressed by this sort of statement elaborating Child's hypothesis: "certain external factors set up quantitative differentials in the egg and embryo, as a result of which qualitative differences of structure ultimately ensue" (Huxley and de Beer, 1934). Such statements really tell us very little about cellular differentiation, though at least this one had a measure of scientific validity, in contrast to the absurd preformationist hypothesis of Charles Bonnet, the idea of embryonic emboitement, or the theory of entelechies or supraphysical forces, by H. Driesch. Yet to be fair there was a great corpus of knowledge on vertebrate embryology, at the level of light microscopy, during the 1930s as Huxley and de Beer showed. Cellular biochemistry was to appear in full force somewhat later, stimulated by the work of J. Needham. It seems to me, therefore, that any really useful work dealing with amphibian morphogenesis needs to include descriptions not only of morphological change, but also of why they occur and of what causal mechanisms operate during this time. With this in mind a useful approach is to compare and contrast the activities of inducer substances and hormones during development.

Unfortunately, however, the chemistry and mode of action of inducers are still not very clear, notwithstanding the efforts by T. Yamada in Japan, H. Tiedemann in Germany, L. Saxen and S. Toivonen in Finland, and P. D. Nieuwkoop in Holland, who with their colleagues have investigated them.

Section IV is thus devoted to the mechanisms concerned with cellular differentiation arising during ontogeny and considers the modern concepts now prevailing, based on research involving electronmicroscopy and molecular biology. A firm basis is provided by the framework of endocrinology and its relevance to amphibian metamorphosis. Among the many who have worked in this field, probably W. B. Etkin and J. J. Kollros in the USA, J. M. Dodd in Britain, P. G. W. J. van Oordt in Holland and A. A. Voitkevitch in the USSR have contributed more than most, though alas the most commonly quoted complex hypothesis by Etkin, on the mechanisms controlling thyroid hormone synthesis, secretion, and utilization by developing larval anuran tissues still leaves many questions unanswered. But the real crunch in any study of cellular differentiation, and by this we mean new protein synthesis, must be at the level of molecular biology. Here E. Frieden and P. P. Cohen in the USA, J. B. Gurdon and J. R. Tata in Britain, and R. Weber in Switzerland, and their coworkers among the increasing number of investigators, have contributed so much valuable information on nuclear–cytoplasmic interrelationships of amphibian larval differentiating cells with their work on receptor (binding) sites for thyroid hormones. As with research on steroid hormone cellular receptors in mammalian cells, this is an expanding field for elucidating some of the causal mechanisms of cellular differentiation at the molecular level, in a modern interpretation of *"Entwicklungsmeckanik"* of the classical German embryologists. The foundations were laid down by J. Monod and F. Jacob and somewhat earlier, in the 1950s, by J. Watson, F. Crick, and M. Wilkins, with their pioneer work on DNA. Section IV, therefore, attempts, albeit in a limited and generalized way, to provide a synthesis of the principles of classical embryology and modern molecular biology with reference to induction, via the activity of inducing substances, and the thyroid hormones on their target cells. This section, to my mind, deals with the most exciting yet frustrating material in the book, probably because of the complexity and frequently the confusion of the information available, but indeed we are reaching to the very heartland of the molecular interrelationships influencing biochemical synthesis and activity.

The small, apparently insignificant, amphibian tadpole is a wondrous mechanism of organic complexity that is of profound significance in the study of basic biological problems, i.e., how living animals reproduce, grow, and differentiate, and thence die. Classical embryology, which exploited amphibian embryos, appeared to be a relative backwater of biological research barely 50 years ago. It has reemerged with new vigor and importance, diversifying into a range of disciplines now at the core of modern biology.

Acknowledgments

I am pleased to acknowledge and offer my thanks to numerous sources for permission to publish illustrations used in this work. In practically every case the authors who generously allowed their work to be reproduced in addition offered encouragement and good wishes, for which I was deeply touched. The various individuals and the publishers who granted their permission to use copyrights are listed against the actual figures separately, on a succeeding page. I repeat my thanks.

My warmest appreciation goes also to various members of the Department of Zoology, University College, London: Rosina Down, Dave Franklin, Roy Mahoney, Edwin Perry, and Brian Pirie, for their valued help during the preparation of this book. This is a good opportunity to thank them all for much technical assistance and good-humored patience in dealing with my needs during past and present days. I have enjoyed good fortune in this respect. I also acknowledge the assistance of: Carl Gans and Michigan University for supplying a copy of the unpublished Michigan University PhD thesis on *Rana catesbeiana* by Ira D. George; Reiko Makidono, who kindly translated and thus rendered comprehensible to the author the beautiful pictorial script of several Japanese papers; Christine Oates who sent me a copy of her Newcastle University PhD thesis on *Xenopus larval* morphology; Keizo Takata of Nagoya University for so promptly sending me the monograph on *Megalobatrachus japonicus* by Haruo Iwama, which I could not obtain in England; K. Watanabe of Tsurumi University, Yokohama who kindly supplied a copy of Tahara and Ichikawa's paper on *Bufo bufo japonicus,* and the countless kind individuals who have sent me reprints of their original publications.

Thanks are also due to Margaret Keenan for patiently and efficiently typing and retyping drafts of the manuscript and to the numerous librarians, especially Joan Marsh and Jill Bailey at University College and Susan Bevis at the Zoological Society of London, who unfailingly and cheerfully satisfied my requests for innumerable references.

Among colleagues with whom I discussed different aspects of the subject of amphibian structure and development, Mary Whitear here at University College, and Stanley Turner at the Portsmouth Polytechnic, were particularly helpful. I recall with pleasure, albeit tinged with sadness, the long talks I enjoyed with the late Dvorah Boschwitz of the Hebrew University, Israel, when she visited our department. So often I was made to think again when I became too complacent, or made to return to the original source of reference for verification and correction. Of course the faults in this work are mine alone, as doubtless I will be informed in due course.

In a work such as this the number of new and often relevant publications that appear literally every week is growing at an alarming rate. I hope, therefore, that readers will forgive me for omitting any aspect of the subject that might have been significant to them. Perhaps in this instance, however, one is fighting a losing battle, and the prize of being right up to date will always be just out of reach.

Finally, it is a pleasure to record my appreciation for the helpful cooperation and courtesy from all those involved at the Humana Press.

Permission from the following authors and publishers for the use of original Illustrations, Tables or their modified versions, reproduced in this book is gratefully acknowledged:

Fig. 1, W. Etkin, Academic Press and Plenum Publishing Corporation; Fig. 2a, Gustav Fischer Verlag; Fig. 2b, R. Cambar, J. D. Gipouloux and the Editors of *Bulletin Biologique*; Figs. 3 and 4, H. Broyles and Plenum Publishing Corporation; Figs. 5 to 9 and Table 1, H. Nanba and the Editors of *Archivum Histologicum Japonicum;* Fig. 10, Eliane Larras-Regard and Academic Press; Figs. 11a and 11b, M. E. Rafelson and S. B. Binkley and Macmillan Publishing Company; Figs. 12, M. H. I. Dodd and J. M. Dodd and Academic Press; Fig. 13, W. Etkin and Plenum Publishing Corporation; Fig. 14, D. H. Copp and the Editors of the *Journal of Endocrinology;* Figs.15 to 29, The Company of Biologists; Figs. 33 and 34, J. Hourdry and M. Dauca, and Academic Press; Fig. 35, P. P. Cohen, R. F. Brucker, and S. M. Morris and Academic Press; Figs. 36 and 37, J. R. Tata and the Editors of the Biochemical Society; Figs. 38A to 38E, American Institute of Biological Sciences; Figs. 39 to 46, The Company of Biologists; Figs. 48 and 49, N. Yoshizaki; Fig. 50, J. J. Picard and J. Gilloteaux and the Editors of the *Archives de Biologie;* Fig. 51, J. J. Picard and Alan R. Liss Inc.; Figs. 52 to 54, H. L. Kaung and Alan R. Liss Inc.; Fig. 59, K. Sterling and the Editors of the *Bulletin of the New York Academy of Medicine;* Fig. 60, C. C. F. Blake and J. .S. Oatley and Macmillan Journals; Fig. 61, C. C. F. Blake and Pergamon Press; Table 2, U. M. Spornitz and Springer Verlag; Table 3, Plenum Publishing Corporation; Table 4, T. Yamada and Alan R. Liss Inc.

CONTENTS

Section IV. Cellular Differentiation, Ontogeny, and Molecular Biology

Section 1

Amphibians

1. The Origin of the Amphibians

The term amphibian refers to the dual mode of life that members of this group of animals enjoy. Although some modern toads spend most of their lives on land, most amphibians remain near ponds and streams; others remain wholly in water throughout life. In the majority of cases amphibians lay eggs in water, where they are fertilized, such eggs thenceforth developing into water-dwelling, gill-breathing larvae, essentially similar in form to their fish ancestors. Only later, after metamorphosis, do amphibians assume a terrestrial life, though they are never completely independent of water, owing to their mode of respiration and their inability to avoid desiccation. A number of modern amphibians have evolved various methods to obtain such independence, but few, if any, have really succeeded in developing a life style only first achieved among the vertebrates by reptiles. Some amphibians that have no terrestrial existence at all stay in water permanently as pedogenic forms *(vide infra)*. The first amphibians, distinguishable from fishes by structural differences that enabled them to survive for periods of time out of water, are found as fossils in the Devonian period of the Paleozoic era, about 350 million years ago. Mainly based upon his investigations on the paired fore and hind limb extremities, Holmgren (1933) concluded that urodeles evolved from dipnoan lungfishes and all the remaining tetrapods evolved from crossopterygian lobed-fin fishes. Though this view is now discredited, one still cannot be other than impressed by the close similarity in appearance of urodele embryos and larvae and those of dipnoans such as *Neoceratodus forsteri*, from Australia. Furthermore, *Neoceratodus* and urodele larvae both develop fore extremities before the hind ones appear; in contrast the opposite occurs in anurans. Probably such affinities are the result of convergent evolution. It is generally believed by most paleontologists that amphibians evolved from rhipidistian crossopterygian fishes with lobed fins (Watson, 1926, see Schaeffer, 1965; but *vide infra*). During the prolonged dry conditions of the late Devonian, individuals able to traverse short distances between partially dried out rivers or lakes to those less stagnant or more favorable ecologically, would have the better chance of survival. The late A. S. Romer (1945) at Harvard University elegantly posed and answered the question why tetrapod amphibians originated at all during evolution. Many of the earliest primitive aquatic amphibians, the Labyrinthodontia, were fairly large and carnivorous, and they lived for most of the time alongside their close relatives, the crossopterygians. The latter differed from them mainly by having median fins and a lesser

development of paired limbs, indeed many features of early labyrinthodonts were closely comparable with those of the crossopterygians. Labyrinthodonts could breathe air at the pool surface, like crossopterygians
they could obtain food in the aquatic environment, for there was little for
them on land, and they did not need to escape predators. Paradoxically,
tetrapody was an adaptation and an evolutionary advantage to enable labyrinthodonts to remain in water, for by progressing overland, using simple limbs, they could reach other pools and lakes if their original one
dried up. Crossopterygians, having a lesser ability to travel, would die if
fresh water did not return to a muddy or stagnant marshy lake. The subject has been reviewed in some detail by Thomson (1969) in terms of reliable available data.

The ancestral labyrinthodonts (named because of folding in the en-
amel of their teeth), are mainly distinctive by virtue of the fact that they
had transformed the crossopterygian lobed fins into primitive tetrapod
limbs. They respired by breathing air through simple lungs, evolved from
hinder gill pouches. Other modifications concerned the skull and skeleton, closure of the fish gill slits, and changes in the first (hyoid) gill cleft
and skeletal components of the hyoid arch and jaw, to form an auditory
apparatus. During the changing ecological conditions of the Devonian period, natural selection favored these modified crossopterygians, which
gradually evolved into the labyrinthodonts, who were better able to survive, at least for short periods, in a terrestrial environment. Ultimately the
short periods involved in traveling from one pond to another presumably
extended into longer periods of terrestrial existence, and by basking and
feeding at the water bank, a land fauna would originate.

During the warm, moist period of the Carboniferous era, about
280–350 million years ago (lasting about 60–80 million years), the earth
was covered by swampy forests and giant ferns, which later fossilized
into coal. Conditions were favorable for amphibian development and a
great variety of them evolved in this great age of the "Amphibia," for
they required warmth to maintain a high enough body temperature, the
prevailing moist atmosphere hindered desiccation and food was plentiful.
Some labyrinthodonts were large creatures reminiscent of crocodiles; others were small, somewhat like salamanders. The majority, however, were
still mainly aquatic. Most of these groups were eliminated steadily during
the succeeding Permian and Triassic periods, and by the end of the Jurassic period (130–150 million years ago) ancestors of the modern amphibians were in existence, though their origins are still obscure. Other forms
gave rise to the reptiles, for, as found in the Permian *Seymouria,* a very
early form intermediate between an amphibian and a reptile has been described.

The earliest known true amphibian, *Ichthyostega,* from fresh-water
beds of the Mississipian period, doubtless had primitive tetrapod limbs,

though for most of its life it was probably mainly aquatic, for it retained a tail fin and vestigial gill covers. Apart from its superior limb development, its skull had many similarities with those of crossopterygians, especially osteolepiformes, and it could well have been derived from them. Other skull features are highly comparable to those of labyrinthodonts of the Carboniferous and Permian periods. The labyrinthodonts did not survive the Triassic period, but an offshoot, the small scaly aquatic lepospondyls of the Carboniferous period, gave rise to salamander-like microsaurs, probably a vaguely generalized ancestral group of the modern amphibians: Branchiosaurs were probably larval labyrinthodonts, many transforming during life into adults, without gills of the *Eryops* type. Frogs are the most successful of these modern forms and their phylogenetic history can be traced back to the Jurassic period, though remains are fragmentary. In 1940 the late D. M. S. Watson of University College, London, published descriptions of *Miobatrachus* from the Pennsylvanian period of Illinois, a rare small-tailed amphibian that could well be in the ancestral stream of frogs. Likewise *Protobatrachus* of the Triassic period of Madagascar has a skull similar to that of frogs and toads. Urodeles appear after the Triassic period, though some of the more generalized microsaurs have a skull pattern that suggests that they have an ancestral relationship. Apodans, the legless amphibians, alone among the living amphibians have retained body scales. Romer (1946) believed that *Lysorophus,* a microsaur, could well represent an ancestral form of this order. Whether tetrapods are monophyletic in their derivation, that is, that they arose as a single group from the rhipidistian crossopterygian assemblage, or are diphyletic (or polyphyletic) and that there were at least two separate and distinct sources of amphibians, is still widely debated by experts in this field. The Swedish paleontologist Erik Jarvik (1960) concluded from a vast series of investigations, which still continue, that porolepiforme rhipidistians are the progenitors of the urodeles (and possibly apodans), and osteolepiforme rhipidistians of the anurans and ultimately of the amniotes including man. The importance of Jarvik's view is that land-inhabiting tetrapods, with their pentadactyl limbs, would thus have originated from more than one source, each one independently during evolution. Furthermore, the reptilian–avian and reptilian–mammalian assemblages are derived solely from osteolepiformes. At present it would seem likely that the supporters of the view of monophyletic tetrapody (see Szarski, 1962), strongly canvassed among others by Thomson (1968) at Yale University, are in the majority. However, previous evidence and methodologies of long standing have been questioned by Rosen et al. (1981), and from their results they argue in favor of lungfishes, and not any of the rhipidistians, as being the sister group to the tetrapods. Their work and its significance in terms of phylogenetic theory has been considered by Thomson (1981). Indeed, though the conclusions of Rosen et al.

(1981) were utterly rejected by Jarvik (1981), it is possible, nevertheless, that some at least of our long-established concepts of the evolutionary relationships of the tetropods and the 'rhipidistian barrier' may have to be reevaluated.

2. Living Amphibians

The class of vertebrates termed Amphibia includes three living subclasses (or orders) of animals, among which are the tailless frogs and toads, the Anura and the tailed newts and salamanders, the Urodeles. These groups are also known as the Salientia (from the Latin *saliens* to leap) and the Caudata (from the Latin *cauda* a tail). The third lesser known group is the Gymnophiona (from the Greek *gymnos* naked and ophioneos serpent-like), often called Apoda (from the Greek footless) or Caecilia. This group of about 20 genera and 50 species comprises the elongate, legless, wormlike creatures of the tropics (see Taylor, 1968) considered to be more primitive in some ways than the other two groups, though indeed they are structurally specialized for burrowing beneath the soil. The longest is probably *Caecilia thompsoni* from Columbia, which is about 4.5 ft long; the smallest species are *Siphonops hardyi* of Brazil and *Gymnophis parviceps* of Panama, both about 7 in. long (Cochran, 1972).

There are about 2500 known species of anurans, typically represented by the common frog *Rana temporaria*, whose characteristic shape is of a head grading into the body; there is no neck movement. The elongated hindlegs provide an efficient mechanism for leaping; indeed "champions" can jump many feet. Probably some of the best jumpers 'pound for pound' are the small arboreal members of the Hylidae. The absolute and relative (jumping distance/snout–vent length) jumping ability of 80 species of frogs were investigated in detail by Zug (1978). Ranking of relative jumping performance in order of the weakest to the strongest is as follows: Bufonidae, Pelobatidae, Discoglossidae, Microhylidae, Ranidae, Leptodactylidae, Dendrobatidae, Ascaphidae, Miobatrachidae, Hylidae, and Pelodryadidae. The best jumper *Acris gryllis* jumped an average 31.1 × body length, with a maximum of 61.7 × body length (equivalent to a 6-ft human, allowing for leg length, jumping over 180 ft). Zug (1978) quotes the recorded "Guiness Book" world record of 5.21 m for a single leap by a *Rana (catesbeiana?)* and 9.675 m for three consecutive leaps by a *Ptychadena oxyrhynchus*.

There are 300 species of known urodeles. Large adults of some genera, like *Cryptobranchus japonicus*, may reach several feet in length, though the majority of adult newts and salamanders are usually not longer than 9 in. In contrast to anurans, in addition to their tail the four legs are of equal size, though some wormlike American urodeles, like *Amphiuma,* have only tiny limbs and *Siren* has lost its hind limbs completely.

Adult anurans and urodeles usually spend most of their lives on land and only return to water during the breeding season, lasting a month or so. A few anurans, such as *Xenopus laevis,* the South African clawed toad of the tongueless Pipidae, are permanently aquatic though they breathe air using lungs, and like their larvae retain lateral line organs in the skin. Of particular interest is the fact that the adult *Xenopus* still excretes high quantities of ammonia, like its larvae (Deuchar, 1975), in contrast to the urea excreted by terrestrial amphibians. The larva of the spotted newt, *Notophthalmus viridescens,* is aquatic, but after metamorphosis the newt is terrestrial for 3 yr. as a red eft. It thence returns to water, loses its tongue, and reacquires lateral line organs (Dent, 1968).

Some African species of the genus *Arthroleptis* (e.g., *A. poecilonotus*) deposit highly yolked eggs on the ground that develop directly into fully metamorphosed froglets 2–3 wk later; this is an example of amphibian oviparity. *Nectophyrnoides tornieri* and *N. vivipara* demonstrate ovoviviparity, for fertilized yolky eggs are retained within the female's oviduct for about 2 months until the birth of the froglet. The small eggs of *N. occidentalis* (4–20) are retained in the oviduct for up to 9 months, and the embryos absorb nutrient secretions ("lait uterin") through the mouth by pharyngeal contractions. Together with the internal fertilization, this type of embryonic 'gestation,' though there is no placenta, constitutes a primitive form of viviparity (Lamotte and Xavier, 1972).

Some species of *Eleutherodactylus* (e.g., *E. nubicola*) also deposit eggs that develop directly within the egg through a non-aquatic 'tadpole' stage, into froglets (Lynn, 1942; Lynn and Lutz, 1946). There is enough yolk to provide food for development throughout ontogeny. Such forms have large highly vascularized tails used for gaseous exchange; the external gills have disappeared or are vestigial, and there are no gill slits, or internal gills (see Dent, 1968).

Among the Gymnophiona some species develop as aquatic larvae, while others show direct development within the egg or may be ovoviviparous.

3. Metamorphosis in Urodeles

Among urodeles metamorphosis of the aquatic larva to a terrestrial juvenile is a far less dramatic event, especially in terms of bodily change, than in anurans *(vide infra).* Superficially, at least, young urodelans before metamorphosis are in general similar in appearance to older metamorphosed forms. In contrast the difference in appearance between larval anurans and the postmetamorphic froglet is remarkable. Included among the more obvious external morphological changes generally recognizable during urodelan metamorphosis are the involution of the external gill filaments and fusion of the gill slits, resorption of the tail fin and folds of the lower jaw, and differentiation of the eyelids. Different species or urodeles develop definitive coloration, usually specific in different sexes, and coloration changes at breeding times are well known. Internal changes that occur during metamorphosis may include the development of narial valves, to prevent access of water if the adult is submerged, and alterations in the skeleton and musculature. In the skin the larval epidermal Leydig cells typically disappear.

Urodeles are of particular interest in that a number of genera are neotenous, that is, they retain a variety of larval characteristics during later life, but as such they can reproduce and are pedogenic. Noble (1954) among others has described these animals. The adult *Siren* retains three pairs of external gills and gill slits, and *Cryptobranchus* and *Amphiuma* each have one pair of gill slits and do not develop eyelids. Dent (1968) has summarized our information on the relationship of the thyroid to metamorphosis in neotenic urodeles. On the whole the failure of some neotenic forms to fully metamorphose seems to occur because of a failure of the larval tissues to respond to thyroid hormones, rather than to a deficiency of thyroid tissue or of its function. Administration of thyroid hormones to some neotenic genera may elicit specific, albeit modest, metamorphic changes, which Dent listed in category 1. These changes include gill reduction in *Siren* and skin shedding in *Cryptobranchus* and *Amphiuma* but the effects are variable and do not significantly alter the neotenic condition. Other neotenic forms of the Plethodontidae and Ambystomidae, listed in category 2, include possibly *Haideotriton wallacei, Typlomolge rathbuni,* and species of *Eruycea, Gyrinophilus* and the axolotl *Siredon*. These forms may be induced by relatively high dosages of thyroxine to show greater metamorphic change. There may be loss of tail fins, labial folds, lateral line organs, and gills. In *Gyrinophilus palleucus,* changes in the skull also occur and eyelids develop. Kezer (1952) brought about complete metamorphosis in *Eurycea tynerensis* and *E. neotines* within 18 d by immersing them in 1:500,000 thyroxine solution.

Specimens listed in category 3, which are occasionally neotenic in nature, metamorphose more easily in relatively low concentrations of thyroxine than do those in category 2. This group includes species of the Ambystomidae, Salamandridae, Plethodontidae, and Hynobiidae. Neotenic specimens of *Notophthalmus viridescens* readily metamorphose in 1:2,000,000 thyroxine solution. It is of interest that Ducibella (1974a,b) showed hemoglobin and serum protein in *Ambystoma mexicanum,* whose composition normally changes from the larval to that of the neotenous metamorphosed adult at 125 and 210 d of development, respectively, can be induced to change prematurely by (4.5 or 9.0 × $10^{-8}M$)T_4. Other normally metamorphosing anatomical features did not change, and it was concluded that such blood biochemical transition is more responsive to a lower threshold of T_4 than other responsive tissues.

It is probable that some at least of the facultative neotenic species of Dent's (1968) categories 2 and 3 fail to metamorphose owing to the hypofunction of the pituitary; thus the failure of the thyroid is the result of insufficient stimulation by inadequate secretions of thyroid stimulating hormone (TSH). Larsen (1968) suggested that the total absence of meta-morphosis in the perennibranch *Necturus maculosus* may have evolved in a series of steps. First, there was reduction in the output of thyroxine be-cause of reduced thyroid activity, possibly caused by inadequate secretion of TSH; thence there is a loss of tissue sensitivity, followed by increased thyroid activity to subserve, in some manner, other new function(s).

4. Metamorphosis in Anurans

Anuran post-embryonic development has been separated by Etkin (1964, 1968) into three specific phases: premetamorphosis, prometamorphosis, and metamorphic climax. In *Rana pipiens,* premetamorphosis, which lasts about 7 wk (at 22–25°C), is a period of larval growth, but little change in external form (approximately TK St. I–XI, Taylor and Kollros, 1946). Prometamorphosis, lasting about 3 wk, begins when the ratio hindlimb length/body length (HL/BL) exceeds 0.2 and it ends when the ratio is about 1.0 (approximately TK St. XI–XX). During pro-metamorphosis the hind legs grow extensively in length from about 2 to about 20 mm, while the larva itself grows in length from 55 mm (body length 18 mm + tail length 37 mm) to about 65 mm (body length 21 mm + tail length 44 mm). Near the end of prometamorphosis the anal canal piece (basal lobe of the ventral fin, which includes the anal opening on the left side of the tail), is resorbed. Soon afterwards the opercular skin covering the gill chambers thins and degenerates, and on the left side the tubular wall of the spiracular opening is reduced. True climax begins at about TK St. XX with the loss of both horny beaks. Thence within 24 h the left and right forelegs emerge, the mouth progressively widens, the head is remodeled, and the tympanum is recognizable. Soon after the foreleg emergence the larval tail begins to involute; for example, in *R. temporaria* it has disappeared in less than a week (approx. 20°C), which episode marks the end of climax, the completion of metamorphosis, and the creation of a small fully formed terrestrial froglet (Fig. 1).

The extensive recognizable external larval body changes are accompanied by profound alterations to the internal anatomy, which probably involves to a greater or lesser extent all the larval organ systems, in particular those of the skeleton, musculature, intestine and associated glands, kidneys, circulation, and skin. Such profound changes during metamorphosis bring forth parallel changes in larval physiology, biochemistry, and behavior in order to anticipate the transition from an aquatic to terrestrial life.

5. General Consideration of the Endocrinal Mechanisms Responsible for Anuran Larval Metamorphosis

Based upon the results of his own extensive investigations over many years and those of his coworkers and others, Etkin (1964, 1968) proposed an elaborate theory, within the framework of the endocrinal system, to account for the mechanism controlling anuran metamorphosis. The theory, very broadly, is as follows: During the phase premetamorphosis, when there is a large increase in larval body size but little hindlimb growth, the pituitary–thyroid axis is maintained at a steady low level of activity and the pituitary thyroid-stimulating hormone cells (TSH cells) are highly sensitive to negative thyroid hormonal feedback. In this way TSH cell secretion is minimal and the thyroid hormone level is thus so low as to hardly differ from that of thyroidectomized larvae.

Prometamorphosis is marked by an increase in thyroid activity with a substantial storage of hormone within enlarging thyroid follicles. Immersion in thyroxine of normal or thyroidectomized larvae showed that maximal leg growth occurred in 10–20 parts/billion (ppb) of T_4, while tail resorption at climax required about 200 ppb T_4. Furthermore, in the premetamorphic *R. pipiens,* the hypothalamo–hypophysial portal blood capillaries are poorly developed. During prometamorphosis, when the hindlimbs commence growth, this blood capillary system differentiates in the floor of the hypothalamus and the hypothalamic neurosecretory fibers secrete a thyroid releasing factor, or hormone (TRF) into the capillary bed, which drains from the differentiated median eminence into the pituitary. TSH release from the TSH cells of the pituitary is thus activated. The median eminence does not develop in thyroidectomized larvae, but does so under the influence of graded concentrations of exogenous thyroxine (Etkin, 1963, 1965). Just before the onset of prometamorphosis, the TRF mechanism becomes sensitive to the initial level of circulatory thyroxine and a positive T_4 feedback occurs to elicit prometamorphosis. The progressive increase in the production of TRF stimulates TSH, which in turn stimulates thyroid hormone secretion. This leads to a spiralling

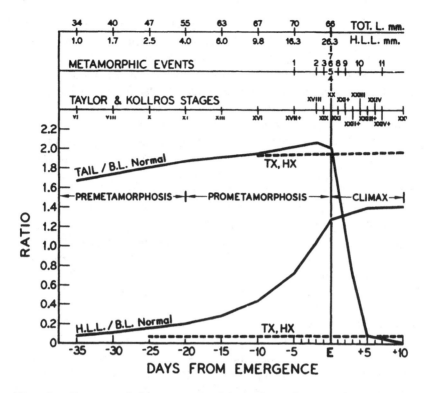

Fig. 1. Pattern of the metamorphic cycle in larvae of *Rana pipiens* (23 ± 1°C) shown in solid line. Broken line refers to thyroidectomized (TX) and hypophysectomized (HX) specimens. Numbered events are: 1, Definite reduction of anal canal piece; 2, completion of anal canal piece reduction; 3, recognizable foreleg skin window; 4, loss of both beaks; 5, emergence of first foreleg; 6, emergence of second foreleg; 7, mouth widened to reach nostril (events 4–7 in 24 h); 8, mouth widened to reach between nostril and eye; 9, mouth to anterior edge of eye; 10, mouth post-middle of eye; 11, recognizable tympanum. Climax at E. (After W. Etkin, 1964, 1968.)

effect and further raises the levels of production and a secretion of TRF, TSH, and thyroid hormones, with concomitant characteristic and sequential larval morphological changes. Ultimately there is maximal activation of the pituitary–thyroid axis with the resulting climax at TK St. XX. Etkin further postulated that TRF desensitizes the TSH cells of the pituitary to the effects of negative feedback by a high circulatory level of thyroid hormones. However, from the beginning of climax the hypothalamus ceases to be sensitive to positive feedback by circulatory T_4, and thenceforth the secretion of TRF, TSH, and thyroid hormones declines. At this time the TSH cells presumably reassume their sensitivity to negative feedback by thyroid hormones.

During early premetamorphosis prolactin, which has an antithyroid action in larvae (Gona, 1967, 1969; Etkin and Gona, 1967), is produced in the pituitary at a high rate. At this period there is substantial larval growth in size, but little organic differentiation. As the hypothalamus is activated, prolactin production is inhibited and its circulatory level drops, further decreasing as the level of TSH rises during prometamorphosis. The overall larval growth rate falls and the restraint exercised by prolactin on metamorphosis diminishes. Thus, during premetamorphosis, prolactin predominates over TSH; the reverse is true during prometamorphosis and climax. It is of interest that the action of thyroid hormones antagonizing growth in size of anuran larvae was postulated by Steinmetz (1952).

Nevertheless, in order to fit the available vast array of evidence into a comprehensive whole, Etkin was obliged to speculate and make assumptions not as yet verified, or indeed possibly not true (see Frieden and Just, 1970). Thus the circulatory level of thyroid hormone does not appear to increase to the high levels described by Etkin (see Just, 1972). Protein bound iodine (PBI) of plasma samples of tadpoles of *R. pipiens,* during TK St. V-XVII, is between 2.0–4.0 µg PBI/100 mL plasma; at about TK St. XX (the beginning of climax) it is about 10 µg PBI/100 mL plasma. Only a small percentage of larvae attained levels of 20–30 µg PBI/100 mL plasma, and thus about a 10–15-fold increase of T_4 between premetamorphosis and the highest level recorded at climax. This increase is much lower than that recorded from immersion experiments by Steinmetz (1954) and by Etkin (1964, 1968) of almost 200-fold. Recently in support of Just's results Regard et al. (1978) reported on thyroid hormone blood plasma-binding in *R. catesbeiana* tadpoles, and T_4 showed a 10-fold and T_3 a 15-fold increase at climax, compared with premetamorphic levels, during normal development. Again, there is little information available on larval circulatory levels of TRF, TSH, and prolactin and there are no data to support the view of a positive feedback mechanism on hypothalamic TRF secretion, or that TRF desensitizes the pituitary TSH cells to a negative T_4 feedback. Nor is it known whether the larval pituitary contains prolactin or that it really has antimetamorphic properties.

An alternative theory to explain anuran metamorphosis, in the light of present and in some cases more reliable data, has been proposed by Dodd and Dodd (1976). This is based on the view that the avidity of peripheral tissues and thus their utilization of thyroid hormones continually increases during prometamorphosis and early climax, with the result that hormones are removed from the circulation so quickly that their concentration level rises only relatively modestly (see Frieden and Just, 1970). The speculative negative–positive–negative and thus changing feedback mechanism need not be involved, and negative feedback only would operate on the appropriate endocrinal system throughout larval life. Thus the prime control of the activity of the TRF–TSH–thyroid hormonal axis resides in the metamorphosing tissues themselves. Low temperatures would operate at the tissue level to reduce their utilization of thyroxine. The resulting higher circulatory levels of thyroid hormone decrease the activity of the endocrinal axis by negative feedback.

This hypothesis explains the mechanisms controlling metamorphosis more simply than others because its basic assumption denies Etkin's claim of a relatively high and rising circulatory thyroid hormonal level occurring during metamorphosis compared with that at the premetamorphic stage. Nevertheless a number of features still require explanation. For example, what influences switch on the metamorphic mechanism? How does the thyroid begin and thence improve its secretory performance? And which processes are essential to stimulate the growth and differentiation of the median eminence? Perhaps the incipient origin, development, and secretion of the thyroid are independent of the pituitary and hypothalamus. If these endocrine organs exerted tissue priorities by virtue of higher avidity to the originally prevailing low threshold level of thyroid secretion than other tissues, then this could well get the endocrinal system on the move, so to speak. Thereafter the ever-rising production and secretion of thyroid hormones stimulated by the hypothalamus and the pituitary would be controlled by tissue utilization and negative feedback.

Within the above theories on the endocrinal control of metamorphosis [see discussion by White and Nicoll (1981)], the relationship of prolactin or growth promoting hormone (STH) and indeed whether they have any antimetamorphic effect at all, is not clear though it does appear likely that mammalian prolactin does have an antithyroid effect on prometamorphic anuran larvae (see Dodd and Dodd, 1976). T_4-induced metamorphosis of *R. catesbeiana* larvae is accelerated after injection of antisera to ovine prolactin (Eddy and Lipner, 1975; Yamamoto and Kikuyama, 1982). Perhaps the growth-promoting hormones influence tissue utilization of thyroid hormones, as envisaged by Dodd and Dodd. Again, mammalian adenohypophysial somatotropin appears to be the more potent stimulator of growth (expressed in terms of body size and weight) in larvae of *R. temporaria* (Enemar et al., 1968; Enemar, 1978) and *Xenopus laevis* (Turner, 1973), whereas prolactin only is really significantly effective in *R. catesbeiana* and *R. pipiens* (Berman et al., 1964; Brown and Frye, 1969; Etkin and Gona, 1967). Snyder and Frye (1972) considered that in *R. pipiens* prolactin is the larval growth hormone and somatotropin stimulates growth in postmetamorphic juvenile adults.

It is of interest to find that in anuran larvae (e.g., Ranidae) porphyropsin (based on 3,4-dehydroretinol, Vitamin A_2) is the major visual pigment, and in terrestrial forms rhodopsin (based on retinol, Vitamin A_1) predominates (Wald, 1981). T_4 induces rhodopsin deposition in the pigment epithelium of tadpole eyes by directly reducing the activity of the enzyme 3,4-retinol dehydrogenase, which is necessary for the synthesis of porphyropsin. In this case ovine prolactin does not antagonize T_4 induction, though it does so, for example, with T_4-induced tail regression. Crim (1975) concluded from these investigations that T_4 and prolactin are antagonistic in larval metamorphic development when: (a) each hormone exerts a specific morphological and/or physiological effect when given alone, and (b) when the effects are opposite. Antagonism may thus result from differential gene activities influencing opposite parameters in development. Again bovine prolactin was found to inhibit T_4-induced tail resorption, but did not influence shortening of the alimentary tract, nor the development of the hindlimbs and the median eminence in larval bullfrogs. Liver orginase activity, but not that of ornithine transcarbamylase, was likewise blocked by prolactin (Noguchi and Kikuyama, 1979). It may be mentioned, however, that Wright et al. (1979) did find that bovine prolactin significantly retarded hindlimb growth and development in young TK St. V-VIII larvae of *R. pipiens*. Bovine prolactin enhances collagen synthesis in the tail of the bullfrog larva 40-fold compared with that of untreated controls (Yoshizato and Yasumasu, 1970). The hormone may act through the agency of cAMP, which similarly inhibits tail regression in *Bufo bufo japonicus* (Yamamoto et al., 1979); but see also Stuart and Fischer (1978), who found cAMP to stimulate tail regression. It may be noted that recently a growth-promoting, prolactin-like, antimetamorphic hormone has been isolated from the pituitary gland of larval and adult *R. catesbeiana* (Kikuyama et al., 1980). The subject has been discussed in some detail recently by White and Nicoll (1981).

Clearly more information about the origin, secretion, and circulatory levels of various hormones other than of the thyroid, and the behavior of peripheral tissues in these cases, is required to support the theory of Dodd and Dodd. Other hormones including ACTH of the interrenals and of the pineal (see Rémy and Disclos, 1970; Rémy and Bounhiol, 1971), and external factors such as light and temperature doubtless are also involved.

The sequence of metamorphic changes that occurs during larval development is elicited by the gradual increase in the circulatory level of thyroid hormones. The stoichiometric view of Etkin (1964, 1968) proposes that any effective concentration of thyroxine can induce metamorphic changes if allowed to operate long enough, and that each tissue has a specific total requirement or amount of thyroid hormones needed to undergo metamorphic change. This is satisfied either by a higher concentration acting for a shorter period, or by a lower concentration acting for a longer period. In contrast, Kollros (1961) has advocated the view that each tissue has a minimum hormone requirement or threshold level for its response, and that the various tissues have different thresholds that vary according to the temperature. Thus hypophysectomized tadpoles immersed in different concentrations of thyroxine proceed to develop only to certain stages of the metamorphic sequence, however long the duration of the treatment; to proceed further to a higher stage requires a higher threshold concentration of hormones.

It is likely that larval tissues change in their capacity to respond to thyroid hormones during development, thus that there are changes in tissue sensitivity (Derby, 1968). Older larvae of *R. pipiens* (TK St.X) responded to T_4 treatment by having better-developed skin glands than did those larvae similarly treated (in time and T_4 concentration) but at TK St.VII, and the latter larvae in terms of skin glands are superior to those of larvae treated at TK St.V (Heady and Kollros, 1964). Likewise the capacity of the larval tail to respond to T_4 stimulation by involution improves during development. The latent period of this response was considerably lower at TK St. XII than in younger stages, when immersed in the same concentrations of T_4 (Chou and Kollros, 1974). Frieden and Just (1970) itemized a substantial list of biochemical and morphological changes that occur from the beginning of climax (TK St.XX), in response to an abrupt rise in the circulatory level of thyroid hormones (see Just, 1972). They suggest that during earlier larval stages, presumably in different tissues the competence of these various processes to react to the significant increase in the level of thyroid hormones increases.

Nevertheless it is probable that most if not all larval tissues have some capacity to respond to thyroid hormones from their beginning, and this property improves as the larva develops. Various organs acquire thyroid hormone sensitivity early in larval development (see Moser, 1950; Prahlad and Delanney, 1965). In *Xenopus laevis,* biochemical and other metamorphic events respond to T_3 2–4 d after fertilization, at NF St.36–41 after hatching (Tata, 1968b). Most responses described reached maximal level 15–18 d after fertilization, at NF St.50–51. However, the originally median thyroid Anlage is only apparent at NF St.33–34, and the definitive paired Anlagen are seen at NF St.41. A few diffuse lobes are recognizable on each side of the pharynx at NF St.48 and the first follicles arise soon after at NF St.48–49. Presumably there is little thyroid hormone present in the larval circulation before NF St.50, before premetamorphosis, though Hanaoka et al. (1973) have claimed thyroid hormone synthesis to begin in *Bufo bufo japonicus* and *Xenopus laevis* almost from the beginning of the thyroid Anlage, with an "appreciable" amount of ^{131}I-labeled thyroglobulin formed at a stage equivalent to NF St.45 of *Xenopus.*

It is likely that the degree of development of various cellular structures, including their biochemical synthetic activities, after the beginning of premetamorphosis (around NF St.53) is independent, or practically so, of the thyroid, as demonstrated experimentally in larvae hypophysecto-mized, thyroidectomized, or treated with chemical thyroid inhibitors, or goitrogens, at appropriate stages of development.

White and Nicoll (1981) reviewed the most recent evidence on hor-monal control of amphibian metamorphosis. Since plasma protein-bound iodine and thyroid hormone concentration appear to be highly variable during climax, they considered that perhaps thyroid hormone and TSH levels are pulsatile; indeed, that diurnal variations may possibly occur. They emphasize the importance of tissue sensitivity to T_3 and T_4, and possibly to prolactin. Perhaps there is a synergistic effect with growth hormone (somatotropin). Adrenal steroids may also be involved by al-tering tissue sensitivity. There is quite some way to go! The subject is dealt with in more detail, at the molecular biological level, in Section IV.

Section II

Staging of Embryonic and Larval Amphibians

6. Anurans and Urodeles

The ontogenetic development of amphibians from the fertilized egg to the completion of metamorphosis has been widely investigated. Details of the topography, anatomy, and histology of the various organs and tissues are now available in the literature, most of them examined in lesser or greater detail to the level of electronmicroscopy. Analyses of ontogeny have also been made at the gross visual level, for a variety of reasons (*vide infra*), and like a series of photographic stills, successive stages of development of individual species have been arbitrarily tabularized and numbered, though various authors frequently have included the specific levels of internal anatomical development that have occurred in the referred stages. Any examination of the gross external features of amphibian ontogenetic change during development clearly is relevant to a study of morphogenesis especially in anurans; indeed such changes provide a working basis for more detailed research in this field. The following account, therefore, deals with the subject of staging in some detail, but of more importance, it brings up-to-date and lists the relevant publications in the field.

Among amphibians the rate and in many cases pattern of embryogeny and larval development usually differ between genera and species. Such variations are mainly intrinsic and are influenced at the genetic level. For example, two of the shortest periods of ontogenetic development in anurans, 9–11 d in *Bufo pentoni* from Senegal (Forge and Barbault, 1977) and 12–13 d from the egg to the end of metamorphosis, found in *Scaphiophus holbrooki hurteri* from N. America (Bragg, 1966), are adaptations to utilize breeding ponds that are only briefly available during arid conditions. Likewise the Indian bullfrog *Kaloula pulchra*, from Thailand, has a brief developmental cycle to complete metamorphosis in 15 d in order to make use of the temporary ponds in the breeding habitat (Schmidt, 1978). Minor genetic differences may, to a lesser degree, be responsible for modest differences in ontogeny between varieties of the same species, but here the significant factors are mainly ecological, such as the ambient temperature and the conditions in the breeding ponds, i.e., pH, O_2, and CO_2 levels, feeding, and overcrowding. Indeed in *Ambystoma mexicanum*, for example, the iodine level in lakes where development occurs can determine the rate of metamorphosis, or even whether it occurs at all. The neotenic *Pseudotriton palleucus* can be made to metamorphose within 3 months when a few drops of potassium iodide are added to the ambient water of its tank (Blair, 1964). Thus information in terms of larval age or size, available to workers in different fields of amphibian research, could well be of doubtful value.

The need for accurately described, arbitrarily determined, standard development stages of amphibian embryos and larvae, of different species, for reference or comparison, has long been apparent to embryologists; this applies equally for developmental stages of all vertebrates including humans. The importance of standardized developing morphological stages in vertebrates, including amphibians, has been emphasized by Just et al. (1981) and they have prepared lists of partially completed comparative tables of stages (based on limb growth and tail and gill resorption) of some of the better known anuran and urodelan species reported in the literature.

In amphibians it has been found most useful to list the series of developmental stages in terms of the external morphology and the age at a known constant temperature (usually 18°C) (Pollister and Moore, 1937). One of the earliest series was by Adler (1901), who described 16 stages of *Bufo vulgaris* from the gastrula to the end of metamorphosis. Keibel subsequently published a number of "Normentafeln," of various amphibians, one of the early ones was by Eycleshymer and Wilson (1910) on *Necturus maculatus* probably the first complete ontogenetic series of stages of a urodele (Fig. 2a), and another by Gläesner (1925) on *Triton vulgaris*. Eventually, a series of stages (numbered in Arabic numerals) of various amphibian species, from the fertilized egg through larval development to the newly metamorphosed froglet (or newtlet), became available though the number of complete series known is limited and most of the earlier published series were incomplete. Thus, 25 stages of *Rana pipiens* comprised the egg leading only to the larva with a completed operculum (Shumway, 1940). Likewise the 23 stages of *R. sylvaticus* by Pollister and Moore (1937) terminated with even younger larvae having visible external gill filaments. Again the well-known series of stages of *Ambystoma maculatus,* unpublished by Harrison but illustrated by others (see Hamburger, 1950), includes 46 stages. However, the series terminated abruptly with larvae whose balancers are fully developed at H St.45 but absent at H St.46, when three pairs of external gill filaments and merely three-toed forelimbs are recognizable. A comparable incomplete series for *Triturus pyrrhogaster* by Anderson (1943) includes 25 stages (I–XXV), and only hindlimb buds have appeared at the three-toed forelimb stage, as occurs in the fully staged *Pleurodeles waltl* (Gallien and Durocher, 1957).

Complete series (from the end of embryogenesis through metamorphosis) of *R. pipiens* (Roman numerals I–XXV were used for the postfeeding stages) by Taylor and Kollros (1946), of *Bufo valliceps* (46 stages) by Limbaugh and Volpe (1957), *B. regularis* (66 stages) by Sedra and Michael (1961) and *B. melanostictus* (43 stages) by Khan (1965), among others, became available. They compared similar though differently numbered stages of different species of anurans by various authors. From the middle of the century the widespread and convenient use of *Xenopus* embryos in a variety of research fields on developmental biology demanded a comprehensive series of stages, and the detailed 66 stages of *X. laevis* were published by Nieuwkoop and Faber (1956). They also listed equivalence in staging with other anurans.

In 1953 at the XIV Zoological Congress in Copenhagen, Emil Witschi proposed a scheme for international agreement by embryologists on staging normal vertebrate embryos: Their ontogeny should be divided arbitrarily into a number of periods. Thus Period I would include stages 1–6; Period II the blastula, St.7; Period III the gastrula, St.8–11; Period IV the primitive streak, St.12; Period V the neurula, St.13–17; Period VI the tail bud, St.18–24; Period VII the embryo, St.25; Period VIII metamorphosis, St.26–33; Period IX the fetus, St.34–36. However, though Sven Horstadius at the Congress voiced a general concern about the difficulties of applying such an all-embracing system because of the profound differences in the patterns of development between different classes of vertebrates, Witschi notwithstanding elaborated the scheme in 1962. Some workers have used a somewhat comparable scheme of Roman numerals to subdivide amphibian premetamorphosed ontogeny into Periods, with Arabic-numbered stages for each one (Cambar and Gipouloux, 1956, for *Bufo bufo,* and Cambar and Martin, 1959, for *Alytes obstetricans)* (Fig. 2b), but the added complexity would seem to be a doubtful benefit. On the whole Witschi's proposals have not been adapted and, for each species, Arabic-numbered stages alone would still appear to be the most satisfactory and commonly used arrangement. A relatively simplified general table of developmental stages, which with minor modifications may be applied to pelobatids, bufonids, hylids, and ranids, has been prepared by Gosner (1960). Thus stages 1–25 include the embryonic and prefeeding stages, the transition to a feeding free-swimming larva occurring during stages 21–25; stages 26–40 are based on hindlimb development (a component very commonly used by other workers also); stages 40–46 include the period of metamorphic climax when there are profound larval bodily changes in shape, the tail regresses, mouth parts break down, and the forelimbs break through. As Gosner (1960) rightly claims, staging tables provide a useful shorthand for distinguishing different ontogenetic levels of development of an amphibian species.

Schemes of anuran and urodelan larval development from the fertilized egg to completion of metamorphosis, of representative forms, using numbered stages, are shown in Figs. 3 and 4. Most species could well fit into the numerical arrangements that are illustrated. Many workers, especially those from the USA, concerned particularly with anuran ecology and breeding behavior, refer their species to Gosner's stages. Others deal-

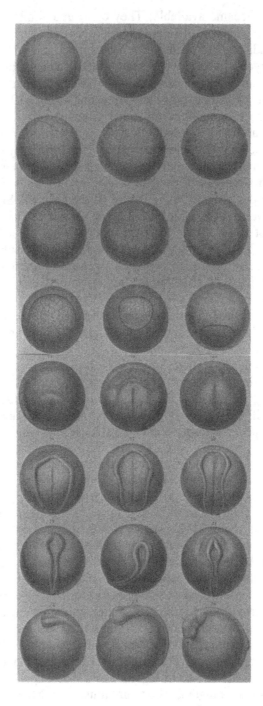

Fig. 2A.   The series of stages of *Necturus maculatus* published in 1910
by A. C. Eycleshymer and J. M. Wilson, of its ontogeny from the egg to

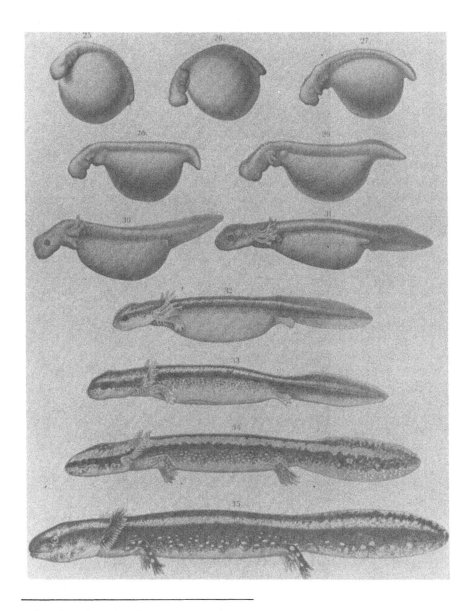

Fig. 2A. (*cont.*) the metamorphosed animal: a classic example of a Keibel series and probably the first complete one published for a urodele. There are nearly 600 references included in the work from the earliest in 1766 by J. Ellis and J. Hunter on *Siren lacertina* (*Phil. Trans. Roy. Soc. B* **56,** 189–192; 307–310) and C. Linnaeus (Uppsala) again on *Siren lacertina*, to the latest reference in 1909.

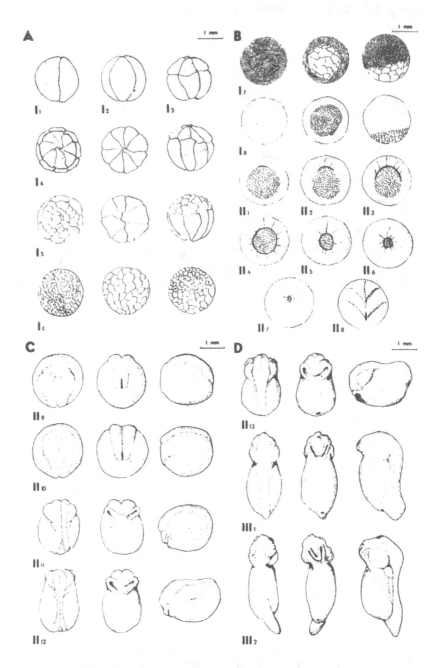

Fig. 2B. Parts I-III. Example of a staged series of an anuran *Bufo bufo* from the first cleaved egg to the completion of metamorphosis and formation of the tiny toadlet. In this series. Period I includes egg segmentation; Period II the formation of the gastrula and the neurula; Period III organogenesis of the embryo and larva; and Period IV larval metamorphosis, from the first origin of the hindlimbs. (After Cambar and Gipouloux, 1956.)

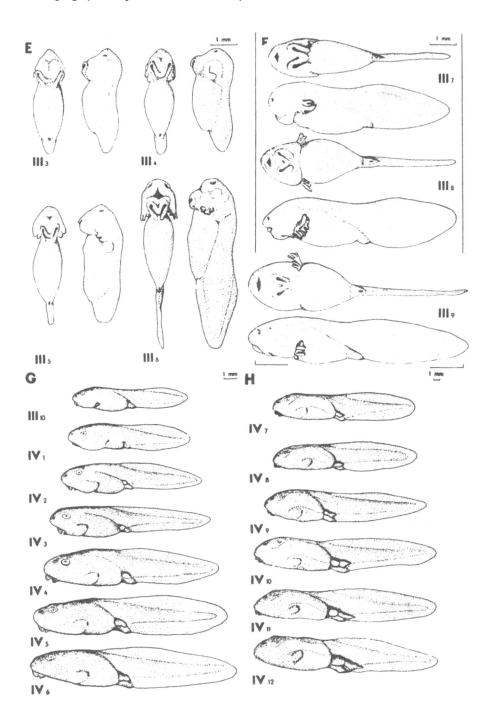

Fig. 2B (*continued*)

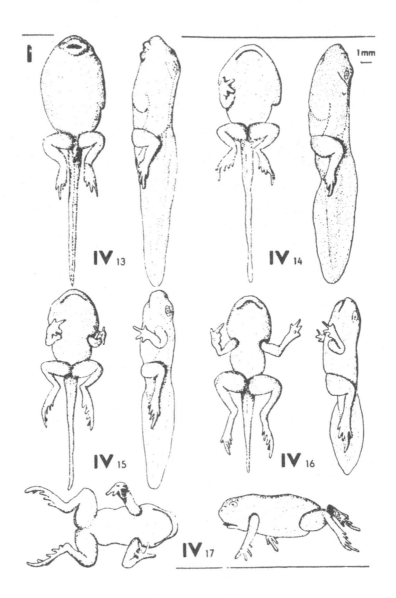

Fig. 2B (*continued*)

ing with problems of fundamental cellular biosynthetic mechanisms, now more often at the level of molecular biology, either use the series of *Xenopus laevis* by Nieuwkoop and Faber (1956) or that of *Rana pipiens* by Taylor and Kollros (1946), still probably the most popular one, though not complete for developmental stages of the genus *Rana*. This preference may well reflect the high quality (notwithstanding the Roman numerals), but also perhaps the conservatism of so many excellent biologists.

The amphibians here listed for staging, or for relevant information on their larvae, are grouped in Families, of the Suborders of Anura and Urodela (mainly after Noble, 1954). The author has no knowledge of any existing comprehensive series of stages of embryos and larvae of any caecilian (apodan) species. In the case of this Order, only a brief summary of some relevant references to developmental stages is given. Different species, where listed, have the author reference and short comments describing their results. Those species that are completely staged from the egg to the fully metamorphosed individual, are distinguished by a superscript [a] against the author(s). Species incompletely staged have a superscript [b] against the author(s). Larvae usually of random stages, when referred to Gosner's (1960) series have a superscript [c] against the author(s).

The completed list doubtless has omissions, though probably most of the important references to staged specimens (complete or incomplete) are included. Information on larvae of various amphibians (in particular of anurans and urodeles, even when they are not staged), is given when it has some relevance to the general scheme. In many cases there is little information available on larvae of relatively obscure species—indeed, there are very many species of amphibians in this category—and in such cases they are listed. If *all* references on different aspects of amphibian larval ontogeny were included the list would in practice be endless.

7. Caecilia (Apoda or Gymnophiona)

The heart and associated arteries, external gills, and developing vertebral column of young larvae of *Ichthyophis glutinosus* and *Gegenophis carnosus,* were studied by Ramaswami (1944, 1954, 1957). Inger (1954) described larvae of *Ichthyophis monochrous,* 65–231 mm long from the Philippines. Cochran (1972) provided some brief information on embryos and larvae of various caecilians. Some genera have species that lay eggs (e.g., *Rhinatrema, Siphonops,* and possibly *Caecilia*), since no developing embryos have yet been found in the female's oviduct. Other genera have viviparous species (e.g., of *Gymnophis, Chthonerpeton,* and *Typhlonectes*). A free-living larva of a *Rhinatrema* species had three pairs of external gills. Usually, however, these structures are present only in

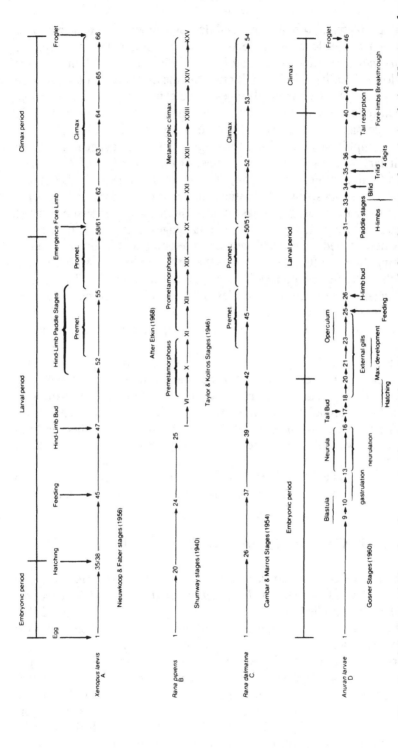

Fig. 3. Scheme of the development of anuran larvae represented by the staged series of A. *Xenopus laevis* (Nieuwkoop and Faber, 1956); B. *Rana pipiens* (Taylor and Kollros, 1946); C. *Rana dalmatina* (Cambar and Marrot, 1954); and D. stages applicable to common species in general (Gosner, 1960). (Modified after Broyles, 1981.)

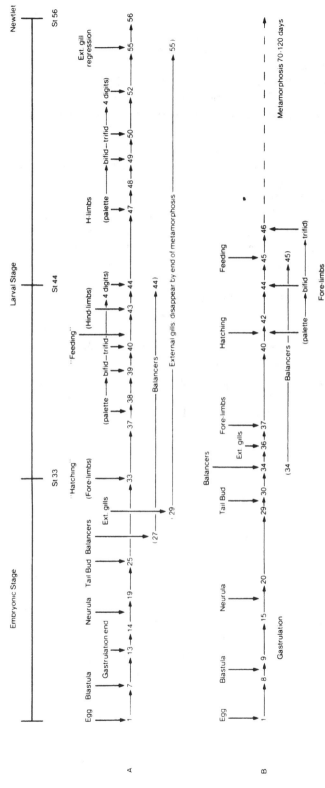

Fig. 4. General scheme of development of urodele larvae represented by the examples in A. *Triturus helveticus* Razoumowsky (= *Triturus palmatus* Schneider) (Gallien and Bidaud, 1959) and *Pleurodeles waltl* Michah (Gallien and Durocher, 1957); and B. *Ambystoma maculatum* (Harrison, after Hamburger, 1950). Though feeding appears to commence at a slightly earlier stage in *Pleurodeles* (St. 38) than in *Triturus* (St. 42), and though there are variations in stage numbers in the case of limb development, overall their patterns of development are extremely similar in terms of staging. At 18°C metamorphosis of *Pleurodeles* is achieved in 110 ± 10 d, and in *Triturus* 90 ± 10 d. *Triturus taeniatus* (Gläesner, 1925; Rotman, 1940) has a chronology generally similar to these species, especially of *Triturus helveticus* (Gallien and Bidaud, 1959). (Prepared from an idea by Broyles, 1981.)

29

uterine embryos or in those in the egg of other genera. Taylor (1968) listed taxonomic descriptions of adult Apoda and in the case of some genera briefly referred to their larvae, mainly in terms of size, dentition, or the presence of gill slits. These include: *Ichthyophis mindanaoensis* (the neuromast system is retained until transformation occurs), *I. supachii, I. youngorum* (larvae live two years before transformation); *Epicrionops petersi noblei* (a 63-mm long specimen had a gill slit and external gill fragment present); *Typhlonectes obesus* (a female had an embryo in the uterus); *Nectocaecilia petersii; Chthonerpeton indistinctum* (larvae 108–113 mm long at birth); *Caecilia pressula; Dermophis mexicanus mexicanus* (oviducal broods may contain up to 10 young), *D. oaxacone; Gymnophis proxima* (embryonic forms 132 mm long in the uterus of an adult female); *Grandisonia alternans* (a pair of circular gill slits present in two larval specimens); and *G. larvata.* Information on gill structure (10.5 mm larva found in the uterus of a female *Gymnophis multiplicata proxima,* and of other larvae 45–84 mm long) and on fetuses (including their dentition) of other Caecilia including *Typhlonectes obesus,* has been published by Wake (1967, 1969, 1976, 1978). The lateral line system of three species of *Ichthyophis* has been described by Hetherington and Wake (1979). Badenhorst (1978) in a study of the development of the Organ of Jacobson and the tentacular apparatus of *Ichthyophis glutinosus* listed St.I (25 mm larva in the egg capsule); St.II (35 mm larva); St.III (4 specimens 6–11 cm long); St.IV (2 older larvae 23.6–24 cm long). Exbrayat et al. (1981) likewise described Stages I–IV in the ontogeny of *Typhlonectes compressicauda,* from the beginning of organogenesis to the final larva 150 mm long.

Summary

We can briefly summarize the data on the published series of developmental stages of amphibian larvae. Of the 2500 or so known species of anurans there are complete tables for fewer than twenty. Apart from the fairly brief series of Sampson in 1904 for *Hylodes martinicensis* (*Eleutherodactylus nubicola*), all the other complete series of anurans through metamorphosis were prepared after 1950. Members of the Liopelmidae of the Suborder Amphicoela are not represented in the list. Of the 300 known species of urodeles, even fewer have been completely staged, most of these from about 1940 onwards, and there are no staged species of the Plethodontidae (though there is an incomplete one of *Plethodon cinereus* by Dent, 1942), Ambystomidae, and Meantes. There seems to be no complete ontogenetic table from the egg to the juvenile stage for any caecilian species. This is indeed a sorry state of affairs and clearly great gaps exist in our knowledge in this field of amphibian developmental biology. Probably most of the tables now available were prepared for use in experimental investigations on cellular biofunction, exemplified in particular by *Rana pipiens* (Taylor and Kollros, 1946) and *Xenopus laevis* (Nieuwkoop and Faber, 1956). Perhaps this dismal record will stimulate future workers to do something about it. There is need for it.

8. Tables of Staged Specimens

ORDER ANURA (Salientia)

SUBORDER AGLOSSA (flattened aquatic frogs)

Family Pipidae

Species	Reference	Description
Hymenochirus boettgeri	Sokol, 1962	General information on larval growth and development, including metamorphosis
Pipa carvalhoi and *P. parva*	Sokol, 1977	External characters and organogeny of larvae, stages as Nieuwkoop and Faber (1956) for *Xenopus*
Xenopus gilli	Rau, 1978	Compared *X. gilli* and *X. laevis* larvae
Xenopus laevis	Bles, 1905	Classical description of larval development from the egg through metamorphosis
	Peter, 1931	Description of Bles' material from larvae (4.7 mm long, unhatched with gill buds, cement glands, and frontal glands) to metamorphosed froglet
	[b]Gasche, 1944	15 Stages including metamorphosis.
	[b]Weisz, 1945a	1st form larvae: hatching to St.21 (6–9 mm); 2nd form larvae: St.22 (9–30 mm); 3rd form larvae: St.23 (30–58 mm)
	Weisz, 1945b	General development: anatomy
	[b]Bretscher, 1949	Study of hindlimb development. St.I–VII (equivalent to NF St.50–56/57)
	[a]Nieuwkoop and Faber, 1956	Complete series of 66 stages

SUBORDER AMPHICOELA (primitive frogs, aquatic or terrestrial)

Family Liopelmidae

Species	Reference	Description
Ascaphus truei	Stephenson, 1951	Eggs and larval stages described
	[c]Brown, 1975a	20 Stages described
Leiopelma archeyi	Stephenson, 1951	Eggs and larval stages described
Leiopelma hochstetteri	[b]Archey, 1922	7 Stages: eggs to metamorphosed froglet; from specimen to 35 d (hatching) and 1 month post-hatching stage

SUBORDER OPISTHOGLOSSA (aquatic and terrestrial frogs)

Family Discoglossidae

Alytes obstetricans	[a]Cambar and Martin, 1959	44 Stages
Bombina orientalis	[b]Sussman and Betz, 1978	25 St; after Shumway (1940) to formation of spiracle; no hindlimb buds
	[b]Michael, 1981	Early development in relation to different temperatures
Discoglossus pictus	[b]Gallien and Houillon, 1951	32 St; egg to larva 8.5–9 mm long (formation of the spiracle)

SUBORDER ANOMOCOELA (spade foot frogs)

Family Brachicephalidae

Atelopus spp.	[c]Duellman and Lynch, 1969;	Described eggs and larvae (illustrated) of *Atelopus spumarius* (St.37); *A. certus* (St.25–44); *A. ignescens* (St.25); *A. minutes* (St.31, St.40)
Rhinoderma darwini	[a]Jorquera et al., 1972, 1974	Species from Valdivia (Chile), St.1–26 (St.1–11 embryonic, St.1–15 metamorphic, direct development). Sub-spp. from Concepcion. St.1–27 (St.1–12 embryonic, St.1–15 metamorphic, partially aquatic development)

Family Pelobatidae

Pelobates cultripes	[c]Busak and Zug, 1976	General information on larvae
Pelobates fuscus	[b]Zabroda and Il'enko, 1981	St. I–19 illustrated, including mouth parts, of larvae from the Ukraine
Scaphiophus bombifrons	Bragg, 1966	General information on larvae
	[b]Trowbridge, 1941	Eleven Stages to end of cleavage illustrated
	[b]Trowbridge, 1942	St.12–25 (64 h), hatching at St.19; data on larvae through metamorphosis (32 d)
Scaphiophus holbrooki holbrooki	[b]Gosner and Black, 1954	Information on embryos and larvae. Staging after Rugh (1951)
	Voss, 1961	General information on larval growth

(*continued*)

8. Tables of Staged Specimens (*continued*)

Family Dendrobatidae *Phyllobates femoralis*	[c]Lescure, 1976	Both larval spp. carried on back of adult males. Described larvae of *P. femoralis* 11 mm long (tiny hindlimb buds) and *P. pictus* 11.4 mm long (stage just before first hindlimb bud). Good reference list for Dendrobatids)

SUBORDER DIPLASICOELA (typical frogs)

Family Microhylidae (small mainly upland frogs)

Cophylinae (Subfamily)	[c]Blommers-Schlösser, 1975a	General descriptions
Dyscophinae (Subfamily)	[c]Blommers-Schlösser, 1975a	St.25-46.
Dasypops schirchi	[c]da Cruz and Peixoto, 1978	Described larva 52 mm long (St.37)
Kaloula pulchra	Schmidt, 1978	Larval development 15 d (26°C)
Otophryne robusta?	Pyburn, 1980	Eggs described; illustr. larvae of unknown frog have unusual mouth parts and spiracle
Family Ranidae *Rana afghans*	Hora, 1933	Illustrated unhatched stages and hatched larvae minus external gills. Described suctorial disc
Rana alticola	Smith, 1924	Illustrated a well-developed larva
Rana aurora	[c]Brown, 1975b	Information on embryos and larvae
Rana berlandieri (and *R. sphenocephala*)	[c]Hillis, 1982	Compared larval morphology of the 2 Spp. at St.28-40 of Sosner (1960)
Rana brevipoda porosa	[a]Iwasawa and Morita, 1980	45 Stages. St.1-40 (several stages subdivided). Illustrated with tables (in Japanese: brief English summary)

Species	Reference	Description
Rana catesbeiana	George, 1940	Information on eggs and larvae, through climax of specimens from Louisiana. Life cycle to froglet 4 months, in contrast to specimens from New York, which remain as tadpoles through two winters
	Hammerman, 1969	Described an arbitrarily selected series of Stages A1–A12, B-L from larva with small hindlimb buds to metamorphosed froglet: 39 mm → 125 → 125 → 45 mm (total length)
	[c]Hedeen, 1975	St.20-40 described
	Viperina and Just, 1975	General information after Taylor and Kollros (1946) stages
	Bruneau and Magnin, 1980a,b	General information on larvae and data on their growth
Rana chalconata	[b]Hing, 1959	Egg → gastrula → St.A,B,C (neurulation) → St.I–X (left foreleg through spiracle)
Rana clamitans	Ting, 1951	Larval development, 3 months
Rana crassipes	Lamotte and Zuber-Vogeli, 1954	Description of older larvae (57 mm: body, 18 mm; tail, 39 mm; well-developed hindlimbs)
Rana curtipes	Lobo, 1961	Brief information on larvae in India
Rana cyanophlyctus	[b]Ramaswami and Lakshman, 1959	St.1–25 to completion of the formation of operculum (10 mm larva, 11 d old, 18–20°C)
Rana cyanophlyctus	Mohanty Hejmadi and Dutta, 1979	Description of breeding and of four stages of larval development; from a prelimb to the metamorphosed stage
Rana dalmatina	[a]Cambar and Marrot, 1954	Complete series of 54 Stages
Rana esculenta	[b]Michniewska-Predygier and Pigon, 1957	St. 1–25 to closure of operculum. Compares with *R. temporaria*, *R. terrestris* and *Bufo bufo*. Illustrated
	[a]Manelli and Margaritora, 1961	St.1–50 egg to end of metamorphosis. St.1–30 (embryogeny); St.31–32 (larvae); St.33–50 (metamorphosis). St.31 incipient hindlimb bud; St.44–50 tail regression

(continued)

8. Tables of Staged Specimens (continued)

Rana fusca	[a]Kopsch, 1952	30 Stages, detailed organogenesis: gastrulation (1–6); neurulation (7–9); embryonic period (10–17); larvae (18–26); metamorphosis (27–30)
Rana japonica	[b]Tahara, 1959	25 Stages to completion of operculum; no hindlimb buds developed
Rana johnstoni inyangae	[c]Channing, 1976a	Described 24 tadpoles (39–65 mm long) (= St.26–37). Compared larvae with those of Rana j. angolensis in terms of keratodonts
Rana lessonae	Günther, 1978	Developmental stages of larvae described
Rana occidentalis (= R. occipitalis)	Lamotte and Zuber-Vogeli, 1954	Described larvae 40–55 mm long
Rana palmipes	Volpe and Harvey, 1958	Described larvae and mouth parts; large tadpoles up to 80 mm long. Staged according to Taylor and Kollros (1946)
Rana pipiens	[b]Emmet and Allen, 1919	11 Stages similar to those of Tschernoff (1907) for R. temporaria, with respect to the hindlimb development
	[b]Miller, 1940	23 Stages; operculum closing over gills (12 mm larvae); compared with R. sylvatica (Pollister and Moore, 1937)
	[b]Shumway, 1940	25 Stages to completion of operculum
	[b]Taylor and Kollros, 1946	End of embryonic period (St.25), then St.1–XXV (based mainly on limb development)
	[b]Rugh, 1948	Photographic series after the Shumway stages
	[a]Witschi, 1956	St.1–33. Egg to froglet with small tail stub
Rana pretiosa	Licht, 1975	Life histories in high and low elevations
Rana ridibunda	Günther, 1978	Developmental stages described
Rana septrionalis	[c]Hedeen, 1971	General information on larvae
Rana sierramadrensis	Webb, 1978	Described tadpoles; compared them with those of Rana sinaloae
Rana sinaloae	[b]Webb and Korky, 1977	Stages described after Taylor and Kollros (1946)
Rana spp?	[b]Terent'ev, 1950	Stages described and compared with other anurans by Just et al. (1981)
Rana sylvatica (= R. pallustris)	[b]Pollister and Moore, 1937	23 Stages; egg to larva with first indication of hindlimb; external gills still recognized
	Herreid and Kinney, 1967	Temperature and larval development. After Pollister and Moore (1937) and scheme of Witschi (1956): St.26–33 metamorphosis

Rana tarahumarae	Webb and Korky, 1977	Tadpole variation; after Taylor and Kollros (1946)
Rana temporaria (and *R. arvalis*)	[b]Tschernoff, 1907	11 larval stages used for the development of the hindlimb (10 mm larva to onset of climax)
Rana temporaria (and *R. viridis*)	[b]Wintrebert, 1905a,b	Hindlimb development, 10 larval stages. St.I (incipient hindlimb bud) through ontogeny (climax St.X). No illustrations
	[b]Moser, 1950	13 Stages to larva with hindlimb rudiment
	[b]Michniewska-Predygier and Pigon, 1957	Egg to formed operculum: St.1–15. Illustrated
	[b]Claes, 1964; 1965	Embryonic and larval development: St.1–17 (177 h). Staging after Kopsch (1952)
Rana temporaria	Angelier and Angelier, 1968	Embryonic and larval development in relation to temperature (in the Pyrenees)
Rana temporaria	[a]Dabagyan and Sleptsova, 1975	St.0–54: From egg through embryogeny, larval development and climax. St.39–54 are like those for *Rana dalmatina* (Cambar and Marrot, 1954)
Rana temporaria	Koskela, 1973	Information on larvae in Finland
Rana tigrina	[a]Agarwal and Niazi, 1977	St.1–55 (unfertilized egg to metamorphosed froglet) (8 wk, 26–30°C). St.1–26 to hatching; St.32 feeding; St.47–55 limb growth and metamorphosis
Rana vibicaria	Zweifel, 1964b	Described and illustrated larvae and mouth parts. St.20–38 according to Limbaugh and Volpe (1957). Compared with *R. palmipes* and *R. warschewitschi*
Tomopterna delalandei cryptotis	Channing, 1976b	Information on eggs and hatched larvae
Ranidae of Malagasy	[c]Blommers-Schlösser, 1979b	Described known tadpoles of various genera: *Mantidactylus ulcerosus, M. opiparus, M. wittei, M. tornieri, M. punctatus* (all illustrated); and *M. betsileanus, M. curtis, M. lugubris, M. albofrenatus, M. aerumnalis, M. aglavei, M. blommersae, M. depressiceps, M. pulcher, M. liber, M. bicalcaratus, Aglyptodactylus madagascariensis, Manella aurantiaca*

(continued)

8. Tables of Staged Specimens (*continued*)

Family Brevicipitidae (narrow mouthed toads)

Phrynomerus annectens	Channing, 1976b	Information on breeding and hatching larvae
Pseudohemiscus granulosus	[c]Blommers-Schlösser, 1975a	Stages 25–46
(Subfamily Scaphiophryninae)		

Family Rhacophoridae (tree frogs: some return to terrestrial life)

Philautus carinensis	[c]Wassersug et al., 1981	Described larva at St.37. From Thailand
Rhacophorus arboreus	[a]Iwasawa and Kawasaki, 1979	Complete series St.1–46; egg to completion of metamorphosis (44 d), climax 7 d (22°C)
Rhacophorus leucomystax	[b]Ting, 1970	St.1–25; after Shumway (1940) for *Rana pipiens*.
Malagasy Frogs of Genus *Boophis*	[c]Blommers-Schlösser, 1979a	*B. rappiodes, B. luteus, B. madagascariensis, B. hillenii, B. williamsii* (all illustrated); and *B. mandraka, B. erythrodactylus, B. spp. unknown, B. majori, B. gondoti, B. understeini, B. tephraeomystax, B. idae, B. granulosus,* and *B. microtympanum*.
Theloderma stellatum	[c]Wassersug et al., 1981	Described larva at St.28, from Thailand

Family Ceratophrydidae

Odontophrynus carvalhoi	[c]Carmaschi, 1979	St.36 and 43 larvae described: after Limbaugh and Volpe (1957) and Gosner (1960)

Family Hyperoliidae

Heterixalus betsileo and *H. tricolor*	[c]Blommers, Schlösser, 1982	Described larvae of *H. betsileo* (St. 25–41) and *H. tricolor* (St. 25–40) of Madagascar
Family Polypedatidae		
Gephyromantis methueni	Razarihelisoa, 1974	Described and illustrated larvae and their growth
Gephyromantis liber	[c]Blommers-Schlösser, 1975b	Described eggs, embryos, and larvae (no external gills) to end of metamorphosis

SUBORDER PROCOELA

Family Bufonidae (toads and their allies)

Ansonia muelleri and *A. penangenis*	Inger, 1954	Information given on larvae

Species	Reference	Description
(=*Bufo muelleri* and *B. penangenis*) *Bufo andersonii*	[a]Shivpal and Niazi, 1979	St.I–XXII, from hatching to the end of metamorphosis (30–32°C). From Jaipur
Bufo arenarum	[b]del Conte and Sirlin, 1952	St.1–15 according to *R. pipiens* (Shumway, 1940): 11 d development to completion of operculum
Bufo bufo	[a]Cambar and Gipouloux, 1956	Complete series, St.1–48: 47–48 d (20°C).
	[b]Michniewska-Predygier and Pigon, 1957	St.1–25: egg to completion of operculum
	[a]Rossi, 1958	St.1–25 (embryonic; zygote to disappearance of external gills), St.I–XV (larval; hindlimb bud stage to end of metamorphosis). Development 66 d (18–19°C)
Bufo bufo japonicus	[a]Tahara and Ichikawa, 1965	St.1–45: complete series. In Japanese; no English summary
Bufo cavifrons	[b]Korky and Webb, 1973	St.33–42 according to Limbaugh and Volpe (1957) for *Bufo valliceps*
Bufo coccifer	[c]McDiarmid and Foster, 1981	Described breeding behavior and larvae in Costa Rica. The life cycle to the end of metamorphosis is 5 wks.
Bufo cognatus	Bragg, 1936	General information
Bufo mauritanicus	[a]Siboulet, 1970	St.1–47: from 1st cleavage to completion of metamorphosis (comparisons from 17–26°C)
Bufo melanostictus	[a]Khan, 1965	St.1–43: egg to completion of metamorphosis (29–32°C)
	[a]Sit and Kanagasuntheram, 1973	Stages according to schemes of Shumway (1940) and Taylor and Kollros (1946)
Bufo pardalis	Channing, 1973	Four tadpoles 25–27 mm long (= NF St.52 of *Xenopus*)

(continued)

8. Tables of Staged Specimens (continued)

Bufo pentoni	Forge and Barbault, 1977	In Senegal: hatching 30 h, larval development 9–11 d in the field
Bufo pusillus	[b]Channing, 1972	Six tadpoles 14–17 mm long (= NF St.52 of *Xenopus*)
Bufo quercicus	[b]Volpe and Dobie, 1959	Stages according to Limbaugh and Volpe (1957) for *Bufo valliceps*: general information
Bufo regularis	Lamotte and Zuber-Vogeli, 1954	Described a few larval stages, 12–23 mm and metamorphosed froglet 11 mm long
	[a]Sedra and Michael, 1961	St.1–66; egg to completion of metamorphosis
Bufo rubropunctatus	[c]Formas and Pugin, 1978	General information
Bufo valliceps	[a]Limbaugh and Volpe, 1957	St.1–46; complete series; also statistics of larval growth, and development of mouth
Bufo vertebralis hoeschi	Channing, 1976b	General information on eggs and larvae
Bufo vulgaris	[b]Adler, 1901	16 Stages: St.I–VII (embryogeny from gastrula inside the egg membranes); St.IX–XII (larval development outside egg membranes); St.XIII–XVI (metamorphosis)
	[b]Schreiber, 1937	8 Stages: St.1 (premetamorphosis), St.2–8 (metamorphosis)
Bufo woodhousei: fowleri	[b]Gosner and Black, 1954	Information on embryos and larvae. Staging after Rugh (1951)
Nectophrynoides malcolmi	[c]Wake, 1980	Described and illustrated the free-living, nonfeeding (till metamorphosed) tadpoles: 6 stages from 12–33 d post-laying (equivalent to Gosner's St.24–44 approx.)
Nectophrynoides occidentalis	[a]Lamotte and Xavier, 1972a	8 Stages: general information
	Lamotte and Lescure, 1977	General information on larvae including illustrations of those of the following spp. *Acanthixalus spinosus*; *Adenomera hylaedactyla*; *Arthroleptis poecilonotus*; *Chiromantis rufescens*; *Discodeles opisthodon*; *Eleutherodactylus johnstoni*; *Hyperolius zonatus*; *Myersiella microps*; *Nectophrynoides viviparus*; *Phyllobates femoralis*; *Phyllomedusa tomoptera*; *Pipa parva*; *Platymantis hazelae*; *Rhacophorus reticulatus*

Species	Reference	
Nectophrynoides tornieri	Angel and Lamotte, 1944	St.1 (larva, hindlimb buds) → St.4 (froglets). In NW Africa, near Liberia
	Lamotte and Xavier, 1972b	General information
Stephopaedes gen-nov.	bOrton, 1949	St.1–5. Descriptions of larval anatomy.
	cChanning, 1978	Zimbabwe. 9 tadpoles and 2 metamorphosing young: larvae 14–22 mm long (St.21–38 of Gosner, 1960)
Family Hylidae (tree frogs, usually very small)		
Aplastodiscus perviridis	cCaramaschi et al., 1980	Stages 38 (illustrated), 40–42, and metamorphosed froglet described. Stages after Limbaugh and Volpe (1957) and Gosner (1960)
Gastrotheca christiani	Barrio, 1976	6–10 Eggs in female pouch. Incubation, 60 d. Larva with umbrella-like external gills illustrated
Gastrotheca riobambae	Fitzgerald et al., 1979	Information on breeding and the tadpoles
G. riobambae (and other hylid spp.) (*G. argenteovirons*; *Flectonotus pygmaeus*)	bdel Pino and Escobar, 1981	St.1–25 for incubation periods of embryos in the pouch. Series can be used for embryos of other egg-breeding hylids
	cDuellman and Maness, 1980	Information on breeding and the tadpoles
Hyla avivoca	aVolpe et al., 1961	St.1–46; complete series according to Limbaugh and Volpe (1957) for *Bufo valliceps*
Hyla brunnea	bSchreckenberg, 1956	St.I–IV (late embryo to end of premetamorphosis; hindlimb 3 mm long), St.V–VII (metamorphosis and climax), St.VIII (frog with tiny tail stub)
Hyla burrowski	cMartin, 1967b	General information
Hyla californiae	cGaudin, 1965	General information
Hyla carnifex	cDuellman, 1969	Described tadpoles (St.25–41) and illustrated a typical St.38 and mouth parts
Hyla carvalhoi	cPeixoto, 1981	Described and illustrated larvae (from Rio de Janeiro and their mouth parts; related to *H. circumdata*

(continued)

8. Tables of Staged Specimens (*continued*)

Species	Reference	Description
Hyla cinereus	Garton and Brandon, 1975	Hatching 2–3 d (larva 4.5–9.5 mm long); largest tadpole 60 mm long; then 4–6 wk to completion of metamorphosis
Hyla claresignata	Lutz and Orton, 1946	Tadpoles through metamorphosis and some organogeny. Rio de Janeiro
Hyla crucifer	°Gosner and Rossman, 1960	Various eggs and larvae to metamorphosis
Hyla elkejungingerae	°Henle, 1981	Described St.25–45 from Peru: illustr. aSt.33 larva and mouth parts
Hyla fimbrimembra	°Savage, 1980b	Described and illustrated a tadpole at St.38 from Costa Rica
Hyla jervisiensis	°Martin and Littlejohn, 1966	Various larvae between St.18–40
Hyla lesureri	°Martin et al., 1966	General information
Hyla ocularis	°Gosner and Rossman, 1960	Various eggs and larvae to metamorphosis
Hyla regilla	bEakin, 1947	St.1–24; larva has no hindlimb buds; according to Pollister and Moore (1937) and Shumway (1940) (see also illustrations in Rugh (1948))
Various Hylidae (from Melbourne, Australia)	°Gaudin, 1965	General information
	°Brown, 1975c	Embryonic temperature adaptation
	°Martin, 1965	Illustrated *H. aurea raniformis*, *H. ewingi*, and *H. verreauxi*; according to stages of Gosner (1960) and Limbaugh and Volpe (1957)
Litoria chloris, *L. fallax* and *L. gracilenta*	°Watson and Martin, 1979	Described and illustrated (St.20, 27–29) embryos, larvae, and mouth parts. These spp. show typical ontogenetic patterns of Australian hylids. The reference list is useful for this group
Litoria verreauxi	°Anstis, 1976	General information
Phrynohyas venulosa	°Zweifel, 1964a	Described and illustrated larvae to metamorphosis and mouth parts: Stages acc. to Pollister and Moore (1937), Limbaugh and Volpe (1957), and Gosner (1960)
Phyllomedusa trinitatus	bKenny, 1968	Blastrula → larva with maximum external gill development; no limb buds. 10–13 mm long, 184 h
Pseudacris brachyphona	Bayard-Green, 1938, 1952, 1964	Descriptions of eggs and tadpoles
Pseudacris clarkii	bEaton and Imagawa, 1948	Illustrated 23 stages from egg and compared developmental age but not stages, with that of *R. pipiens* (Shumway, 1940)

Family Leptodactylidae (frogs with only slight webbing between toes)

Species	Reference	Description
Batrachyla antartandica	[c]Formas, 1976	St.30–33; from Valdivia, Chile
Batrachyla leptopus	[c]Formas, 1976	St.27–31; from Valdivia, Chile
Batrachyla taenata	[c]Formas, 1976	St.27–32; from Valdivia, Chile
Calyptocephalella caudiverbera = *C. gayi* (Jorquera and Izquierdo, 1964)	[c]Jorquera and Pugin,	St.1–14; egg to larva with operculum, no hindlimb buds
Crinia signifera, C. victoriana	[c]Martin, 1965	St.28–34 of Limbaugh and Volpe (1957)
C. tasmaniensis	[c]Martin, 1967b	General information on larvae
Cyclorana australis	Tyler and Martin, 1975	General information on larvae
Cyclorana cryptotis	[c]Tyler et al., 1982	Information on embryos and larvae. Larval life span 24 d. Morphology identical with *C. cultripes*
Eleutherodactylus augusti latrans	Valette and Jameson, 1961	General embryology
Eleutherodactylus guentheri	Lynn and Lutz, 1946	Series of embryos to hatching and fully metamorphosed froglet. Also some internal organogeny
Eleutherodactylus jasperi	Wake, 1978	Small oviducal froglets
Eleutherodactylus nubicola	Lynn, 1942	Classic study of organogenesis. Compared prehatching stages with those of Sampson (1904)
(=*Hylodes martinicensis*)	[a]Sampson, 1904	Two-celled St. → blastula (St.I) → hatching (St.XIV). St.XV (tail reduced to small knob, specimen otherwise same as previous stage. Also internal organogeny
Eleutherodactylus ricordii	Hughes, 1959	Embryology, especially of spinal cord. Stages in days before specimens in eggs hatched (after usage by Lynn (1942) for *E. nubicola*
Eupsophus vittatus	[c]Formas and Pugin, 1978	General information
Heleioporus eyrei	[c]Packer, 1966	Described and illustrated stages 1–40 (but not climax) of Gosner (1960), from area of Perth, W. Australia

(continued)

8. Tables of Staged Specimens (*continued*)

Hylorina sylvatica	cFormas and Pugin, 1978	General information
Lepidobatrachus asper	Cei, 1968	Various developmental stages briefly described. Well formed tadpole illustrated
Lepidobatrachus llanensis	Cei, 1968	Same as *L. asper.*
Leptodactylus albilabris	Dent, 1956	Neurulae to end of metamorphosis. Related them to Taylor and Kollros stages
Leptodactylus melanotus	Orton, 1951	General information
Leptodactylus pentadactylus	cHeyer, 1979	Larval characteristics (teeth formula, maximum larval length, eye diameter relative to body length, etc.). Described five of the 11 spp. represented; *Leptodactylus flavipictus, L. labyrinthicus, L. pentadactylus, L. rhodonotus, L. rugosus*
LEPTODACTYLID FROGS AND LARVAE (AUSTRALIA)	cWatson and Martin, 1973	13 Genera, St.30–38
	Tyler and Martin, 1975	General information
	Martin, 1967a	Various species; general information St.30–36
	cMartin et al., 1966	
Pleurodema cinerea	cHulse, 1979	Eggs laid in foam nests. Hatching 36–48 h; maximum larval size, 35–40 mm; fully metamorphosed 31–36 d
Proceratophrys boiei	cIzecksohn et al., 1979	Staged after Gosner (1960) and Limbaugh and Volpe (1957). Stages 34, 40, and 43 described and illustrated
Proceratophrys laticeps	cPeixoto et al., 1981	Described and illus. larvae and mouth parts of St.36 larva
Ranidella bilingua	cMartin et al., 1980	New spp. larval mouth parts described. Development of larva in field rapid (13–14 d)

TADPOLES OF NORTH AMERICA	Wright, 1929	Described general information of 38 spp. of nine genera of mature tadpoles, mainly E. and S. USA: mouth parts and larvae (poorly) illustrated. *Acris, Ascaphus, Bufo, Gastrophryne, Hyla, Hypopachus, Pseudacris, Rana, Scaphiophus*
TADPOLES OF MIDDLE AMERICA	Starrett, 1960	Described and illustrated (with mouth parts) tadpoles of the Families Rhinophrynidae (*Rhinophrynus dorsalis*); Leptodactylidae (*Engystomops pustulosus*); Dendrobatidae (*Dendrobates pumilio*); Centrolenidae (*Centrolene prosoblepon, Cochranella fleischmanni, C. reticulata, C. granulosa, Teratohyla spinosa*); Hylidae (*H. elaeochroa, H. moraviensis. H. pseudopuma, H. rivularis, H. uranochroa, H. zeteki, Hylella sumichrasti, Phyllomedusa helenae; Anotheca coronata*); Ranidae (*Rana warschewitschii*)
TADPOLES OF CONTINENTAL USA AND CANADA	[c]Altig, 1970	Of the 72 spp. of anuran tadpoles in the USA and Canada then known many are identified by the key provided. Gosner stages (St.25–40) are used. Useful bibliography
TADPOLES OF NORTH CAROLINA, USA	Travis, 1981	Key to identify tadpoles: Useful throughout SE USA
TADPOLES OF COSTA RICA	[c]Savage, 1980a	Described larvae of Costa Rican frogs and toads and provided a synoptic morphological key of the 61 known tadpoles of the 87 species
TADPOLES OF SOUTH AFRICA	Power, 1926	Described structure and life histories of uncommon *Bufo carens* (Bufonidae), *Phrynomantis bifasciata* (Brevicipitidae), and *Kassina senegalensis* (Ranidae)
FROGS OF S.W. AFRICA	Channing and van Dijk, 1976	Described eggs, larvae and adults. Keys for recognition of larvae and adults from Pipidae, Microhylidae, Bufonidae, and Ranidae
ANURANS OF ETOSHI NATIONAL PARK, SW AFRICA	Jurgens, 1979	Brief, varied information given on eggs, larvae, and breeding of: *Bufo garmani, B. vertebralis hoeschi* (Procoela, Bufonidae): *Breviceps adspersus adspersus, Phrynosoma bifasciatus bifasciatus, P. annectens* (Diplasicoela, Microhylidae): *Pyxicephalus adspersus, Tomopterna delalandei cryptotis, Hildebrandtia ornata* (one specimen only), *Cacosternum boettgeri, Kassina senegalensis* (Diplasicoela, Ranidae)

(continued)

8. Tables of Staged Specimens (*continued*)

AMPHIBIANS OF RUSSIA (USSR)	Terent'ev and Chernov, 1965	Information given on breeding, eggs, and larvae of: *Bombina bombina, B. variegata, B. orientalis, Pelobates fuscus, Pelodytes caucasicus, Bufo viridis, B. raddei, B. calamita, B. bufo, Hyla arborea, H. japonica, Rana ridibunda, R. esculenta, R. nigromaculata, R. terrestris, R. chensinensis, R. temporaria, R. dalmatina*
ANURANS FROM INDIA	Annandale, 1918a,b; Annandale and Rao, 1918; Rao, 1918	Described and in some cases illustrated larvae and mouth parts: Other species were referenced: *Nyctobatrachus pygmaeus; Oxyglossus lima; Rana beddomei; R. breviceps; R. brevipalmata; R. cancrivora; R. crassa; R. cyanophlyctus; R. hexadactyla; R. leptodactyla; R. limnocharis; R. semipalmata; R. sternosignata; R. tigrina; R. tyleri; R. verrucosa* (Diplasicoela: Ranidae): *Rhacophorus maculatus; R. malabaricus* (ref.) (Diplasicoela: Rhacophoridae): *Bufo fergusoni; B. melanostictus; B. microtympanum; B. stomaticus; B. viridis* (Displasicoela: Bufonidae): *Microhyla achatina; M. berdmorei; M. ornata; M. pulchra* (ref.); *M. rubra* (ref.) (Diplasicoela: Microhylidae): *Cacopus systoma* (ref.); *Kaloula obscura; K. pulchra* (ref.); *K. triangularis; K. variegata* Diplasicoela: Brevicipitidae)
ANURANS FROM THE PHILIPPINE ISLANDS	[b]Alcala, 1962	Described and illustrated the larvae and mouth parts of eleven spp. of anurans. Stages were referred to Shumway (1940), Limbaugh and Volpe (1957), and Taylor and Kollros (1946). The *Cornufers*, as their developmental pattern is very different, are divided into six stages: cleavage and blastulation; gastrulation; neurulation; limb bud; hind-limb paddle and operculum; metamorphosis. *Ooeidozyga l. laevis; Rana c. cancrivora; R. e. everetti; R. magna visayanus; R. microdisca leytensis; Cornufer guentheri; C. hazeli; C. mayeri* (Ranidae): *Rhacophorus leucomystax quadrilineatus; R. P. pardalis* (Rhacophoridae): *Kaloula conjuncta negrosensis* (Microhylidae)

AMPHIBIANS OF NORTH BORNEO — Inger, 1956

Of the 26 spp. listed, some significant details are given of larvae of nine of them: *Megophrys hasselti* (Pelobatidae); *Kalophrynus pleurostigma pleurostigma* (Microhylidae); *Chaperina fusca, Rana macrodon macrodon, R. macrodisca palavanensis R. nicobariensis nicobariensis* (Ranidae); *Rhacophorus leucomystax linki* (Taylor and Kollros stages for this), *R. otilophus R. pardalis pardalis* (illust) (Rhacophoridae)

AMPHIBIANS OF WESTERN CHINA — Liu, 1950

In this large treatise among the many spp. listed and described, many of their larvae are detailed and most of them, including mouth parts, are illustrated. These include: *Bombina maxima* (Discoglossidae); *Aelurophryne mammata, A. breviceps, A. glandulata, Scutiger pingii, S. rugosa, S. popei, S. schmidti, S. spp.* (Mt. Omei), *S. spp.* (Lungtang), *Megophrys minor, M. oshanensis* (Pelobatidae); *Bufo raddei B. tibetanus, B. bufo wrighti, B. bufo gargarizans* (Bufonidae); *Hyla annectans* (Hylidae); *Kaloula rugifera, K. macroptica, Microhyla ornata* (Microhylidae); *Rana adenopleura, R. boulengeri, R. phrynoides, R. temporaria chensinensis, R. japonica, R. chaochiaoensis, R. andersonii, R. margaretae, R. nigromaculata, R. limnocharis, R. guentheri* (Ranidae); *Nanorana pleskei* Gunther (= *Rana pleskei* Boulenger); *Staurois mantzorum, S. chunganensis, S. lifanensis, S. kangtingensis* (Ranidae); *Rhacophorus leucomystax, R. bambusicola, R. omeimontis* (Rhacophoridae)

Order Caudata = Urodela
SUBORDER CRYPTOBRANCHOIDA
Family Hynobiidae (land salamanders from Asia) — Liu, 1950

Information given on larvae

Batrachuperus karlschmidti, B. pinchonii

Hynobius naevius — Oyama, 1929

Hynobius nigrescens — [a]Usui and Hamasaki 1939

Brief information on eggs and larvae St.1–66 (egg to metamorphosed form). First anterior limb bud (St.39), postlimb bud (St.46). Balancers (St.37–46). In Japanese; no English summary

(continued)

8. Tables of Staged Specimens (*continued*)

Hynobius shihi		
Onychodactylus japonicus	Liu, 1950	Information given on larvae
	[a]Iwasawa and Kera, 1980	St.1–72, egg to the end of metamorphosis (2 yr). Illustrated (In Japanese, with tables and summary in English)
Family Cryptobranchidae (water dwelling forms)		
Cryptobranchus alleghaniensis	[b]Smith, 1912	Described eggs and oviposition: St.1–23 (to hatching in 6 wk). Larvae at 2 yr (80–95 mm long); postlarval forms. Illustrated prehatching stages and larvae
Megalobatrachus japonicus	[a]Kudo, 1938	St.1–33: Complete series illustrated. Detailed organogeny described and there is a comprehensive large bibliography from 1782
	[a]Iwama, 1968	St.1–59: Complete series of stages from egg to metamorphosed larva (150–160 mm long). In Japanese; no English summary
SUBORDER AMBYSTOMOIDEA (aquatic or land forms)		
Family Ambystomidae		
Ambystoma gracile	Snyder, 1956	Described larvae at low and high elevations. Metamorphosis mainly after 1 yr at sea level. T_4 immersion 1:500,000 induces metamorphosis (70°F)
Ambystoma gracile	[b]Brown, 1976	Stages according to Harrison
Ambystoma jeffersonianum	Grant, 1930b	Information on larvae at metamorphosis
Ambystoma maculatum	[b]Hamburger, 1950	Harrison St.1–46; egg to larva with three digits to forelimb; three well-developed external gills, balancers disappeared
	[b]Leavitt, 1948	Redrawn series of Harrison (see Rugh, 1948)
	[b]Hara and Boterenbrood, 1977	Harrison St.6–10 refined
Ambystoma mexicanum	Marx, 1935	Stages I–VIII through climax with loss of external gills
	[b]Schreckenberg and Jacobson, 1975	St.1–40; egg to larva at hatching, with first indication of forelimb bud
Ambystoma mexicanum	[b]Bordzilovskaya and Dettlaff, 1979	Stages described of normal development

Ambystoma opacum	Grant, 1930b	Information on larvae at metamorphosis
Ambystoma ordinarium	Anderson and Worthington, 1971	General information on larvae; metamorphosis when 60–70 mm long
Ambystoma punctatum =*A. maculatum*	Grant, 1930a	Information on larvae at metamorphosis, based on skin-shedding and other external features
	[b]Dempster, 1933	Stages after Harrison. Information on larvae and growth rates, *vide supra* (see *A. maculatum*)
	[b]Leavitt, 1948	
Ambystoma rosaceum	Anderson, 1961	General information on larvae
	Anderson and Webb, 1978	Compared larvae; *A. rosaceum* with *A. ordinarium*
Ambystoma talpoideum	Patterson, 1978	Information on larvae in populations from S. Carolina. Metamorphosis in specimens > 25 mm (SVL)
Dicamptodon ensatus	Nussbaum and Clothier, 1973	Data on population structure growth and size of larvae
Rhyacotriton olympicus	Nussbaum, 1969	Information on eggs and larvae
	Worthington and Wake, 1971	Larvae 17–45 mm (SV length); organogeny, especially skeleton, skin, and teeth

SUBORDER SALAMANDROIDEA (typical salamanders and newts; aquatic and terrestrial forms)

Family Salamandridae		
Diemictylus pyrrhogaster	[b]Oyama, 1930	Egg to larva, St. 1–42 (equivalent to reach H. St. 40–42 *A. maculatum* of Harrison. In Japanese; no English summary
Diemictylus viridescens	Gage, 1891	Classical account with bibliography; describes eggs, larvae, metamorphosed forms, 2½–3-yr-old red efts and newly returned aquatic forms. No Stages
	Jordan, 1893	Early description on embryogeny from egg to the neurula. Illustrated; no staging
Euproctus asper	[b]Gasser, 1964	Staged after Gallien and Durocher (1957) and Gallien and Bidaud (1959). Larval development to feeding; no balancers
Notophthalmus viridescens	Pope, 1924	Described life histories of larvae and adults
	Wilder, 1925	Illustrated a few developmental stages

(continued)

8. Tables of Staged Specimens (*continued*)

Pleurodeles waltl	aGallien and Durocher, 1957	St.1–56; egg to fully metamorphosed larva, 110 ± 10 d, 72 ± 10 mm long (18°C)
Salamandra atra	Wunderer, 1910	139 Illustrations of egg to fully transformed specimen (46 mm long). Detailed descriptions; no stages
Salamandra salamandra	bJuszczyk and Zakszewski, 1981	External morphology through metamorphosis: (St.I–III a and b)
Triturus alpestris	bKnght, 1938	St.1–30 (egg to larva with bifid forelimbs); equivalent to H. St.1–41 and Gläesner St.0–38
Triturus alpestris	bEpperlein and Junginger, 1982	St.1–37; (to feeding); equivalent to H. St.45 and St.41 (Gläesner, 1925) (illustrated)
Triturus helveticus (=*T. palmatus*, Schneider)	aGallien and Bidaud, 1959	St.1–56: complete series from egg to the end of metamorphosis
Triturus pyrrhogaster	bAnderson, 1943	St.I–XXV; equivalent to H. St.1–46
	aOkada and Ichikawa, 1947	St.1–60: egg to fully metamorphosed form. In Japanese; no English summary
Triturus taeniatus	bGlucksohn, 1931	St.36–62 (metamorphosis) from first forelimb buds to metamorphosis; equivalent to H. St.36–46 and Gläesner St.32–52/53. Stages of limb development
Triturus torosus	bRotman, 1940	St.1–36 according to Gläesner St.1–31/32; Harrison St.1–36 (= Glucksohn St.36); to stage when incipient forelimb bud first appears
	bTwitty and Bodenstein, 1948 (in Rugh, 1948)	St.1–40; to forelimb palette stage. Similar to Harrison stages
Triturus viridescens (=*Notophthalmus viridescens* Dent, 1968)	Grant, 1930a Dent, 1968	General information on larvae at metamorphosis based on skin shedding and other external features
Triturus (Molge) vulgaris (=*T. taeniatus*)	aGläesner, 1925	St.1–56: Detailed series from zygote to the end of metamorphosis. Tables of organogeny and comprehensive reference list 1894–1924
Tylotriton andersoni	aUtsunomiya and Utsunomiya, 1977	St.1–60: egg to metamorphosed form without external gills. Compared with *T. pyrrhogaster* (Okada and Ichikawa, 1947). In Japanese; English summary

Species	Reference	
Tylotriton verrucosus	Smith, 1924	Described and illustrated larvae
	[a]Ferrier, 1974	St.1–56 according to Gallien and Durocher (1957) for *Pleurodeles waltl*
Family Plethodontidae (majority of American urodeles; lungless, brook-dwelling, or terrestrial forms)		
Batrachoseps (= *Plethopsis wrighti*)	Stebbins, 1949	Hatching 133 days after laying (12°C): illustrated larvae and hatching process
Desmognathus fusca	Wilder, 1913	Described eggs, hatching, metamorphosis, and organogeny
Eurycea bislineata	Wilder, 1924	Statistical analysis of growth, development, and transformation (begins after 2 yr)
	Wilder, 1925	Organogeny of postembryonic, larval, premetamorphic, and metamorphic stages
Eurycea bislineata	Bruce, 1982	Information on growth and metamorphosis of the larva; comparison with *E. pongicaudor guttolineata* in N. Carolina
Eurycea longicauda longicauda	Anderson and Martino, 1966	General information; 18–42 mm larvae and 51 mm form recently metamorphosed
Eurycea longicauda guttolineata	Bruce, 1970	General information on larvae
Eurycea longicauda melanopleura	Ireland, 1974	Described features of larval development
Eurycea lucifuga	Clergue-Gazeau and Thorn, 1976	General information on larvae
Eurycea multiplicata griseogaster	Ireland, 1976	Described features of larval development
Eurycea neotenes	Sweet, 1977	Information on larvae and pedogenic forms
Plethodon cinereus	[b]Dent, 1942	St.I–XXIV: egg to hatching (resorption of external gills). Direct development (no aquatic stage, open gill slits, Leydig cells, or neuromast organs). Some stages illustrated
Plethodon glutinosus	Highton, 1956	Information on egg laying, hatching, and larval development
Plethodon vehiculum	Peacock and Nussbaum, 1973	Information on larval development: comparison with other spp. of *Plethodon*
Pseudotriton montanus	Bruce, 1974	General information on larvae

(continued)

8. Tables of Staged Specimens (*continued*)

Pseudotriton montanus floridanus	Goin, 1947	Information on eggs and larvae (12–13.5 mm long)
Pseudotriton ruber	Bruce, 1972, 1974	General information on larvae
Pseudotriton (=Spelerpes) ruber	Dunn, 1915	General information on larvae
Pseudotriton striatus axanthus	Goin, 1947	Information on eggs and larvae (14.5–16 mm long)
Stereochilus marginatus	Bruce, 1971	General information on larvae

SUBORDER PROTEIDA (neotenous forms)

Family Proteidae

Necturus lewisi	Ashton and Braswell, 1979	Hatchling (newly hatched; 20.7 mm long; av. 22.8 mm) and post-hatchling (65 mm long) larvae described and illustrated
Necturus maculatus	[a]Eycleshymer and Wilson, 1910	St.1–35: Complete series from egg to metamorphosed specimen (126 d old, 39 mm long). Illustrated. Detailed organogeny described and there is a large bibliography from 1766 to 1909
Proteus anguinus	Durand, 1971	Describes and illustrates embryos and larvae. Blastula (4–5 mm diameter) → juvenile (30–40 mm long). Hatching at 19 mm; forelimb blastema at 17–18 mm, hindlimb, 20–21 mm. Range 15–120 days

SUBORDER MEANTES (neotenous forms)

Family Sirenidae

Siren lacertina	Goin, 1947	Described larva 16 mm long
	Ultsch, 1973	Brief information on egg laying
AMPHIBIANS OF RUSSIA (USSR)	Terent'ev and Chernov, 1965	Information given on breeding, eggs, and larvae of: *Hynobius keyserlingii, Ranodon sibiricus, Onychodactylus fischeri, Triturus vulgaris, T. montandoni, T. alpestris, T. vittatus, T. cristatus, Mertensiella caucasica, Salamandra salamandra*

[a]Completely staged from egg to fully metamorphosed individual.
[b]Incompletely staged.
[c]Random stage larvae, referred to Gosner's series (1960).

Section III

Organic Origin, Development and Change in Anuran Larvae to the Completion of Metamorphosis

9. Introduction

The preceding account emphasises that in order to prepare for terrestrial life amphibian larvae, especially of anurans, undergo a complex metamorphosis that embraces elaborate morphological, biochemical, physiological, and behavioral changes. During premetamorphosis there is no real effective level of hormonal secretion in the circulation (Myauchi et al., 1977) and thyroid activity is minimal (Etkin and Gona, 1974). From the end of premetamorphosis (at the hindlimb bud stage) through prometamorphosis and climax (Etkin, 1964, 1970) larval development is under endocrinological control of high complexity (Dodd and Dodd, 1976). Thyroxine and triiodothyronine from the thyroid, growth promotion hormone(s) from the pituitary, and possibly glucosteroids of the interrenals, act directly (or indirectly in some way) on reactive larval tissues. These hormones are influenced in their synthesis and release by thyroid stimulating hormone(s) (TSH) and adrenocorticotropic hormone(s) (ACTH) of the pituitary, and by the hypothalamus, which secretes a thyroid releasing factor (TRF) controlling the activity of TSH and possibly the growth promoting hormone(s) of the pituitary (Etkin and Lehrer, 1960). Indeed extirpation of the hypothalamus wholly inhibits metamorphosis (Voitkevitch, 1962; Dodd and Dodd, 1976). These hormones operate in a *milieu* influenced by light, temperature, and various other factors such as diet, iodine, or larval crowding in the ambient water (see Kaltenbach, 1968; Dent, 1968). Whether ACTH is essential (Rémy and Bounhiol, 1971) for complete metamorphosis is not clear. The relationship and influence of ovine somatotropin (Enemar et al., 1968) and thyroid and prolactin hromones (see Etkin and Gona, 1967; 1974; Brown and Frye, 1969; Blatt et al., 1969; Medda and Frieden, 1970, 1972; Jaffe and Geschwind, 1974a,b; Dodd and Dodd, 1976) on larval growth and metamorphosis have been extensively investigated. To some extent it appears that their actions in *Xenopus* can be separated into growth or enlargement (prolactin or somatotropin) and new protein synthesis and tissue differentiation (thyroxine) (see Turner, 1973).

The progressive increase in the rate of synthesis and release of thyroid hormones raises their circulatory level and as a result elicits metamorphic changes. It is likely, however, that the measured level is lower (Frieden and Just, 1970; Just, 1972) than heretofore believed (Etkin, 1970). The lower increased levels of hormone obtained during later stages of metamorphosis are explained by the increased propensity of larval tissues avidly to utilize the ever-increasing amount of thyroid hormones produced by the thyroid (Ashley and Frieden, 1972). Contrasting theories of endocrinological functions and their relationship to metamorphosis are discussed by Dodd and Dodd (1976).

The overt expression of anuran larval hormonal activity is morphological change. Probably all larval organs are influenced to a greater or lesser extent, either directly or indirectly, some to completely disappear and others to be partially or wholly remodeled. A work of this kind cannot adequately deal with the wealth of information, especially on ultrastructure, of all larval tissues during metamorphosis. The thymus, for example, is best treated within the framework of immunology (see review by Cooper, 1976). Its origin, development, and ultrastructure through metamorphosis have recently been described (see Nagata, 1976, 1977; Henry and Charlemagne, 1981). The scheme in Section III is arbitrary, in part determined by the amount and reliability of previous information, but also a result of the author's personal involvement with specific tissues. In general more consideration is given to organs receiving thyroid hormones than to those directly (or indirectly) assisting in the manufacture of them.

Nevertheless, in order to function properly during metamorphosis, the endocrine organs need to grow and differentiate. The thyroid in particular is required to synthesize and release its hormones fairly early in larval development, so that prometamorphosis can be initiated. In relation to the various changes that occur in tissues during larval life through metamorphosis, it would seem appropriate, therefore, to commence the account with the endocrines, and thence subsequently deal with other morphological systems, whose growth, differentiation, and function are dependent upon them.

10. The Thyroid Gland

The organo-morphological changes that occur during larval metamorphosis are closely related to the functional changes in cellular activity required for terrestrial existence. Morphological development of a larva, certainly from prometamorphosis onwards in anurans, is elicited at the primary level by the endocrines, which likewise grow and differentiate during metamorphosis. Some endocrinal glandular development, however, could well be short-term, necessary to regulate the metamorphic cycle. Thus the thyroid, for example, shows features of reduction in the size of some of its cellular components at the end of climax, when the circulatory thyroid hormone level is reduced (Regard, 1978). Indeed the recognizably low levels of circulatory T_3 and T_4 in premetamorphosis tadpoles and in adult anurans suggests that thyroid hormones may well be important only during the period of metamorphosis (Regard et al., 1978).

Nevertheless, during larval life the end product of the various and complex endocrinological interrelationships is the synthesis, release, and maintenance in the circulation of the required level of thyroid (and other) hormones, regulated by tissue demand and the hormonal concentration feedback on the hypothalamo-hypophysial axis.

Origin and Development

The paired thyroid glands of larval amphibians are generally similar in appearance to those of other vertebrates (Lynn and Wachowski, 1951). Seen by light microscopy (see Schreckenberg, 1956) they comprise colloid-filled follicles supported by connective tissue. In *Xenopus laevis* a median thyroid Anlage originates at about NF St.33–34, which has split into paired definitive Anlagen by NF St.41. At NF St.46–47 the paired thyroid lobes each consist of a small mass of cells about 20 μm long. Incipient follicles form at NF St.48–49 and at NF St.49–50 there are 13–16 follicles per lobe. Chromophobe droplets first appear before the beginning of premetamorphosis (about NF St.53), at NF St.51 (Turner, 1973). During anuran larval development the thyroid steadily increases in overall size (Schreckenberg, 1956; Fox, 1966; Fox and Turner, 1967; Michael and Adhami, 1974). Thus in *R. temporaria* from CM St.41–53 (equivalent to the stages for *Rana dalmatina* by Cambar and Marrot, 1954), the measurements of the thyroid nuclear population and of its tissue volume increase by about 50 times, the overall volume of the thyroid by 60 times, and the volume of the colloid substance by over 200 times. The calculated individual cell volume (volume of thyroid tissue/nucleus) is about 700–800 μm^3. At CM St.54 (the end of climax) when the thyroid is reduced in size, cell volume is also reduced to about 400 μm^3. Peripheral chromophobe droplets are largest and most plentiful at the height of climax (Fox, 1966). The thyroid glands of *Bombina bombina*, *B. variegata*, *Pelobates fuscus*, *Hyla arborea*, and *Rana temporaria* were investigated qualitatively and quantitatively during larval development by Francois-Krassowska (1978). She found that the maximum activity of the thyroid in terms of volume, secretory surface area of the follicles, and highly vacuolated colloid coincides with the period of the emergence of the forelimbs at climax.

Follicular cell height increases during metamorphosis (Saxen, et al., 1957a,b; Coleman et al., 1968). Though the follicle individual cell volume of *Rana* is relatively constant until the end of climax, cell height, as with the thyroid of the larval *Xenopus*, probably does increase at climax though it is still not reduced by the end of metamorphosis (compare Figs. C and J, Fox and Turner, 1967). Thus presumably follicle cells increase their biosynthetic activity during metamorphosis, among other things, by increase in their number and change in shape.

During the period of prometamorphosis to climax, the thyroids of larvae of *Arthroleptella bicolor villiersi* (the entire life cycle to the froglet is 7–10 d), and of *Bufo angusticeps* (the life cycle to the froglet is 3–4 months), increase in volume by hyperplasia and there is increase in epithelial cell height. Thyroid vascularity becomes more extensive. Colloid accumulates in the follicles of *Bufo,* but not in those of *Arthroleptella,* where it appears to be continuously released during the rapid period of larval development and metamorphosis. Furthermore, in thyroxine-treated *Bufo* larvae thyroid development is repressed (Brink, 1939). Indeed, thyroid size is generally inversely proportional to the concentration of thyroxine in the external fluid medium of the larva, as demonstrated in *Rana* and *Xenopus* by Fox and Turner (1967). Immersion of older larvae in goitrogens such as phenylthiourea and thiourea results in the inhibition of thyroid hormone secretion and the consequent development of enlarged thyroids, which are hyperplasic, with large follicles filled with colloid. Tajima (1977) described the effect of the goitrogen propylthiouracil on the fine structure of the thyroid of the larval urodele *Hynobius nigrescens.* Compared with controls, within 6 months the RER of the follicle cells was highly dilated, electron dense granules and homogeneous vacuoles increased in amount, and the apical microvilli were elongated and more profuse. After treatment of goitrogen for 9 months these organelles were reduced in number and amounts. The 'goitrous' enlarged thyroids of *Rana* and *Xenopus,* obtained under such experimental conditions, were interpreted in terms of 'feedback' control mechanisms, involving the reduced thyroxine circulatory level and hence the raised hypothalamo-hypophysial endocrinal secretory stimulation of the thyroid (Fox and Turner, 1967).

Fine Structure of the Thyroid

The ultrastructure of the early developing, premetamorphic thyroid gland of *Xenopus* has been described by Jayatilaka (1978). After the first recognition of two longitudinal strands of the thyroid Anlagen at NF St.40, incipient follicles with colloid vesicles in the cells and in small lumina have appeared by NF St.46–47. At NF St.48–50 well-developed follicles are lined by mucous and cuboid cells that, among other things, contain a RER in stacks, mitochondria, and a Golgi apparatus; colloid vesicles are situated near the lumen and microvilli project into large intercelluar spaces. The follicle lumen is filled with colloid. Where applicable the *Xenopus* NF Stages related to thyroid development generally agree with those of Turner (1973) (*vide supra*). Between the basement membrane and the follicle cells lightly staining cells were described by Jayatilaka, which may be equivalent to the 'light' cells of the rat thyroid (Young and Leblond, 1963). Other cells, similar in appearance, but situated outside the basement membrane of the follicle cells may be related to the true parafollicular cells (Nonidez, 1932) or C cells, which produce thyrocalcitonin in mammals (Busolati and Pearse, 1967). At St.51–52 the follicles contain large osmiophilic membrane-bound cytolysomes and vesicles. The fine structure of the thyroid of *R. japonica* throughout its development has been described by Nanba (1972) (see Figs. 5–7 and 9; Table 1).

The well-developed thyroid cells of the prometamorphic *Xenopus* larva generally appear to be similar to those of *Rana* and of other vertebrates (Coleman et al., 1968; Regard and Mauchamp, 1971; Neuenschwander, 1972; Regard, 1978), and there are increased amounts of RER and of the Golgi complexes, changes presumably concerned with the synthesis and secretion of thyroid hormones (see Regard, 1978). Intracellular colloid droplets increase in number and dense lyosomal-like vacuoles are recognizable (Fig. 10). During late prometamorphosis the serum thyroxine level of *R. pipiens* has increased 10-fold over that of younger larvae, which suggests that at climax there is hydrolysis of colloid droplets and thyroxine release into the circulation (Kaltenbach and Lee, 1977). The lightly-dense, rounded parafollicular or C cells of the mammalian thyroid are only rarely found in the thyroids of

Table 1

Ultrastructural Features of the Developing Thyroid Gland of the *Rana japonica* Larva[a]

Premetamorphosis (TK St.24/25–XI) (approx.)

The paired thyroid Anlagen situated below the pharynx are first recognized as solid masses of immature cells containing pigment, yolk, and lipid droplets. An occasional follicle lumen lined by a few short microvilli initially appears between adjacent cells. There are abundant lysosomes (positive for acid phosphatase) early on in this stage; probably they are responsible for digesting yolk. Midway through this period (at about TK St.VI) follicle units are developed, and by TK St.X there is an increased amount of granular endoplasmic reticulum (RER) and a well developed Golgi apparatus in the follicle cells.

Prometamorphosis (TK St.XI–XX)

The cuboidal or columnar-shaped follicle cells have a well-developed RER and Golgi apparatus and there are numerous apical secretory granules. Synthesized secretory material is stored in the follicle lumen.

Metamorphic climax (TK St.XX–XXV)

Numerous dense lysosomal granules of variable size and secretory granules are present in the apical region of the follicle cells. Presumably the lysosomes are active in the hydrolysis of reabsorbed colloidal thyroglobulin during the formation of thyroid hormones.

[a]Compiled from the work by Nanba (1972). Pre.-prometamorphosis and climax periods after Etkin (1968) and Stage numbers after Taylor and Kollros (1946).

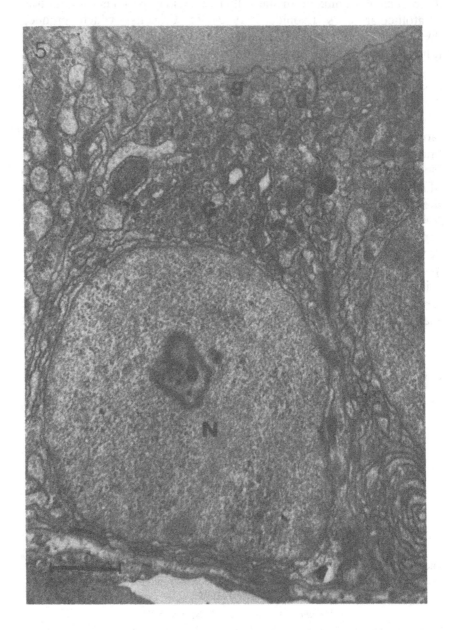

Fig. 5. Thyroid follicular cell of *Rana japonica* larva at TK stage X (near the beginning of prometamorphosis), which is similar in appearance to that in higher vertebrates. Scale mark, 1 μm. Lettering: g, granules; G, Golgi complex; lysosomes (arrows); N, nucleus. There is a well-developed granular endoplasmic reticulum. (After Nanba, 1972.)

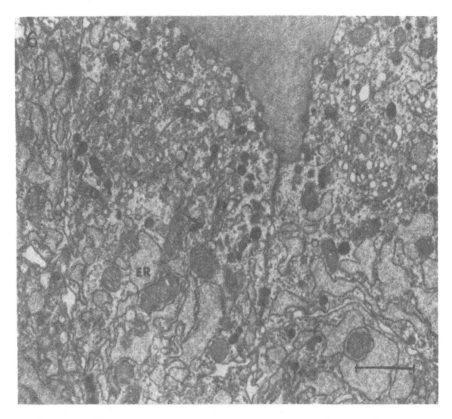

Fig. 6. Follicular cell at stage XX (beginning of climax). There is an enlarged granular endoplasmic reticulum (ER), a well-developed Golgi complex (G), and many granules and lysosomes (arrowed). Scale mark, 2 μm. (After Nanba, 1972.)

metamorphosing or young toads of *Xenopus*. An occasional "parafollicular cell" is located basally in the epithelium and never has a free luminal surface. Their role in *Xenopus* is unknown (Coleman, 1970). Whether such cells are equivalent to either (or both) of the light-type cells reported by Jayatilaka (1978), or to the true mammalian parafollicular C cells, is not clear.

Biosynthetic Activity and Enzymology of the Thyroid

Acid phosphatase has been identified in large lysosomes in the apical regions of the follicle cells of *Xenopus laevis* larvae. The Golgi complex also reacts strongly for this enzyme, which is active in proteolysis during the release of thyroid hormones from colloid (Coleman et al., 1967; Regard and Mauchamp, 1973) (see Fig. 8). Likewise alkaline phosphatase

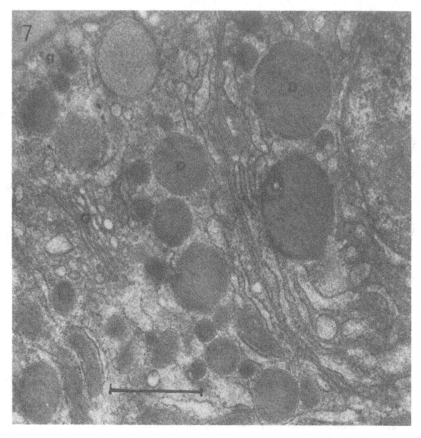

Fig. 7.   Apical region of a follicular cell at TK stage XXIV (end of climax). There are numerous dense droplets (D) and granules (g) and a well-developed Golgi complex (G). Scale mark, 1 μm. (After Nanba, 1972.)

and ATPase are localized in the cell surfaces and intercellular spaces of the follicle cells (Coleman et al., 1967). During anuran metamorphosis as the thyroid becomes more functionally active, there is a parallelism between thyroid endocrinological activity and its ultrastructure, reflected in the high development of the RER and thyroglobulin synthesis, which reach a maximum at climax. During this time variation in peroxidase activity, responsible for iodide oxidation, leads to the incorporation of iodide into the thyroglobulin molecules located in the colloid. The follicle cellular distribution of peroxidase changes during larval life, and its activity is recognized in the nuclear envelope endoplasmic reticulum (ER), Golgi saccules, and small apical vesicles; there is wide distribution of peroxidase during prometamorphosis and climax when it shows maximum activity (Nanba, 1973), and its site of utilization is at the apical microvilli. After metamorphosis peroxidase activity in the thyroid cells is reduced, and it is recognized only in some RER formations and microvesicles. At this time the thyroid cellular RER is reduced and the amount of thyroglobulin synthesis is less. Hypophysectomy of *Xenopus* larvae at the end of prometamorphosis leads to ultrastructural changes in the thyroid cells, which regress to a condition similar to that of premetamorphosis. Only a few RER cisternae remain and within 15 d thyroglobulin is no longer recognizable within the cells, for there is a decrease in thyroglobulin synthesis and of iodide utilization (Regard, 1978.)

The rate of uptake of radioactive iodide by the anuran thyroid is low during premetamorphosis, but this increases to reach a maximum near the beginning of climax (Kaye, 1961; Hanaoka, 1966). Thyroglobulin biosynthesis also increases during prometamorphosis of *Xenopus* (Regard and Mauchamp, 1971). The chemistry of amphibian thyroglobulin is not known exactly, but immunologically it is unrelated to mammalian thyroglobulin, differing from it in several chemical and physicochemical properties (Dodd and Dodd, 1976). Amphibian thyroglobulin appears to synthesize diiodotyrosine (DIT) residues more efficiently than does that of mammals; again thyroglobulin of *Xenopus* differs from other thyroglobulins in its higher T_4 content and lower DIT/T_4 ratio (Sorimachi and Ui, 1974). Monoiodotyrosine (MIT) and DIT are synthesized by the thyroid before thyroxine (Shellabarger and Brown, 1959; Flickinger,

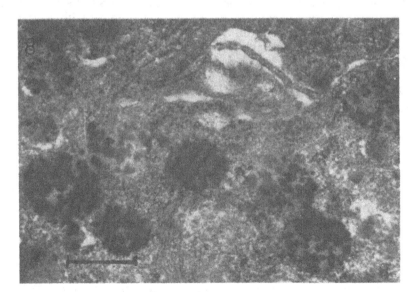

Fig. 8. A well-developed acid phosphatase reaction is obtained in the dense droplets of thyroid follicular cells at TK stage XXIV. Scale mark 0.5 μm. (After Nanba, 1972.)

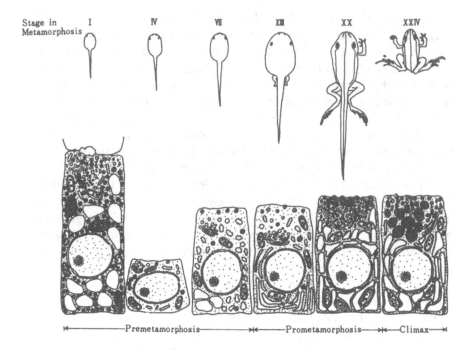

Fig. 9. General scheme of the major changes in ultrastructure of thyroid follicular cells during the larval metamorphic cycle in *Rana japonica*. Premetamorphosis (to TK stage XI) is a period of differentiation of the thyroid cells. Prometamorphosis is a period of synthesis and storage of the thyroid hormones in thyroid cell lumina with a gradual increasing level of release. Climax (TK stages XX–XXIV) is the period of maximum secretion and release of thyroid hormones. (After Nanba, 1972.)

1964) in anuran larvae. Indeed MIT is already present in tail bud stages of *R. pipiens* (Flickinger, 1964), presumably, therefore, before the origin of the thyroid Anlage. The initiation of thyroid hormonal biosynthesis, in relation to the degree of morphogenesis of the thyroid, is claimed by Hanaoka et al. (1973) to occur extremely early in larvae of *Bufo bufo japonicus* and *Xenopus laevis*. Thus small, but significant, amounts of iodothyronines were found to be present at the time of the first origin of the thyroid, when the Anlage was merely a thick lining of the ventral pharyngeal epithelium. Nevertheless, the level of any thyroidol Iodocompounds that may be present in embryos before the thyroid develops and becomes functional must be considered infinitesimal, relative to the level that occurs during the metamorphic cycle. Labeled thyroglobulin appeared in appreciable amounts during premetamorphosis (after NF St.46 of *Xenopus*), when thyroid follicles start to develop. Before NF St.43–44 the main iodotyrosine was MIT and the main iodothyronine was triiodothyronine (T_3). Afterwards DIT and thyroxine (T_4) predominated.

A high proportion of the thyroid hormones T_3 and T_4, however, is always present near and at climax (Regard, 1978; see also Cohen et al., 1978). The chemistry of the thyroid is described in Figs. 11a and 11b.

An active transport system in the thyroid gland concentrates iodide from the plasma which is oxidized probably involving peroxidase enzyme(s). Tyrosine residues in thyroglobulin (a glycoprotein) are iodinated to form protein-bound 3-iodotyrosine (MIT) and 3,5-diiodotyrosine (DIT). Two molecules of DIT or one molecule each of MIT and DIT thence unite to form thyroxine (T_4) or triiodothyronine (T_3) respectively, still bound to thyroglobulin as peptide-linked residues.

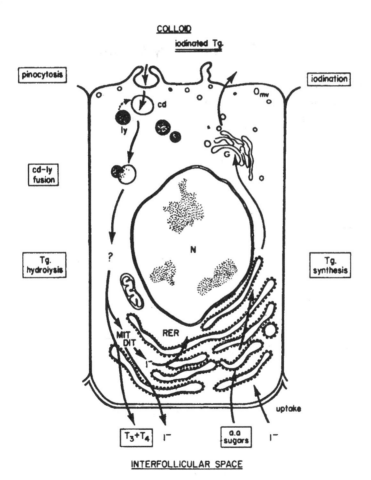

Fig. 10. Specific reactions of anuran thyroid tissue depicted grammatically: aa, amino acid; cd, colloid droplets; G, Golgi complex; I, iodide; ly, lysosome; mv, microvesicle; N. nucleus; RER, granular endoplasmic reticulum; Tg, thyroglobulin; MIT, monoiodotyrosine; DIT, diiodotyrosine; T_3, triiodothyronine; T_4, thyroxine (after Regard, 1978).

Other iodothyronines formed by mixed or homologous condensations of MIT and DIT are 3,3'5'-triiodothyronine (reverse T_3) or 3,3'-diiodothyronine, which have little or no thyroxine-like activity.

T_3 and T_4 are released from thyroglobulin by proteolysis to diffuse into the blood and thence bind reversibly to alpha-globulin and prealbumin, the main thyroid hormone plasma binding proteins (the protein-bound iodine, PBI). Probably only small quantities of free T_3 and T_4 are metabolically active, whose level depends upon the concentration of the PBI and the rate of utilization by the target tissues (tissue avidity). Any free MIT and DIT released at the same time as the T_3 and T_4, are deiodinated by an NADP enzyme located in the microsomes of thyroid cells.

11. The Pituitary Gland

The vertebrate pituitary gland (see Holmes and Ball, 1974), originates during embryogenesis from a downgrowth (infundibulum) of the brain, which joins an upgrowth (hypophysis) from the roof of the mouth. These

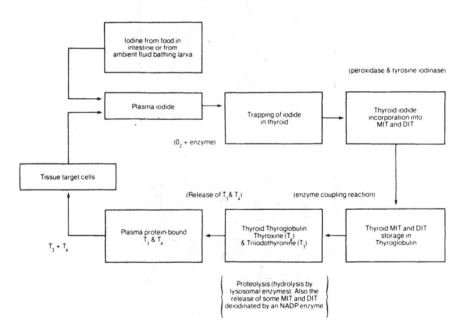

I. Iodide is trapped by the thyroid; an oxygen-dependent process probably enzymic in nature.

Fig. 11A, B. Chemistry of synthesis and liberation of thyroid hormones: mainly from the rat, though probably the mechanism is similar in anurans. (From Rafelson and Binkley, 1968, and Cohen et al., 1978.)

$$H_2O_2 + 2\,I^- + 2\,H^+ \xrightarrow[\text{peroxidase}]{\text{iodine}} 2\ \text{"Active I"} + 2\,H_2O$$

II. Oxidation of trapped iodide and organification.

$$2\ \text{"Active I"} + \text{L-Tyrosine} \xrightarrow{\text{enzyme 3}}$$

L-3-Iodotyrosine (MIT)

+

L-3,5-Diiodotyrosine (DIT)

III. Coupling of iodine to tyrosine to form MIT and DIT.

$$2\ \text{Diiodotyrosine} \longrightarrow$$

+ Alanine

3,5,3',5'-Tetraiodotyrosine (T_4)
thyroxine

$$\text{Monoiodotyrosine} + \text{Diiodotyrosine} \longrightarrow$$

+ Alanine

3,5,3'-Triiodothyronine (T_3)

IV. Formation of T_3 and T_4 by enzyme action coupling MIT + DIT or two molecules of DIT with extrusion of alanine.

two main components enclose a small area of mesoderm which includes a network of blood vessels, the so-called mantle plexus. The hypophysis differentiates into (a) a pars distalis or adenohypophysis, whose various diversely differentiated cells (such as thyrotrophs, gonadotrophs, and somatotrophs, etc.) each synthesize and secrete their individually distinct hormones (b) a pars intermedia whose cells secrete a melanoctye-stimulating hormone (MSH), and (c) the pars tuberalis which, situated next to the median eminence, includes blood vessels connecting the two regions. The caudally situated infundibulum, the pars nervosa or neural lobe (Eakin and Bush, 1957) or, neurohypophysis, is in continuity with the brain via an infundibular stalk and in amphibians (as with other tetrapods), secretes oxytocin-like hormones concerned, together with the kidney tubules, with water conservation. Glial cells and pituicytes of unknown function occur in the neurohypophysis, whose most prominent region is the pars nervosa. The main features of the pituitary of the three groups of amphibians and its overall similarity to that of dipnoans have been described by Wingstrand (1966).

The pituitary gland of *Xenopus laevis* starts its development in the larva at about NF St.28. An infundibular wall gradually thins, and in front of the notochord the hypophyseal plate segregates from a stomodeal-hypophyseal anlage. Subsequently demarcation of the pituitary body ensues and by NF St.37–38 the main portion of the hypothyseal plate is situated below the caudal two-thirds of the chiasmatic ridge. By NF St.42 the Anlage of the adenohypophysis reaches the caudal end of the optic chiasma. Differentiation of the pars intermedia occurs at NF St.47, and that of the cells of the neurohypophysis by NF St.49, when a small number of fibers appear in the base of the infundibulum. The first indication of the pars tuberalis is also apparent at this time. All the main components of the pituitary gland are well-developed at NF St.53 during the late period of premetamorphosis, and later stages are mainly concerned with increase in size, particularly of the infundibular process at NF St.57; there is also further cytological differentiation of various cell components (Nieuwkoop and Faber, 1956) (see Fig. 12).

These earlier described gross morphological developmental changes of the pituitary gland of the *Xenopus* larva were subsequently investigated in more detail, especially in the case of the adenohypophysis. Using selective staining techniques and light microscopy Kerr (1965, 1966) described the disposition and histology of at least three types of cells in the larval, and five types in the adult adenohypophysis of *Xenopus laevis*. The larval-type cells differentiate soon after the early differentiation of the pars intermedia and the pars nervosa. Type I basophils (TSH cells) appear ventromedially in the adenohypophysis at about NF St.48 (though subsequently they were recognized at about NF St.45 by Hemme (1972) using electronmicroscopy); at NF St.52 smaller Type I basophils appear in the posterior region and Type III basophils (ACTH cells) are recognized in the dorso-anterior region of the lobe. The granulation of the Type I basophils increases during prometamorphosis, and through climax they were found to have the appearance of solid dense cells. Type III basophils increase in number and extend throughout the anterior third of the gland. Type II basophils of adults are probably gonadotrophs (Holmes and Ball, 1974). The acidophils slowly increase in number and extend throughout the gland dorsal and also posterior to the Type I basophils. Acidophils stain a distinctively clear red with Mallory's stain and are dark green in colour with Luxol-PAS-orange G stains. At NF St.60 they round off and reduce in size, shrinking to their minimum with the loss of granulation at

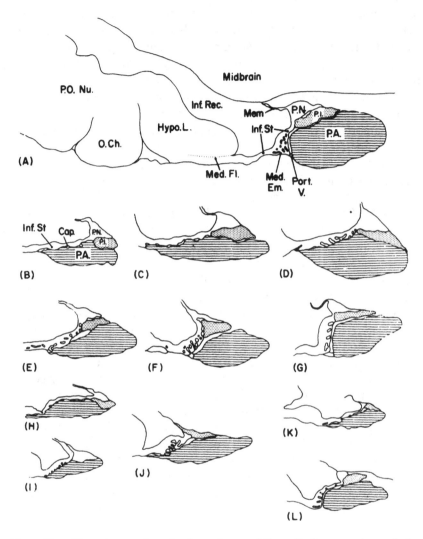

Fig. 12. Drawings of sagittal sections of the pituitary region of: A, adult frog; B, larvae in premetamorphosis; C, early prometamorphosis; D, late prometamorphosis; E, beginning of climax; F, mid-climax; G. post-climax (*Rana pipiens*); H, early prometamorphosis; I, late prometamorphosis; J, 1 day before beginning of climax (*R. sylvatica*); K, mid-prometamorphis; L, early climax (*Bufo americanum*); Cap., capillaries; Hypo. L., hypothalamic lobe: Inf. Rec., infundibular recess; Med. em., median eminence; O.ch., optic chiasma; P. O. nu., preoptic nucleus; P. A., pars anterior; P. I., pars intermedia; P. N., pars nervosa; Port. V., portal vein. (After W. Etkin, 1968.)

NF St.63–64. Regranulation occurs at NF St.65 when there are two types of acidophils differing in size and distribution. Turner (1973) described definitive thyrotrophs in the *Xenopus* adenohypophysis from NF St.47, situated dorsally in the pars anterior, which stain differentially with PAS/ alcian blue. At NF St.50 thyrotrophs occupy a dorsoventrally oriented tract between the median eminence and the lower surface of the pars anterior. These cells are restricted to the ventromedial region between NF St.52 and 54 and their granulation increases. After the onset of prometamorphosis when premetamorphosis has been completed, at NF St.57 or thereabouts, thyrotroph number increases and they occupy the ventral half of the pars anterior to become the most numerous type of cell throughout climax. Watanabe (1969) previously found that the Type I basophils increase in number most rapidly between NF St.55–60 (compare with Turner's 1973 results), decrease their number at about NF St.63–64 and thence increase them again. Likewise their granules increased in size and numbers until reduction at NF St.59, to increase again in size at NF St.61, reflecting granule discharge and new formation (Hemme, 1972). Perhaps the large dense bodies that develop in these cells near the end of climax (Watanabe, 1971a) are lysosomal and are related to the TSH basophil decrease in size. The larval Type I acidophils manufacture growth hormone; Type II acidophils appear after climax (Doerr-Schott, 1968). Whether Type I acidophils secrete prolactin and Type II acidophils somatotropin (growth hormone of adults, Holmes and Ball, 1974), is not clear. The fine structure of different cellular components in the anuran larval adenohypophysis has been described (see Watanabe, 1966, 1971a,b; Hemme, 1972; Mira-Moser, 1972; Pehlemann and Hemme, 1972; Pehlemann et al., 1974). For further information on the amphibian pituitary, reference should also be made to work by Wingstrand (1966), van Oordt (1974), Hanke (1974), Bentley (1976), and Dodd and Dodd (1976). It is of interest that in normally developing tadpoles of *R. temporaria* the disappearance of the aminergic nerve fibres of the pars distalis at climax is dependent upon thyroxine (Aronsson, 1978).

12. The Hypothalamus and Median Eminence

Information on the structure and function of the anuran larval and adult hypothalamus and median eminence has been reviewed (see van Oordt et al., 1972; Holmes and Ball, 1974; Hanke, 1974; Bentley, 1976; Dodd and Dodd, 1976; Oksch and Ueck, 1976; Oota, 1978). These accounts describe neurosecretory nuclei of the hypothalamus as secreting specific substances along axons that terminate in the neurohypophysis and the median eminence. The hypothalamo-hypophyseal portal capillary system transmits releaser substances, the thyroid releasing factor (TRF, probably chemically similar to mammalian TRF, a tripeptide, pyroglutamyl-histidyl-proline amide), to the adenohypophysis; extirpation of the hypothalamus, and thus abolition of sources of hypothalamic neurosecretion, leads to the slowing or cessation of metamorphosis (Goss, 1969; Guastalla and Campantico, 1979) owing to failure of the production of pituitary thyroid stimulating hormone (TSH), which is normally stimulated by the hypothalamic TRF (see Dodd and Dodd, 1976).

During prometamorphosis the median eminence develops between the pars distalis and the floor of the hypothalamus and accumulates neurosecretory material; the capillaries of this region form the hypothalamo-hypophyseal portal system (Fig. 12). There is considerable increase in size and differentiation of the median eminence at climax (Etkin, 1965; Oota, 1978). Neurosecretory fibers develop in the median eminence (see Doerr-Schott, 1968; Bartels, 1971; Aronsson, 1976). The monoamine-containing neurones of the diencephalon are located in the preoptic recess organ (PRO), paraventricular organ (PVO), and the nucleus infundibularis dorsalis (NID). The PRO contains dopamine, and the PVO and NID contain dopamine and 5-hydroxytryptomine and/or 5-hydroxytryptophane. The axons terminate in the median eminence (Kikuyama et al., 1980b), and also probably the pars distalis, which they reach from the fiber tract in the floor of the tuber cinereum (Aronsson and Enemar, 1981). Four types of nerve terminals are recognizable, mainly determined by their granules (Belenky et al., 1973a,b; Belenky and Chetverukhin, 1973); their osmiophilic material probably originated in the parent cells of the preoptic nucleus. The monoamines of the B-type terminals probably originate from the cells of the paraventricular organ and/or the nucleus infundibularis dorsalis (see also van Oordt, 1974; Peute, 1974). During prometamorphosis and climax the numerical proportions of the difference terminals change with increase in the B-type (the dominant type in the adults), and the granular content decreases at climax, which suggests that there is discharge of monoamines and neurohormones into the portal capillaries when the pituitary is most active (Belenky et al., 1973a,b).

Regions of the hypothalamus of larval anurans, especially of *Xenopus laevis*, described by Dodd and Dodd (1976), include the dorsal and ventral preoptic nuclei and the more posterior paraventricular organ and the nuclei infundibularis dorsalis and ventralis, the three latter regions each made up of two types of cells with processes contacting the cerebrospinal fluid (Fig. 13). The paraventricular organ and the nucleus infundibularis dorsalis are aminergic monoamine-containing nuclei, and the nucleus infundibularis ventralis is more typical of a peptidergic nucleus, with its two cell types reflected in granules of differing size. The paraventricular organ appears to regulate the release of a melanophore

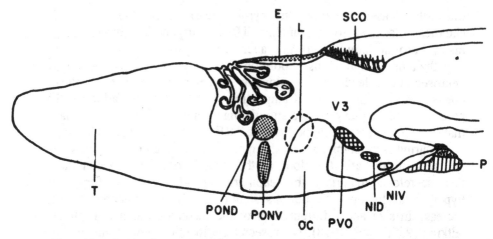

Fig. 13. Sagittal section (diagrammatic) through the brain of a larva of *Xenopus laevis* at NF stage 58 (late prometamorphosis): E, epiphysis; L, region where paired electrolytic lesions inhibit metamorphic climax; NID, nucleus infundibularis dorsalis; NIV, nucleus infundibularis ventralis; OC, optic chiasma; P, pituitary; POND, nucleus preopticus dorsalis; PONV, nucleus preopticus ventralis; PVO, paraventricular organ; SCO, subcommissural organ; T, telencephalon; V₃, third ventricle (After Dodd and Dodd, 1976.)

stimulating hormone (MSH) in the pars intermedia, possibly the synthesis and release of a melanophore inhibiting factor; the nucleus infundibularis ventralis may be concerned with the release of gonadotropic hormones from the adenohypophysis (see Goos, 1978). The peptide growth hormone release-inhibiting hormone (somatostatin or GH–RIH), luteinizing hormone-releasing hormone (LH-RH), and thyroid releasing hormone (TRH or TRF), have been reported to occur in the hypothalamus and some other regions of the body (rest of the brain, skin, gut, and so on) of various adult mammalian and submammalian vertebrates (Jackson, 1978). Immunoreactive forms of these hormones also occur in similar locations in tadpoles of *Xenopus laevis* throughout metamorphosis (King and Millar, 1981). Though their function is not clearly understood the fact that TRF was found initially (presumably in the hypothalamus of the brain tissue) at NF St.54–55 (beginning of prometamorphosis), and its concentration rises steadily throughout metamorphosis, led these authors to suggest that TRF is the causal factor that first triggers off thyroid hormone production and the metamorphic program. However, thyroid hormone biosynthesis appears to begin quite early in larval development, during premetamorphosis (*vide supra*: section on the thyroid), and thus the 'initiator' of the metamorphic cycle is still far from being understood. Indeed there is evidence that at least the preoptic recess of the hypothalamus is dependent upon T_4 for its origin and differentiation. Extirpation of the thyroid primordium of larval *Bufo bufo japonicus* resulted in the monoaminergic (fluorescent) neurones of the preoptic recess organ failing to develop, whereas those of the paraventricular organ and nucleus infundibularis dorsalis were unaffected. T_4 treatment of the thyroidectomized specimens resulted in the differentiation of fluorescent neurones in the preoptic recess organ. After thyroidectomy monoaminergic axon terminals do not develop around capillaries of the median eminence, which process normally ensues in prometamorphic larvae. T_4 treatment to such operated specimens stimulated fluorescent terminal development and penetration amid capillaries in the median eminence. After hypophysectomy, however, in tail bud stages, T_4 failed to elicit these changes (Kikuyama et al., 1979).

Little appears to be known about the larval pineal organ. It manufactures melatonin, which contracts melanophores, and appears to inhibit MSH secretion from the pars intermedia. Yet it is of interest that pinealectomy of the larval *Alytes obstetricans* accelerates metamorphosis. Perhaps the pineal gland partially inhibits the pituitary and thus the thyroid through the hypothalamus (Rémy and Duclos, 1970).

13. The Ultimobranchial Body

The paired ultimobranchial bodies of anurans (see Fig. 14) are believed to produce a hypocalcemic hormone calcitonin. Ultimobranchialectomy of bullfrog tadpoles leads to hypercalcemia in a high calcium medium (Sassayama and Oguro, 1976). Whether it has any other function is uncertain. Together with parathormone—a hypercalcemic hormone of the parathyroids—and perhaps with some involvement from the pituitary gland (Uchiyama and Pang, 1981), it controls calcium metabolism (Roberts, 1972; see Bentley, 1976, and Dacke, 1979).

The ultimobranchial body of *Xenopus laevis* shows features of degeneration at climax and is reduced in postmetamorphic stages, though doubtless it is still functional (Saxen and Toivonen, 1955). Indeed in young *Xenopus* toads the unifollicular gland rarely exceeds 75 μm in diameter and is difficult to locate (Coleman, 1970). Judged by its histological structure, the ultimobranchial body of *Rana japonica japonica* shows reduced activity after climax (Sasayama et al., 1976). However, according to Robertson and Schwartz (1964) the ultimobranchial body is actively secretory in the adult *Rana pipiens,* and is essentially a multifollicular gland of 2–4 connecting follicles, each with a large lumen. The ultimobranchial body of *Hyla arborea, Bufo viridis, Rana ridibunda,* and *Pelobates syriacus* is present for at least 2 yr after metamorphosis

(Boschwitz, 1960a). Perhaps its reduction in the fully aquatic postmetamorphic *Xenopus,* in contrast to some other terrestrial anurans, is bound up with the claim by Boschwitz (1969) that the gland is involved in water drive phenomena during the breeding season.

Each component of the two ultimobranchial bodies of the larval *Hyla, Bufo, Rana,* and *Pelobates* is situated on either side of the pharynx, near the posterior branchial pouches. Before climax they are each composed of one follicle surrounded by a single layered epithelium; or in *Pelobates* it is a coiled tube with parafollicular cells. A vascularized capsule invests the organ. During climax the follicle enlarges and the epithelium becomes pseudostratified. The coiled tube of *Pelobates* divides into several follicles (Boschwitz, 1960b). The follicles contain a small quantity of eosinophilic coagulum. In *Rana pipiens* this includes alpha (acid-mucopolysaccharide with a carbohydrate complex), beta (mucoprotein), and gamma (sudanophilic) components (Robertson and Schwartz, 1964). After metamorphosis the ultimobranchial body further enlarges. There is a large follicle with a pseudostratified epithelium and shallow folds in *Hyla,* 1–2 large follicles and deeper folds in *Rana,* many small follicles each with a single epithelial layer in *Bufo,* and the gland is ovoid and concentrated in *Pelobates* (Boschwitz, 1960b). The origin and development of the ultimobranchial body of *Rana japonica* has been described by light microscopy (Sasayama et al., 1976).

Examination of the fine structure of the ultimobranchial body of *Rana temporaria* larvae revealed a cell type believed to be a parafollicular cell, characteristic of the C cells of the mammalian thyroid. They secrete granules 8 d after larval hatching (Coleman and Phillips, 1974). The gland enlarges by cellular mitosis. A single ultimobranchial gland cell type appears either as an electron-dense "dark" form or a less dense "light" form. The numerical ratio of "dark"–"light" cells varies from gland to gland at any one stage of development. By the end of metamorphosis all the cells are "light" in *Bufo* and *Hyla,* whereas "dark" cells persist in *Rana* (Coleman, 1975). These grandular cells contain secretory granules, granular endoplasmic reticulum and free ribosomes, tonofilaments, microtubules, Golgi bodies and lipid droplets. Probably there is apocrine secretion into the central lumen of the gland. Coleman (1975) considers that the "dark" cells may be precursors of the "light" cells since "dark" cells show no indications of degeneration during larval life.

The ultimobranchial body is present in thyreotic larvae of *Pelobates syriacus;* in *Hyla, Rana,* and *Bufo* at the beginning of climax; increasing hormone concentration inhibits the rate of growth of the ultimobranchial body (Boschwitz, 1960b). Probably prolactin influences the activity of the gland (Boschwitz and Bern, 1971).

14. The Parathyroid Glands

The parathyroid glands of anuran larvae, as typically demonstrated in *Bufo viridis,* originate from the ventral regions of the 3rd and 4th pharyngeal pouches (Boschwitz, 1961; Copp, 1969) (see Fig. 14). Each of the four glands consists of a small group of ovoid cells with sparse cytoplasm but prominent nuclei. The glands are vascularized externally, though in *Bufo,* capillaries are not found between the cells in contrast to the intercellular vascularity of those in *Rana* and in mammals also. Eventually the gland is encapsulated. After metamorphosis the paired glands on each side of the larynx are located near the vena jugularis externa and the nervus hypoglossus. The glands include rounded or whorl-shaped cells, which differ in their staining response to Mallory's hematoxylin, perhaps indicating a difference in function. The maximum size (diameter) of the parathyroid gland may vary between 350 and 1000 μm (Boschwitz, 1965, 1967), usually reached within 2 yr after metamorphosis: thereafter there is a slight but significant involution (see also Sasayama and Oguro, 1974).

During the year there is a cycle of degeneration, and thence regeneration, of cells of the gland, and lumina appear in the regenerated glands. The extent of this cycle of growth and cytolysis is less in animals living in natural conditions and the factors responsible for this feature in captive specimens are not understood (Boschwitz, 1967).

Information on the function of the parathyroid glands and of its hormone in amphibians is limited (see Dacke, 1979). The glands secrete a hypercalcemic hormone (parathormone) and parathyroidectomy in amphibians (anurans) is usually followed by tetanic convulsions. However, the larger the animal the greater the delay in this response, possibly a feature related to the seasonal activity of the glands (see Robertson, 1977). Tetany is doubtless a result of plasma hypocalcemia, probably caused by a decreased rate of calcium mobilization from bone (Cortelyou, 1967). Parathormone appears to be necessary to mobilize calcium carbonate from the lime sacs; calcitonin of the ultimobranchial bodies seems to inhibit mobilization of calcium carbonate (Robertson, 1972).

In urodeles inconsistent effects have been reported after parathyroidectomy. Some species (*Cynops pyrrhogaster*) show tetany, but others (*Megalobatrachus davidianus*) are unresponsive (Oguro, 1973; see Bentley, 1976; Wittle and Dent, 1979). In *Megalobatrachus,* parathyroidectomy has no effect on the serum calcium concentration. Administeration of its parathyroid glandular homogenate to parathyroidectomized *Cynops pyrrhogaster* resulted in an increase in serum calcium concentration within 2 h. Presumably the target tissues of

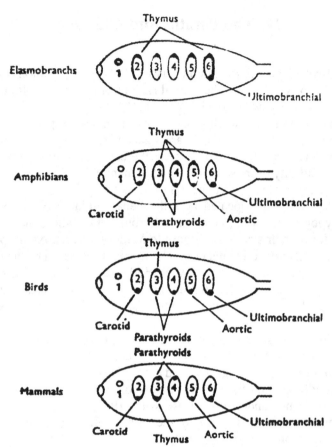

Fig. 14. Diagrammatic illustration of the phylogeny of the branchial derivatives in vertebrates. The small paired parathyroids originate from the endoderm of the 3rd and 4th branchial pouches, ventrally in amphibians and birds and dorsally in mammals, when the hinder parathyoid IV lies adjacent to or is embedded in the thyroid. The ultimobranchial gland from the ventral region of the 6th branchial pouch, and present in all vertebrates except cyclostomes, is paired except in elasmobranchs and reptiles, when the right one is vestigial. In most mammals the ultimobranchial gland (C cells) is embedded in the thyroid. The arrangement is doubtless an oversimplification (see, for example, the variability in the branchial origin of the thymus of amphibians described by Cooper, 1976), but it does provide a useful generalization of the branchial phylogeny of these structures. (After Copp, 1969.)

Megalobatrachus are insensitive to a hypercalcemia factor(s) present in its parathyroids (Oguro, 1973).

Impairment of respiration by relevant respiratory muscles, owing to calcium deficiency after parathyroidectomy, is compensated for by absorption of oxygen through the skin.

15. The Larval Skeleton—Specific Features During Metamorphosis

Origin and Development of Skeletal Components

The extensive changes in the anuran larval skeleton and associated musculature that occur during metamorphosis are well known, and have been documented (see among others de Beer, 1937; Sedra, 1950; Smit, 1953; Witschi, 1955; Sedra and Michael, 1957–1959; Chacko, 1965a,b; de Jongh, 1968; Swanepoel, 1970; Branham and List, 1979; Sokol, 1981). It is of interest, at least for students of evolutionary relationships, that in anurans phylogenies based on anatomical features of the larval chondrocranium resemble at least in major features those arrived at based on adult characters (Sokol, 1981).

In *Xenopus,* for example, the cartilaginous basal plate, and the neural crest derivatives, Meckel's cartilage, and hyobranchial skeleton appear at about NF St.43, the parachordals at NF St.45, and the auditory capsules at NF St.46. At NF St.47 neural arches appear as condensations amid mesenchymal cells opposite the myocommata, and chondrification begins in them at NF St.50; other components chondrify slightly later. The skull shows incipient ossification to form the parasphenoid bone at NF St.51, and the goniale at NF St. 53–54 at about the same stages as perichondrial bone is deposited in the neural arches. Most of the other ossifications arise at climax, simultaneously with the extensive degeneration of the chondrocranial, hyobranchial, and vertebral cartilage. The forelimbs are chondrified at NF St.56, ossified perichondrially at NF St. 57, and endochrondrially by NF St.66. Most of the hindlimb is completely chondrified by NF St.54, periochondrial ossification has commenced in different regions by NF St. 55–56 and in the phalanges at NF St.57. The tarsalia ossify at the end of climax (see Nieuwkoop and Faber, 1956).

Recently, Saint Aubain (1981) described the ontogeny of the chondrified elements in limbs of various amphibian larvae. Limb morphotypes are generally similar in anurans and urodeles, and she concluded that they may be derived from the same type of crossopterygian fin.

Hormonal Control of Skeletogeny

Calcareous material for skeletal ossifications is stored in the endolymphatic sacs (Guardabassi, 1959; Pilkington and Simkiss, 1966). Its deposition in the skeleton is presumably influenced by the hypocalcemic hormone calcitonin of the ultimobranchial bodies, and hypercalcemic parathormone of the parathyroids, which, acting in concert, reduce the level of calcium in the endolymphatic ducts (see Boschwitz, 1960a,b, 1961; Robertson, 1970, 1972; Coleman, 1975; Bentley, 1976).

Skeletal growth, differentiation, and degeneration are directly or indirectly controlled by hormones of the thyroid, the parathyroid, and the
ultimobranchial bodies; presumably the entire process is regulated via
hormone control by the hypothalamo-pituitary axis, that acts mainly at
the level of the thyroid gland. Thus thyroidectomy or the use of chemical
goitrogens inhibit skeletal changes in the larva (Allen, 1918; Terry, 1918;
Steinmetz, 1954) and treatment of larvae with thyroxine hastens skeletal
growth and differentiation (Kaltenbach, 1953a). The ossification of the
femur of young tadpoles of *Rana pipiens* is stimulated by treating them
with triiodothyronine and the osteoclasts increase their amount of granular endoplasmic reticulum (Kemp and Hoyt, 1969).

Intraperitoneal injection of bovine prolactin into *R. pipiens* larvae increases the rate of incorporation of ^{14}C-proline into collagen in the thigh
bone during prometamorphosis (TK St.XII–XVI), but not at climax.
Growth hormone, however, did so during climax, but elicited only a
slight increased rate of incorporation during prometamorphosis. Likewise
T_4 immersion ($5 \times 10^{-8}M$ T_4 solution) of prometamorphic larvae hastened the rate of collagen synthesis compared with controls, a process enhanced by growth hormone though not by prolactin. The sensitivity of
thigh bones to prolactin seemed to be reduced by thyroxine treatment
(Yamaguchi and Yasumasu, 1977a).

Thigh bones from larvae of *R. pipiens* immersed at TK ST.XIV in a
similar thyroxine solution as above for up to 2 wk, or those from larvae
injected intraperitoneally with prolactin, show a higher rate of
^{14}C-leucine and ^{14}C-proline incorporation into protein than in controls.
The effect was similar for leucine with either thyroxine or prolactin treatment, but the rate of proline incorporation was lower in the case of prolactin compared with that of thyroxine. Prolactin injection of T_4-treated
specimens failed to further increase the rates of aminoacid synthesis into
protein. At climax prolactin injection was ineffective, whereas the rate of
incorporation of amino acids was slightly higher in T_4-treated specimens
than in controls. Thigh bones of specimens treated with thyroxine seem to
be insensitive to prolactin in this particular feature (Yamaguchi and
Yasumasu, 1977b). In general it appears that thyroxine stimulates increase in the rate of collagen degradation in the tail fin of tadpoles during
metamorphosis, which process is inhibited by prolactin injection. In the
thigh bone the rates of collagen synthesis and degeneration are increased
by thyroxine treatment during metamorphosis: presumably there is a high
turnover of collagen at this time (Yamaguchi and Yasumasu, 1978).

For further details the reader is referred to the account of the development and differentiation of the anuran tadpole limbs, including chon-
drogenesis, osteogenesis, and myogenesis (origin of the myofibrillar proteins: heavy myosin chain, tropomyosin, α-actinin and M-line protein),

together with the influence of thyroid hormones (which also stimulate enhanced limb vascularity), given by Atkinson (1981)

16. The Notochord

The notochord persists in adults of Amphioxus and Cyclostomata, but in higher vertebrates it is only present in early embryonic (or in larval) stages. It exists merely during the first 10 d of the chick embryonic development (Jurand, 1974). In anuran larvae by the end of climax the notochord has disappeared, or nearly so, in the body and has completely disappeared with the involution of the tail. In *Xenopus laevis* incipient notochordal cellular vacuolation has occurred at NF St.26 and full vaucolation by NF St.37–38. The elastica interna and externa, which envelop the notochord, are recognizable at NF St.32. Degeneration of the notochord begins anteriorly in the body at NF St.46 and there is substantial reduction by NF St.55. Considerable intervertebral compression occurs and by the end of climax little notochordal tissue remains (see Nieuwkoop and Faber, 1956). Aspects of the development of the notochord of *Ambystoma tigrinum, Triturus alpestris* and *T. palmaticus* (Mookerjee, 1953; Mookerjee et al., 1953), *Rana pipiens* (Leeson and Threadgold, 1960), *Rana temporaria* (Claes, 1965), and *Triturus pyrrhogaster* (Takaya, 1973) have been described by light microscopy.

The fine structure of the larval notochord of the urodeles *Triturus, Pleurodeles,* and *Ambystoma* was described by Waddington and Perry (1962), that of the anuran tail of *Rana catesbeiana* by Bruns and Gross (1970) and of *Rana temporaria* by Fox (1973a). The tail notochord sheath fibrils of *Xenopus laevis* (Weber, 1961) and the elastica interna of *Rana rugosa* (Nakao, 1974) have also been described.

More recently, Yoon et al. (1980) investigated by scanning electronmicroscopy the development of the notochord of anuran and urodelan embryos, from the early gastrula stage to the neural tube stage. They noted differences in the notochords of the two groups; that of the anuran (*Xenopus laevis* and *Rana pipiens*) is generally uniform in diameter along its length, but the urodelan notochord (*Ambystoma mexicanum* and *Pleurodeles waltl*) is wider posteriorly and narrows in front. Such differences may be the result of different cellular movements during gastrulation.

The relationship of the notochord to the somites during amphibian embryogeny seems to be a passive one: in normal development the notochord prevents the somites of either side fusing ventrally (Yoon and Malacinski, 1981a).

The preclimactic anuran tail notochord tapers distally. The intracellularly vacuolated cells are enveloped at the periphery of the notochord by a basement membrane and a thin elastica interna (possibly a product of the notochordal cells, Nakao, 1974), and more proximally there is also an elastica externa. Between the two notochordal elastic layers there is a circumferentially disposed collagenous sheath that disappears near the tail tip, and more proximally in the tail it bounds the elastica externa. There is also a connective tissue sheath of fibroblasts and layers of collagen in a structureless matrix (Bruns and Gross, 1970). According to Leeson and Threadgold (1960), the notochordal sheath contains collagen and reticular and elastic fibers embedded in a matrix of either muco- or glycoprotein or neutral mucopolysaccharides. Nuclei and most of the cytoplasm of the cells are located at the periphery of the notochord, though some nuclei occur more centrally where fine cytoplasmic extensions, joined by desmosomes, closely appose one another. Hemidesmosomes are present at the external surface of outermost cells below the basement membrane. The perikaryon, especially of more peripheral notochordal cells, includes a well-developed RER, prominent ribosomes and polyribosomes, some SER, and a few mitochondria with well developed cristae. Golgi stacks are frequently recognized. There is incipient fibrosity in the cell cytoplasm. Numerous plasmalemmal vesicles are situated near or are in open continuity with the intercellular spaces, and less commonly they open into the space below the basement membrane. These vesicles may well be pinocytotic and supply fluid to the intracellular vacuoles (Jurand, 1962).

Ultrastructural features of notochordal degeneration in the anuran (or urodelan) larval body do not appear to have been described. It would seem reasonable to suppose, nevertheless, that the events are generally similar to those that occur at climax in the notochord of the degenerating anuran tail (Fox, 1973a). At the beginning of climax the hinder tail notochordal cells of *R. temporaria* become extremely fibrous. In addition large denser fibrous bodies are present. Throughout climax degenerating notochordal cells (especially those more distally in the regressing tail) develop small secondary lysosomes, autophagic vacuoles, and large membrane-bound cytolysomes, granular bodies, myelin figures and lamellated structures, lipid, and pigment. Mitochondria typically degenerate either by initial disorganization and loss of cristae, leaving an empty vesicle, or the outer membranes degenerate and partially disorganized cristae spill out and merge with the ambient cytoplasm to subsequently disperse and disappear (see Fox, 1927b, 1973a, 1975). Nuclei are slow to become chromatopycnotic. Areas of necrosis become more widespread as organelle integrity is gradually lost. The notochordal collagenous sheath and elastic membranes are invaded by mesenchymal macrophages, which disorganize and thence ingest partially or highly degraded collagen, and thereafter the necrotic debris of the degenerate motochordal cells, within heterophagic phagosomes.

There is widespread deposition of acid phosphatase amid and within organelles such as mitochondria, and it is often freely disposed within the degenerating notochordal cells. By the height of climax the tip of the tail stub shows only few signs of recognizable notochordal structural profiles. In larval *Rana temporaria ornativentris* and *Rhacophorus schlegeleii arborea,* Niijima and Hirakow (1964) showed marked deposition of acid phosphatase and leucine aminopeptidase, which increased around the notochord at the tip of the tail during its regression. Esterase and acid phosphatase have been reported in the degenerating tail notochord of *R. pipiens* larvae (Fry et al., 1973).

17. The Central Nervous System

Aspects of the development of the anuran larval nervous system during metamorphosis, and its relationship to the endocrines, have been reviewed by Kollros (1968, 1981), Frieden and Just (1970), Dodd and Dodd (1976), and Fox (1981). Changes in the anuran tadpole nervous system induced by the addition or withdrawal of thyroid hormones were most usefully itemized in table form by Kollros (1981). Ultrastructural details of neurulation in amphibian embryos have been described (see Schroeder, 1970; Burnside, 1971; Karfunkel, 1971; Mak, 1978), and among other components of the central nervous system (CNS) the development and maturation of cranial nerves (Davies et al., 1982), neurones (Thors et al., 1982), the descending spinal neural pathways of the larval *Xenopus laevis* (Ten Donkelaar and de Boer-Van Huizen, 1982) and *R. cates beiana* (Forehand and Fareh, 1982a,b), and of the larval bullfrog cerebellum and its stimulation by T_4 have been extensively studied (Gona, 1972; Gona and Gona, 1977; Uray and Gona, 1978).

The larval brain of *Xenopus* is generally well developed at NF St.55. Later development is mainly concerned wit its enlargement and further differentiation, and at climax topographical changes occur. In general the structure and arrangement of the brain and cephalic ganglion of *Rana* larvae are similar to those of *Xenopus*. The spinal cord of *Xenopus* differentiates proximodistally and nerves enter the tail by NF St.33–34. At this period anterior spinal nerves originate from the ventral surface of the spinal cord, to be followed subsequently in space and time by those behind. Incipient dorsal root ganglia, from neural crest, each consisting of about 6–10 cells, are recognized at NF St.39 and at NF St.43, at the level of the hind-trunk, they are about 60 μm long (see Nieuwkoop and Faber, 1965).

A detailed examination of the spinal cord of the late embryo of *Xenopus laevis,* by Roberts and Clarke (1982), has revealed nine classes of differentiated neurones. These are: Rohon-Beard cells; extramedullary cells, both primary and secondary neurones; one class of motoneurones that innervate the segmental myotomes; two classes of motoneurones with decussating axons; three classes of interneurones with ipsilateral axons, and a newly described class of ciliated ependymal cells, whose axons project ipsilaterally to the brain.

The Mesencephalic V Nucleus

Kollros and McMurray (1955) noted that the first mesencephalic trigeminal neurones (MTC), of the young *Rana pipiens* larva, appeared deep in the rostral half of the tectum, the number increasing to about 400 fully differentiated cells before metamorphosis. In *Xenopus,* the first MTC form at NF St.20, the number increasing till NF St.58, though substantial neuronal necrosis occurs, probably early, during larval development. The generation times of the mesencephalic neurones and of those of the trigeminal ganglia are dissimilar, and they may not have a common cell lineage. Possibly those of the ganglion at least are of neural crest origin (see Lewis and Straznicky, 1979).

The normal growth and differentiation of the brain are dependent upon thyroid hormones; typically, for example, the median eminence of thyroidectomized larvae remains underdeveloped, characteristic of premetamorphic stages (Etkin, 1970). During larval development of the mesencephalic V nucleus in the midbrain of *R. pipiens,* the cell bodies become sensitive to thyroid hormone fairly late, at TK St.V or even later (Kollros and McMurray, 1956). Local application of thyroxine implants precociously stimulates their cell number and size, which parameters vary inversely in relation to the distance from the source of T_4 stimulation. The cells reduce in size after the withdrawal of T_4 (Kollros and McMurray, 1956), or after treatment of the larva with thiourea (Kollros, 1957), suggesting that their size depends upon the continuous presence of thyroxine.

Rohon-Beard Cells

Rohon-Beard cells were first discovered by Rohon in 1884 and by Beard in 1889, in the spinal cord of fishes, hence their name. These 'large ganglionic cells' were first reported in amphibians (in a urodele) by Burckhardt in 1889, and then they were described by Studnicka in 1895 to degenerate in anurans though they were retained longer in *Triturus* (see Eichler and Porter, 1981 for a brief historical account and the references). Coghill (1914) described their structure and topography in larvae of *Ambystoma punctatum:* he considered them to be part of the sensory system of the trunk, which supplies the skin and muscle.

Large segmentally arranged Rohon-Beard cells differentiate in the dorsolateral regions of the spinal cord of *Xenopus* at NF St.39. However,

Baccaglini and Spitzer (1977) identified them, by virtue of their size and location, in *Xenopus* at NF St.18, where the neural folds are about to form a neural tube. Indeed from autoradiography using ^{3}H-thymidine, Lamborghini (1980) demonstrated that in *Xenopus laevis* 80% of the Rohon-Beard cells have originated by the completion of gastrulation; the remaining 20% form during neurulation and the early tail bud stage. Likewise she showed that Mauthner cells, neurones of the trigeminal ganglion and of the basal plate of the medulla, and extramedullary and primary motor neurones form independently during gastrulation. In larvae of *Rana pipiens* at TK St.I there are about 250 Rohon-Beard cells in the spinal cord, located dorsally near the midline. They are mainly concentrated anteriorly and are of similar numbers bilaterally. As the cells degenerate to disappear by the end of climax, there is a similar rate of cell loss throughout the nerve cord (Eichler and Porter, 1981).

Rohon-Beard cells give rapidly adapting responses to mechanical stimuli (Roberts and Hayes, 1977). Lamborghini et al. (1979) described, by electronmicroscopy, changes in the abundance and distribution of organelles in the perikaryon of the Rohon-Beard cells of *Xenopus*, in four stages from NF St.22–42, when the cells have attained their electrical excitability. They described the progressive loss of intramitochondrial granules (concentrations of divalent Ca^{2+} ions in these neurones), which decrease in parallel with the loss of the Ca^{2+} ion component of the inward current of action potential.

Rohon-Beard neurones degenerate in Xenopus larvae at NF St.50 and merely a few pycnotic cells remain in the nerve cord at NF St.55, their function being superseded by the spinal ganglia. However, they are retained for long periods in hypophysectomized larvae (Hughes, 1968). In *Triturus cristatus* larvae the Rohon-Beard cells normally disappear before the onset of metamorphosis, but thyroxine treatment elicits their premature involution (Trevison, 1972, 1973). Likewise in *Rana pipiens* larvae the loss of the Rohon-Beard cells depends upon a high level of thyroxine during metamorphosis (Stephens, 1965, 1968), though there is evidence of some measure of their dependence upon the maintenance of a neurotrophic peripheral area of the skin (see Bacher, 1973). Meyer (1974) has described the ultrastructure and histochemistry of the Rohon-Beard cells in larvae of fishes and amphibians.

They are believed to originate from neural crest (Dushane, 1938; Horstadius, 1950), though reliable confirmation of this does not appear to be available.

Mauthner Cells

The paired Mauthner cells were first discovered in anuran larvae of *Rana fusca* and *Bombinator pachypus* by Szepsenwol (1935a,b), later to be confirmed by Willis (1947, 1948). These large neurones are situated one on each side of the medulla, each having a large axon descending through the nerve cord of the trunk to the tail (see Model, 1978), and they are visible under the electron microscope in *Xenopus* at NF St.41–42 (Vargas-Lizardi and Lyser, 1974), or in terms of location and nuclear size somewhat earlier at NF St.31–32 (Billings, 1972).

Mauthner cell differentiation (in *Ambystoma* larvae), in part, seems to depend on the ingrowth of the VIIIth (vestibular) cranial nerve, and it is also retarded at low (5 and 8°C temperatures). The VIIth lateral line root fibers also exert an influence (Piatt, 1969, 1971). Mauthner cells are seen at CM St.41 in *Rana temporaria* (Fox and Moulton, 1968). Tusques (1951) reported Mauthner cells in the adult *Rana temporaria;* others believe they disappear in anurans soon after the onset of the froglet stage (see Baffoni and Catte, 1950, 1951; Stefanelli, 1951), possibly because of the reduction in the level of circulatory thyroid hormones, for in larvae of *R. pipiens* immersed in thyroxine, Mauthner cells do not regress and the thyroxine is necessary to maintain their size and level of differentiation. Mauthner cells are smaller than normal in thyroidectomized specimens, compared with controls (Pesetsky, 1962, 1966). These conclusions are supported by results of Fox and Moulton (1968) in *R. temporaria* larvae. Mauthner cells are found in *R. temporaria* at CM St.54 (the end of climax) (Fox and Moulton, 1968), and in *Xenopus* 2 months after metamorphosis, though they are shrunken. A 3-yr-old female had neither Mauthner cells nor axons (Moulton et al., 1968). Perhaps their retention is variable in different anuran species. They are wholly lacking at any stage of development in *Bufo* (Moulton, et al., 1968).

The ground substance of a late prometamorphic Mauthner cell has a fibrillar matrix and is well-endowed with ribosomes and polyribosomes, an RER, Golgi complexes, smooth vesicles, and lysosomal bodies. Mitochondria are abundant in the perikaryon and dendrites, but less so in the axoplasm. Mauthner cells probably ultimately degenerate by autolysis utilizing lysosomal enzymes, and possible indication of postmetamorphic involution are recognizable in *Xenopus* at climax, for mitochondria become swollen and cristae are poorly developed (Moulton et al., 1968). Billings (1972) studied Mauthner cells of *Xenopus laevis* throughout metamorphosis and showed them to have a high content of RER and ribosomes. There are numerous microfilaments and microtubules and gradually the number of multivesicular bodies, alveolar vesicles, and small secondary lysosomes increases. Mauthner cells disappear at climax and were never found in the adult.

A detailed description of the ultrastructure and synaptic connections of the Mauthner cells of premetamorphic larvae of *Rana catesbeiana* has been given by Cochran et al. (1980). They showed that the amphibian Mauthner cells were similar to those in teleosts. Likewise the ultrastructure of the initial segment of the Mauthner cell axon and the axon cap (the surrounding neuropil) of the bullfrog larva (at TK St. VII–XII, 90–120 mm, long about the beginning of prometamorphosis), has been described by Ito (1980). The axon cap cells possibly inhibit Mauthner cell activity. However, in general it appears that the growth of a postsynaptic neurone is locally influenced by its synaptic contacts. Support for this view was provided by Jacoby and Kimmel (1982), who showed that in the larva of *Ambystoma mexicanum,* at regions of the Mauthner cell surface which were growing and differentiating dendrites most rapidly, synaptic contacts from adjacent cells were the most densely arranged and had accumulated most rapidly. The overall function of the extensive coverage of the Mauthner cell and of its initial segment by synaptic terminals thus, so far, has yet to be clarified.

Lateral Motor Column Neurones

Lumbar ventral horn cells, or lateral motor column (LMC) cells, are first recognized in *Xenopus laevis* at NF St. 50, when the hindlimb rudiment is a roundish bud. The peak number occurs at NF St. 52–53, when the hindlimbs are paddle-shaped, and maximum LMC cell loss is at NF St. 54 (Prestige, 1967, 1973). Brachial ventral horn neurones of *Xenopus* differentiate between NF St. 52–53 and 57. These cells were found to reach a peak in number on each side of the nerve cord at NF St. 55. Many of them degenerate and like LMC cells their number is drastically reduced by up to 80% by the end of metamorphosis (Fortune and Blackler, 1976). LMC cells labelled with tritiated thymidine can be followed laterally and thence more medially in the ventral horn, from their origin in the mantle layer; the posterior ones are younger than those situated more anteriorly (Prestige, 1973). Chu-Wang et al. (1981) described and compared the ultrastructure of migrating motoneurones (identified by horse radish peroxidase) and nonmigrating and differentiated LMC cells of bullfrog larvae that had completed their migration. Both types of cells are bipolar, one process extending to the ependyma the other directed towards the ventral root. Centrally and peripherally directed processes had microtubules, mitochondria, and rosettes of ribosomes. Four of the five types of synapses of adult motoneurones are recognizable in migrating motoneurones of tadpoles. Radial glial cells located in the ependyma were found to differ from the LMC and migrating neurones by virtue of their reduced labeling with horse radish peroxidase and their higher content of microfilaments. As cells of the LMC grow and differentiate to innervate the fore and hindlimbs, up to ¾ of the lumbosacral neurones of *Xenopus* and *Eleutherodactylus* (Hughes, 1961) and ⅔ of the neurones of the spinal ganglia of *Xenopus* (Prestige, 1965) degenerate during prometamorphosis. For every differentiated neurone of the LMC 8–9 neuroblasts degenerate, and the chromatopycnotic cells are phagocytosed by microglial phagocytes (Hughes, 1968). In *Rana pipiens* larvae, by TK St. XIV—XXIV the lumbar plexus includes up to 10,000 cells on each side, which number is reduced to about 2000 on each side by climax (Kollros, 1981).

By the use of horse radish peroxidase label, Bennett and Lai (1981a,b) studied the topographical distribution and selective death of cutaneous sensory neurones of spinal ganglia, of the larval anuran *Lymnodastes peronii,* from TK St.XIV—XXIV. Neuronal death occurs throughout the ganglion, but there is a greater elimination of neurones in the ventral than in the dorsal region.

Up to the early hindlimb bud stage ventral horn neurones develop independently. Afterwards they require sensory stimuli via pathways from the limb, through the dorsal root ganglia, for their maintenance. Ventral horn cells and those of their spinal ganglia degenerate when the developing hindlimb is amputated (Hughes, 1968). Similar phenomena occur in the case of brachial ventral horn cells of *Xenopus* after forelimb amputation (Fortune and Blackler, 1976). The events are, however, extremely complicated. If the amputation is performed in very early limb development there is no immediate effect on the ipsilateral LMC cells; then in *Xenopus* after a short delay there is a rapid loss of LMC cells. In *Rana,* likewise, there is no initial effect, but then after retention of some additional cells on the operated side, several weeks later there is significant cell loss compared with that of the controls. In both *Xenopus* and *Rana,* at equivalent TK St.IX or later, amputation (after a short dip in *Xenopus*) leads quickly to an excess of cells on the operated side, and only later (and much later if the operation is performed in late larval or in postmetamorphic stages), does the excess of cells on the operated side disappear and then there is a further loss to reduce the LMC cell population below that of the unoperated control side (Kollros, personal communication and *vide infra*).

These cellular events were elegantly demonstrated, using electron microscopy, by Decker (1978), for after total hindlimb extirpation (uni- or bilateral) in *R. pipiens* larvae at TK St.V (when the LMC cells are first recognized and whose axons in the limb are forming neuromuscular junctions), the ipsilateral LMC neurones failed to differentiate and 95% of them degenerated by TK St.XVII. The retarded degeneration was by pycnosis, but there was little evidence of lysosomal activity and no change in the levels of acid phosphatase until TK St.XV. With limbs unilaterally extirpated at TK St.X (just before the onset of normal ontogenetic neuronal degeneration), the ipsilateral LMC neurones rapidly degenerate and by TK St.XX they comprise only 20% of the total of the contralateral LMC neuronal population. These degenerating cells include numerous secondary lysosomes and there is a significant increase in acid phosphatase activity. The same operations performed at TK St.XV (just after the normal period of LMC neuronal cell death) and at TK St.XX (beginning of climax) resulted in the postponement of the ipsilateral LMC cell death response. This retrograde effect includes an intervening period of neuronal chromatolysis, nuclear positional eccentricity, peripheral reorganization of the RER, hypertrophy of the Golgi apparatus, accumulation of large secondary lysosomes, and large inclusions of lipid. In all cases of neuronal degeneration, glial cells were found to phagocytose the cell debris. These results support those of Prestige (1970), that the LMC neuronal response to axotomy changes during larval ontogeny, and at least three critial phases can be recognized during larval life (see Decker, 1978).

In contrast to the ipsilateral LMC cell death after limb extirpation, it is of interest to find that if in *R. pipiens* larvae the sensory periphery of the hindlimb available to the 9th spinal ganglion is increased, by extirpation of ganglia 8 and 10, mitotic activities and neuronal numbers increase in ganglion 9 (Bibb, 1977).

The total number of ventral root nerve fibers shows a similar initial rise, peak, and decline during normal ontogeny, as in the case of the ventral horn cells, and dying fibers do not myelinate. Fiber loss is accentuated when the hindlimb bud of *Xenopus* is removed (Prestige and Wilson, 1974). Thus thyroid hormones influence the development and/or degeneration of LMC neurones and nerve roots directly (*vide infra*), or indirectly by stimulating normal limb growth (Beaudoin, 1956). The causal factors influencing directional growth of ventral root nerves are extremely complex. In embryos and tail regenerates of *Xenopus laevis* and *Triturus viridescens,* initially when the myotomes are unsegmented multiple adhesive-type contacts join the neural tube and myotomal tissue. When the latter is segmented, early recognizable small bundles of ventral roots, of 1–5 axons and one bundle per myotomal segment, extend for a short distance. One or more glial cell processes accompany the axons and at the point of exit from the cord they appear to funnel the pathway of growth (Nordlander et al., 1981). In *Rana pipiens* larvae the target stimulus for the nerve fibers that invade the hindlimb bud appears to be mesenchyme, for the invaded area is premuscular at this time. This view is further supported by experiments on cultured explants of spinal cord and mesenchymal tissue at TK St.V. Nerve fiber density and morphological complexity are in inverse relationship to target distance. It is postulated that growth of nerves into the limb bud is stimulated and directed by a diffusable target-oriented growth factor(s), that binds to the attachment mesenchymal substratum as a concentration gradient (see Pollack, et al., 1981; Pollack and Richmond, 1981). Myotomal innervation and the formation and differentiation of contacts in *Xenopus laevis* embryos and larvae have been described by Kullberg et al. (1977), by the use of electrophysiology and electron microscopy.

Thyroid Hormones and the LMC Neurones: Fine Structure and Their Degeneration

Thyroid hormones are essential for the growth and differentiation of the LMC neurones during larval development (Kollros, 1968, 1981), though neurones of the CNS may also be influenced in their development by prolactin and somatotropin (Hunt and Jacobson, 1971). Thyroxine can directly elicit substantial necrosis of neurones that fail to establish viable peripheral connections (Race, 1961; Lamb, 1974). Treatment of anuran larvae by immersion or injection of high thresholds of thyroxine, or implantation into larvae of T_4–cholesterol pellets, or surgical hypophysectomy or thyroidectomy, or the use of chemical goitrogens demonstrated that mitotic activity, growth, and differentiation of neurones of the CNS (and their sequential degeneration) are directly influenced by thyroid hormones at different stages throughout metamorphosis (see Beaudoin, 1954, 1956; Kollros and Pepernick, 1952; Kollros and McMurray, 1956; Kollros and Race, 1960; Pesetsky and Kollros, 1956; Pesetsky and Model, 1969, 1971; Reynolds, 1963, 1966; Prestige, 1965, 1973; Hughes, 1966, 1968; Kollros, 1968, 1972, 1981; Kaltenbach, 1968; Decker, 1976, 1977; Dodd and Dodd, 1976; Gona and Gona 1977). It is likely also that the substantial demyelinization that occurs in the CNS of the larva during metamorphosis (seen in *Bufo bufo*), may be influenced by the thyroid hormones (del Grande and Franceschini, 1982).

The fine structure of LMC neurones of *R. pipiens* larvae has been described by Decker (1974a,b; 1976). Young LMC cells at TK St. VI–VII (equivalent to NF St.52–53 of *Xenopus,* at about the middle of premetamorphosis, Etkin, 1968), resemble those of hypophysectomized larvae ultrastructurally and cytochemically. The columns of neurones include cells separated from each other by a distance of about 15–20 nm, and they possess a high ratio of free to membrane-bound ribosomes, a Golgi apparatus, mitochondria, and microtubules, and occasionally a dense body or autophagic vacuole. Nucleoside diphosphatase, acid phosphatase (the most common enzyme), aryl sulfatase, and cathepsin-like esterase are deposited in GERL (Golgi-endoplasmic reticulum–lysosomal complex) (Novikoff, 1963; Novikoff et al., 1964). Occasionally acetylcholine-esterase is present within the inner Golgi saccules. Administration of thyroxine to larvae results in about 1/3 of the LMC neurones differentiating and becoming bipolar, with a well-developed RER arranged in rows as Nissl substance positive for acetylcholine esterase. There is an accumulation of lipid characteristic of aging cells. The majority of T_4-treated LMC cells of hypophysectomized larvae (as in the case of normally developing LMC cells) (Race, 1961; Decker, 1976) degenerate, probably because they fail to establish viable axonal peripheral connexions with the developing fore and hindlimb musculature (see Decker, 1977). The nuclei of the degenerating cells become chromatopycnotic, their Golgi cisternae dilate and fragment into small vesicles. Likewise mitochondrial cristae dilate and become opaque as they degenerate. Autophagic vacuoles, or secondary lysosomes, positive for acid phosphatase, appear when larvae are treated with low doses of T_4 ($\leqslant 10$ mg/L). Glial cells show acid phosphatase in GERL, Golgi saccules, dense bodies, and in heterophagic vacuoles, which also include aryl sulfatase. They phagocytose the neuronal cell debris. A higher dosage of T_4 ($\leqslant 50$ mg/L) influences the degree of hypertrophy of glial cells and their lysosomal enzyme content. In contrast the degree of lysosomal activity of the neurones seems to be directly related to the state of differentiation of the cell and large lysosomes develop only when low doses of T_4 are administered (Decker, 1976). T_4 administration (by injection into the tail musculature of normal or hypophysectomized larvae of *R. pipiens* at TK St. VII), led to a 4–8-fold increase in lysosomal acid hydrolase activity of LMC neurones within 5–6 d. Such lysosomal enzyme activity elicited neuronal necrosis. During this time the granular membranes of the lysosomes become more liable to treatment that disrupts membranes (see Gahan, 1967); T_4 may not be directly responsible for this change, though Hillier (1970) believes that T_3 and T_4 can bind to lipoprotein membranes. It seems that T_4 influences the synthesis and packaging of enzymes via *de novo* production of RNA and protein. During LMC cell death the distribution and activity of these enzymes are major factors in the process (see Decker, 1977).

The Larval Tail Nerve Cord

It is generally believed that tail degeneration of anuran larvae commences at the extreme tip and thence necrosis gradually extends more anteriorly, simultaneously with tail shortening. Furthermore, available evidence from light and electron microscopy reveals that during climax major necrosis is first recognized distally in the tail, and there is less or none seen on proceeding more anteriorly. Indeed this is especially noticeable in the case of the nerve cord, though muscle and notochordal tissue have a more extensive distal range of necrosis. However, intraperitoneal injection of T_3 into premetamorphic tadpoles of *R. catesbeiana* at TK St. VII–IX resulted in tail shortening, with significant reductions in length of the proximal, middle, and distal regions. The rate of shortening nevertheless was greatest distally (Dmytrenko and Kirby, 1981). Whether these results reflect the process that occurs in normal life is not clear, for the 'telescope effects,' representing sequential morphometric distortions in development when inductions are obtained by abnormally high levels of thyroid hormones in younger larval states, are well-documented.

In contrast to the selective degeneration of some cellular components of the spinal cord and ganglia of the larval body, all those of the tail nerve cord degeneration during climax (Fox, 1973b). A small number of cilia and numerous microvilli line the lumen of the pre- and prometamorphic, nondegenerate tail neural tube of *Rana temporaria* (Figs. 15, 16). Climactic tail degeneration begins at the tip and a small circumscribed region of the distal degenerating nerve cord persists as the tail shortens in the distoproximal direction, until its final disappearance. The reduced nerve cord lumen fills with collapsed neural tissue and lipid and pigment accumulate in the cells (Fig. 17). Autophagic vacuoles and larger cytolysomes, positive for acid phosphatase, are recognized, and a large number of membrane-bound bodies, of variable diameter, appear in the nerve cord cells at this time (Figs. 18, 19). Mesenchymal macrophages and granular cells phagocytose the degraded collagen surrounding the neural

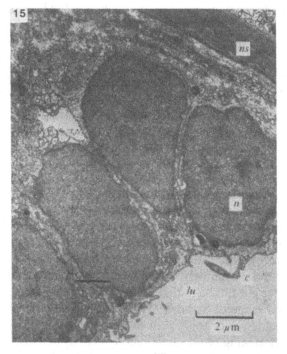

Fig. 15.   Ultrastructure of the tail nerve cord of *Rana temporaria* larvae during its growth and then degeneration at climax. CM stages 48–49 near the onset of climax. The TS section of a low magnification of the nerve cord shows cilia and fine microvilli lining the lumen; the nucleus practically fills the nerve cord cell.

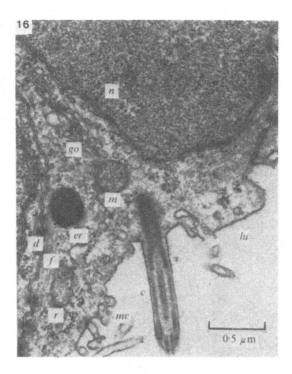

Fig. 16. Ultrastructure of the tail nerve cord of *Rana temporaria* larvae during its growth and then degeneration at climax. T.S. section at higher magnification of the lumen margin of a nerve cord cell at CM stages 48–49. Note the microfilaments extensively distributed in the cytoplasm, particularly near the intercellular junction, the prominent Golgi complex, ribosomes, and the granular and smooth endoplasmic reticulum.

tissue and they infiltrate degenerating cells of the tail nerve cord; presumably they ingest necrotic debris and digest it within their heterophagic vacuoles (Fox, 1973b and 1981).

18. Larval Musculature

Somite Origin and Myogenesis

The first indication of somites and hence of metameric segmentation in *Xenopus*, for example, is at NF St.17 in the neurula, and their segmentation and differentiation continue in the cranio-caudal direction. There are four somites at NF St.18, 4–6 (St.19), 6–7 (St.20), 8–9 (St.21), 9–10 (St.22), 12 (St.23), 15 (St.24), 16 with the first somite reduced (St.25), 17 with the loss of the first somite (St.26), 19 (St.27), more than 20 (St.28), 24–25 with the 2nd and 3rd somites reduced (St.29–30), 22–23

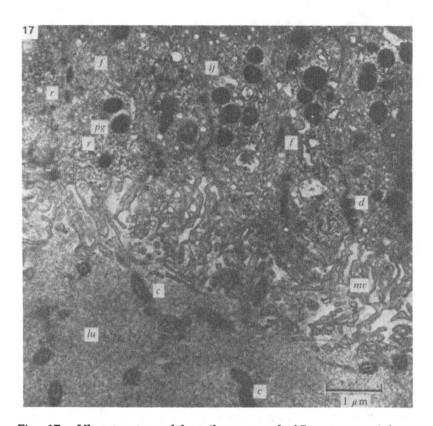

Fig. 17. Ultrastructure of the tail nerve cord of *Rana temporaria* larvae during its growth and then degeneration at climax. CM stages 51–53 during climax. L.S. of a region slightly proximal to the tail tip. A length of about five cells reveals a ragged apical microvillous margin at the lumen and cilia are still present though probably degraded products fill the lumen. Pigment granules are numerous, cytoplasmic microfilaments are extensively distributed, and there are pinocytotic vesicles. Nerve cord degeneration is localized mainly at the tip of the tail as it shortens, and some of the features described are predegenerative phenomena: c, cilium; d, desmosome; er, granular endoplasmic reticulum; f, nerve cell cytoplasmic filaments; go, Golgi complex; ij, intercellular junction; ln, lumen of nerve cord; m, mitochondrion; mv, microvillus; n, nucleus; ns, notochordal sheath; pg, pigment granule; r, ribosome (After Fox, 1973b.)

post-otic somites only; the otic vesicle is the location of the missing 4th head somite (St.31), 26 post-otic somites (St.32), 32 post-otic somites (St.33–34), 36 post-otic somites (St.35–36), and about 40 post-otic somites (St.37–38). Myocoeles begin to disappear in anterior somites at NF St.20 and gradually the somites are separated by myosepta of the connective tissue (see Nieuwkoop and Faber, 1956). The first three somites in the head, which disappear, probably reduce to mesenchyme. Ultimately the extrinsic eye muscles differentiate in this region, where the first recognizable cell masses arise at NF St.39. The ventromedial sclerotome portions of the somite, the presumptive axial mesenchyme, begin to separate off at NF St.24 and they are completely free by NF St.29–30, to completely envelop the notochord as the perichordal tube. Segregated somites are first seen in the tail at NF St.29–30. By NF St.32 myoblasts are spindle-shaped and differentiated myofibrils are present in the anterior tail somites by NF St.35–36.

Myogenesis in anuran and urodelan larvae has been investigated by light microscopy (see Kielbówna, 1966, 1975; Hamilton, 1969; Muntz, 1975), by tissue culture and audioradiography (Kielbówna and Koscielski, 1979) transmission electronmicroscopy (Hay, 1961a, 1963; Blackshaw and Warner, 1976b; Kilarski and Kozlowska, 1981; Peng *et*

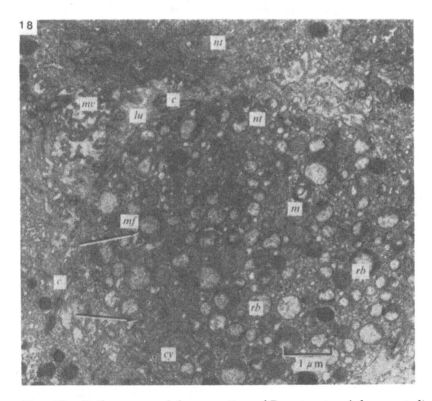

Fig. 18. Tail nerve cord degeneration of *Rana temporaria* larvae at climax. CM stage 52–53 (tail barely one-third of its original length). The nerve cord lumen is reduced and filled with degenerating nervous tissue containing numerous round bodies of varied electron density that may be lysosomal in nature. A cytolysome and myelin figure (both arrowed) are typically present. Cilia are still distinguishable at the lumen margin.

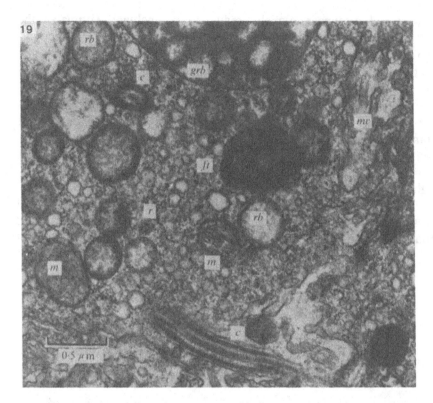

Fig. 19. Tail nerve cord degeneration of *Rana temporaria* larvae at climax. CM stage 52–53. Higher magnification of the lumen margin of a degenerating tail nerve cord showing dense round (lysosomal) bodies, still retained cilia and mitochondria, one of which shows incipient degeneration for its bounding membranes have ruptured and the cristae can disperse into the surrouding cytoplasm: c, cilium; cy, cytolysome; ft, dense fibrous tissue; grb, group of membrane-bound dense bodies, probably a large cytolysome; lu, lumen of the nerve cord; m, mitochondrion; mf, myelin figure; mv, microvillus; nt, nerve cord tissue, r, ribosomes; rb, round dense body (after Fox, 1973b).

al., 1981; Kielbówna, 1981) and by scanning electronmicroscopy (Kordylewski, 1978; Yoon and Malacinski, 1981). A single layer of flattened cells covering the lateral surface of the somites probably represents the dermatome (Kordylewski, 1978), a view in agreement with that of Hamilton (1969) who found the dermatome not to be segmented. In *Xenopus*, somite formation appears to be somewhat aberrant compared with that of other vertebrates. On each side of the embryo, cells nearest the notochord form a horseshoe-shaped arrangement in transverse section; a simple epithelium covers the open end of the horseshoe. Groups of about 10 cells, elongated in the transverse plane, separate from the anterior end of the paraxial mesoderm and rotate through 90°, so that the longer cellular axis becomes orientated anteroposteriorly. In essence a section of the horseshoe-shaped mass partially opens out, and the central ends of the cells move anteriorly and their lateral ends move posteriorly. Initially, therefore, each somite is 1 cell long and about 10 cells wide and is formed in about 1 h (Hamilton, 1969), though Blackshaw and Warner (1967a,b), who confirmed Hamilton's (1969) results, reported quicker formation in 20 min. The somite formation rate doubtless is a variable feature depending at least in part on individual egg clutches and temperature. It is of interest that somite formation in *Bombina* embryos is similar to that in *Xenopus* (Cooke, in Blackshaw and Warner, 1976b).

More recently Brustis (1979) showed in *Bufo bufo* embryos that somites form when unsegmented mesodermal cells make contact by filopodia, with resulting intercellular lacunae. Such cells first differentiate centrally in the somite and thence in the dorsal and ventral zones. "A" cells bordering the intersomitic areas elongate towards the somite center and "B" cells, which are located more deeply, become fusiform in shape. Cells "A" and "B" ultimately fuse to form giant multinuclear cells whose length equals the distance between intersomitic boundaries. The first myofilaments originate near the myoblast nucleus. Myofibrils first form near the middle of the somite and gradually they extend towards the myotomal septa.

Myotubes are first mononuclear and multinucleation, which occurs later on, is probably synchronous. Multinucleation of myotomal muscle fibers of larval *Ambystoma maculatum* is claimed by Muchmore (1962, 1965) to occur not merely by nuclear division of individual myoblasts (Kielbówna, 1966), but by incorporation and fusion of peripheral fibroblasts and interstitial cells, as well as by mitosis of existing myoblast nuclei. However, after middle tail bud stages, myoblast nuclei cease mitosis when cells fuse to form the more elaborate multinucleate fibers. Additional cells to the growing muscle originate by multiplication of peripheral and interstitial cells derived from surrounding mesenchyme. From transplantation experiments between portions of prospective somite tissue (whose nuclei were labeled with tritiated thymidine), of early embryos of *Ambystoma tigrinum* and *A. maculatum,* Loeffler (1969) clearly demonstrated and confirmed in vivo myoblast fusion to give rise to multinucleate myotubes of typical multinucleate muscle. Such myoblastic fusion was previously reported by Hay (1961a) in developing myotomal muscle of *Ambystoma opacum* larvae.

Somite formation and early myogenesis in *Xenopus laevis, Bombina variegata,* and *Pelobates fuscus* embryos were recently described by Kielbówna (1981), who confirmed the rotation of premyoblasts to form somites (see Hamilton, 1969). Mononuclear myoblasts of *Xenopus* and *Bombina* differentiate myofibrils and a sarcotubular system, and multinucleate myotubes form late by fusion of primary myotomal myoblasts with secondary mesenchymal (sclerotome) myoblasts (Kielbówna and Kotcielski, 1979). In *Pelobates,* myogenesis begins with primary myoblast fusion and myofibrils first appear in the multinuclear myotubes. These myotubes eventually fuse with mesenchymal myoblasts.

Other recent work by Yoon and Malacinski (1981b) has described somite formation to be different in urodele embryos compared with that in anurans. In *Pleurodeles waltl* and *Ambystoma mexicanum,* rosettes of presumptive myoblasts first form, but such configurations are not found in *Xenopus laevis* and *Rana sphenocephala* embryos. Again, the shape of the newly formed somites differs in urodeles and anurans. However, the rotation of myoblasts, their fusion and multinucleation during further development takes place in urodeles as in anurans, though unlike other amphibians they found the myoblasts of *Xenopus* apparently not to fuse, at least during their early formative stages of somatogenesis. The time of myoblastic fusion therefore seems to be variable in different amphibian species (see Kielbówna, 1981).

Differentiation of Myoblasts—Fine Structure

Descriptions from electronmicroscopy of the development and differentiation of muscles of tail somites of *Ambystoma opacum* (Hay, 1963), and body somites of *Xenopus laevis* (Blackshaw and Warner, 1976b) are broadly in agreement. In general it appears that myoblasts differentiate from myotomal mesenchymal cells rich in ribosomes and with smooth-surface vesicles, which later on form triads and the rest of the sarcoplasmic reticulum and the T-tubular system. In *Ambystoma,* for example, each muscle fiber (50–200 μm long) extends the length of the somite.

The main features of muscle development, with particular reference to amphibian somitogenesis, can thus be itemized. First, the proximal myotomes are larger and more mature than those that are situated more distally. Second, the earlier myoblasts, which are round or oval-shaped, have incipient scattered myofilaments already developed within them, free ribosomes, but little RER; the latter constituent is extensive in the fibroblastic mesenchymal cells. Third, myoblasts form a syncitium and fuse with adjacent myoblasts by lateral processes according to Hay (1963), or by gap junctions (about 20–40 nm wide) according to Blackshaw and Warner, 1976b), which may be a prerequisite of the full fusion of myoblasts to form multinucleate myotubes. Fourth, the first thick (myosin) and thin (actin) myofilaments are distributed randomly within the cytoplasm of the myoblast. It is of interest that Kilarski and Kozlowska (1981) described four classes of microfilaments, skein, actin, intermediate and myosin, in the developing myoblasts of *Salamandra* larvae, before distinctive actin and myosin myofilaments are established. Fifth, further development of the myoblast results in the heterogeneous combination of elongate orderly arranged myosin and actin myofilaments and the creation of Z bands and, by overlapping, other banding formations, so that the typical recognizable sarcomeres are formed. New myofilaments are added to the lateral and distal ends of existing myofibrils. Diferent stages of sarcomere development, including those of the sarcoplasmic reticulum and the T-tubules, occur within the same myoblast. Sixth, ribosomes are active in the formation of the myofilaments (however, for a detailed account of such subcellular processes, see Watson, 1976; Wilkie, 1976; and Grant, 1978, among others, on protein synthesis and organelle self-assembly). Seventh, it is probable that the synthesis of myosin and actin myofilaments in myoblasts can begin before myoblastic fusion occurs (Kielbówna and Koscielski, 1979). More recent results by Mohun et al. (1980) confirm and extend these observations as did those of Kielbówna (1981, *vide supra*). The former workers showed somitic mesoderm of axolotl embryos to have what they described as a reversible and then an irreversible phase of commitment as it differentiates into striated muscle. The synthesis of alpha-actin and first appearance of thin myofilaments occur at a transition stage intermediate between these phases. At all stages of development and in all tissues, beta-actin and gamma-actin are synthesized, but alpha-actin first appears at the head process stage and only in somites and tail bud where myotomal muscles form. Thin myofilaments originate in the somites near myocoeles and myoblastic fusion occurs afterwards. Well-developed sarcomers are only recognizable some time after the onset of motility.

In summary, therefore: The somitic musculature develops from myoblasts that become multinucleate by mitosis, and that fuse with adjacent myoblasts to form myotubes. There may be some further contribution from mesenchymal cells. Intracellular differentiation of the sarcomeres results in the formation of myofibrils, and increase in muscle volume occurs by the addition of new myoblasts that originate from interstitial and peripheral cells.

Before a description is given of the fine struggle of striated muscle, other kinds of muscle found in the vertebrates should be noted. The smooth, non-striated, involuntary muscles controlled by the autonomic nervous system are usually found in sheets surrounding hollow structures, such as the gut and the blood vessels. Typically, smooth muscle cells have their myofibrils oriented longitudinally with the nucleus at the center. The separate cells are anchored to one another by collagen or reticular fibers of the connective tissue. Cardiac muscle is composed of myoblasts that are apposed end-to-end, the adjacent ends thickening to form desmosomal-like junctions. Such irregular digitation of the cell junction along the long axis are the characteristic intercalated discs. The ontogeny of the larval amphibian heart has been described, among others, by Balinsky (1970) and Deuchar (1975).

The Fine Structure of Differentiated Larval Musculature

During metamorphosis widespread changes take place in the musculature (especially the striated) of anuran larvae, which involve enlargement, differentiation, and degeneration of some muscles, the acquisition of new ones, and changes in location and insertion (de Jongh, 1968). This is particularly true for head and branchial muscles; once they are established in the limbs, however (see Dunlap, 1966), muscle relationships are usually stable (Nieuwkoop and Faber, 1956).

Probably the msot dramatic recognizable change in muscle ontogeny is the complete disappearance of the tail muscles, when the anuran tail involutes at climax (see Fox, 1972c, 1975, 1977b; Watanabe and Sasaki, 1974; Kerr et al., 1974). Apart from enlargement (for thyroprivic larvae can continue to grow larger, though remain at the same stage of development), probably most if not all histomorphological (and topographical) changes that occur in the musculature during metamorphosis are directly or indirectly controlled by thyroid hormones. Muscular development and differentiation are inhibited in larvae treated with goitrogens and accelerated by treatment with thyroxine (Takisawa et al., 1976) and triiodothyronine, including the degeneration of other muscles such as those of the tail, destined to disappear at climax (Weber, 1962; Shaffer, 1963; Fox and Turner, 1967; Turner, 1973; Robinson et al., 1977).

Evidence of thyroid hormone participation in the origin and differentiation of the amphibian larval limb musculature was demonstrated by Dhanarajan and Atkinson (1981). Intraperitoneal administration of T_3 (3 $\times$ 10^{-10} mol T_3/g body weight) to TK St. VII tadpoles of *R. catesbeiana* induces precocious hindlimb development. Indirect immunofluorescence (see also Smidova et al., 1974 on the origin of somitic actin and myosin in urodele embryos), using antibodies against frog skeletal muscle M-line protein and alpha-actin, showed these proteins to be present in the undifferentiated thigh region within 24 h after treatment. Progressively in time myofibrillar protein and fiber protein are stimulated to spread to distal limb regions. The rates of protein synthesis are increased within the initial 24 th period. Whether T_3 (or T_4 for that matter) acts directly on the myoblasts or indirectly via, say, the peripheral nerves, however, is not clear. Investigations of hormone action on denervated limbs or on isolated myoblastic tissue in culture would be of value in this instance.

Muscles from different regions of an adult anuran differ somewhat in fine structure (Eichelberg and Schneider, 1973); this feature also occurs in larval jaw muscles during metamorphosis (Ichikawa and Ichikawa,

1969), and mitochondria of tail muscles of *Rana* differ in size from those of *Xenopus,* which are bigger (Fox, 1975). Sasaki (1974) described tails of species of *Rana, Bufo bufo japonicus,* and *Xenopus laevis* to include superficially situated muscle fibers (pm fibers) classified as red muscle, which had abundant sarcoplasm and mitochondria and deeper muscle fibers (im fibers), or white muscle, with fibers of relatively larger diameter, less sarcoplasm, and fewer mitochondria. There is greater activity of lactic dehydrogenase and succinic dehydrogenase, but less phosphorylase activity in the pm fibers than in the im fibers. Recent further investigations of the enzymology of the larval tail musculature, of various anurans, have been made on lactic dehydrogenase (LDH) (Sasaki, 1979) and ATPase. LDH products were found in the mitochondria, muscle cell membrane, pinocytotic vesicles, T-tubules, vesicular triads, and the sarcoplasm of both the red and white muscle fibrils. The red muscle appears to show higher ATPase activity than the white muscle, and red fibers ar probably active during slow swimming (Watanabe et al., 1978a); however, ATPase activity increases in the white muscle fibers during tail degeneration (Watanabe et al., 1978b). In the latter case the authors suggest various reasons. Conformational change may occur in the enzyme, perhaps because of the binding of myosin to thyroxine, or to one of its metabolic products; or perhaps there is *de novo* synthesis of new myosin; or possibly induction of myosin helix-promoting amino acids such as lysine and aspartic acid; or perhaps there are various combinations of these possibilities.

On the whole the ultrastructure of anuran tail muscle would seem to be generally similar to that of their other larval muscles, and doubtless presumably similar biochemical mechanisms operate if they too suffer degeneration.

The striated tail musculature of an amphibian larva comprises segmental somites, separated by myocommata of connective tissue, situated on each side between the skin and the nerve cord and notochord. Somite size reduces gradually on proceeding distally (Brown, 1946). Like those of other vertebrates, tail muscle tissue of *Rana* and *Xenopus* is composed of sarcoplasm surrounding a large number of parallel, longitudinally oriented, and closely packed myofibrils each about 1 μm thick. The myofibrils have cross-striations of regularly repeating periodicity resulting from the overlapping constituent thick (16 nm) and thin (6 nm) myofilaments of myosin and actin, respectively. Cross-bridges link thick and thin myofilaments at intervals of about 40 nm. In longitudinal section the Z lines are usually straight and the sarcomere is designated as that region between two Z lines. At the center of the A band of thick myofilaments, and H zone includes a central thickened M line where adjacent thick myofilaments join. Thin myofilaments, which originate from the Z line where they tend to be thicker, form the I band and they extend to the edge of the H zone. The dimensions of the I band vary according to the degree of contraction of the sarcomere, which reflects the mutual sliding of overlapping myofilaments whose lengths remain unchanged (see Price, 1969; Fox, 1975; Figs. 20, 21). At CM St.45 and 48–49 of *Rana temporaria*, the mean widths (and standard errors) of the A band were 1.33 ± 0.04 μm and 1.44 ± 0.01 μm, respectively. In *Xenopus laevis* at NF St.45 and 57, these measurements were 1.42 ± 0.04 μm and 1.42 ± 0.09 μm, respectively, and none of these measurements differed significantly. Thus A bands of functional tail muscles of *Rana* and *Xenopus* are stable in width during prometamorphosis, before their climactic degeneration. A bands of twitch and slow muscles of the longus extensor digitorum of an adult *Rana temporaria* were found to be 1.45–1.50 μm wide (Page, 1965), of similar size to those in the larval tail muscles of *Rana* and *Xenopus* (Fox, 1977b).

The sarcoplasm contains ribosomes and polyribosomes, some RER, mitochondria, lipid especially in younger larvae, pigment, and glycogen.

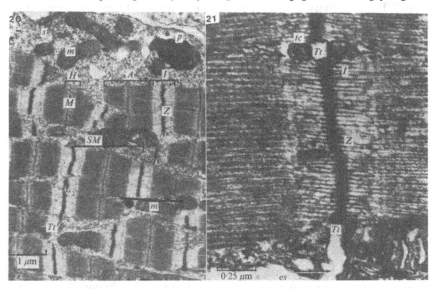

Fig. 20–24. Ultrastructure of tail muscle of *Rana temporaria* larvae during its growth and degeneration at climax.

Fig. 20. CM stage 45 near the onset of prometamorphosis showing the typical banding pattern of nondegenerate striated muscle.

Fig. 21. CM stage 48–49 near the onset of metamorphic climax showing details of the Z-I band region of a nondegenerate-striated myofibril. The triad includes the T-tubule at the base of the Z line and the adjacent terminal cisternae of the sarcoplasmic reticulum. In this instance the T-tubule is open into the extracellular space.

An occasional Golgi complex may be recognized. Large nuclei are located peripherally. A sarcoplasmic reticulum embraces the myofilaments longitudinally, and a closely associated, but separate T-tubular system encircles the myofibrils, at the Z lines, at right angles, and it opens to the exterior at the surface of the cell (see Franzini-Armstrong et al., 1975; Leeson, 1977; Fig. 21). Triads are present at the bases of the Z lines. The trilaminar sarcolemma, bounded externally by a densish basement membrane, is indented by surrounding collagen fibrils which separate blocks of muscles (Fox, 1975). The fine structure of the myotendinous junction at the sarcolemma, which transfers tension from the muscle to the tendon,

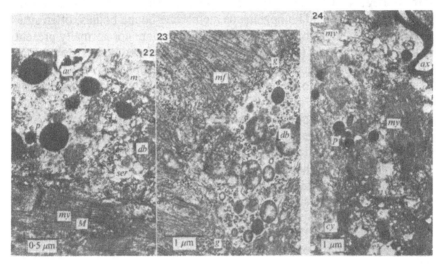

Fig. 22. CM stage 52 at the height of climax (tail half size). The myofibrils are in the process of losing their striations and vestigial Z lines, I bands, and M lines are recognizable. The sarcoplasm now includes numerous dense round bodies that are not present in preclimactic nondegenerate muscle.

Fig. 23. CM stage 53 during climax. The dense round bodies of varied profile in the sarcoplasm of degenerating muscle are comparable to organelles found in degenerating tail nerve cord and they are probably lysosomal.

Fig. 24. CM Stage 52. Myofibrils are variable in appearance as they lose their striation. Extensive areas of necrosis in the sarcoplasm include large cytolysomes and pigment accumulates: A, A band; av, autophagic vacuole; ax, axon to muscle; cy, cytolysome; db, dense round body; es, extracellular space; g, glycogen granules; H, H zone; I, I band; M, M line; m, mitochondria; mf, myofilament; my, myofibrils; p, pigment body; s, sarcoplasm; ser, smooth endoplasmic reticulum; SM, sarcomere; Tt, T-tubule of triad; tc, terminal cisterna of sarcoplasmic reticulum; Z, Z line (after Fox, 1975).

has been described in the tadpole tail myotomal musculature of *Rana rugosa* by Nakao (1976).

Degeneration of the Larval Tail Musculature During Metamorphic Climax

As in the case of other anuran tail tissues, muscle necrosis is more severe distally in the tail as it involutes. Even before the myofibrils show features of degeneration, the sarcoplasm develops various whorled, mottled, or smooth-surface and homogeneous membrane-bound bodies, often situated near the nucleus or between myofibrils, that are not normally present in preclimactic larvae (Figs. 22, 23). Smooth-surfaced bodies are profuse in degenerating tail nerve cord cells and in macrophages (Fox, 1973b). Myelin figures are also present, often near the sarcolemma. Mitochondria eventually lose their cristae to become empty vesicles, or their outer membranes are lost and the contents dispersed amid the cytoplasm, in a similar fashion to the mitochondrial degeneration in cells of tail notochord and nerve cord (Fox, 1973a,b). Small autophagic vacuoles and cytolysomes make their appearance in the sarcoplasm and smooth-surfaced vesicles, lipid, pigment, and glycogen granules increase in amount. As necrotic areas become more widespread, ribosomes and the Golgi complexes disappear and larger cytolysomes are occasionally recognizable (Fig. 24). The nucleus is slow to degenerate, often appearing normal amid regions of sarcoplasmic and myofibrillar degeneration (Fig. 29).

Myofibrils gradually lose their banding pattern and the Z lines often appear wavy in outline and less sharply demarcated. Myofilaments eventually become randomly oriented, less clearly delineated, and blurred in outline, though frequently they are still recognizable when present amid highly degraded myofibrillar tissues. Disorientated myofilaments often merge into electron-translucent homogeneous substance in the sarcoplasm (Fig. 27), and electron-dense fragments of myofibrils, or sarcolytes (Brown, 1946), occur simultaneously with similar but less dense tissues (Figs. 25, 26). Ultimately myofibrils degrade into roundish areas, of variable size, of homogeneous substance, sometimes containing a core of myofilaments or showing spatial continuity with lesser degraded and recognizable myofilaments (Figs. 27, 28). All stages from hardly degenerate striated myofibrils to the fully degraded nonfibrillar tissue can be traced through climax, often in different regions of an individual degenerating tail or in tails of successive larval stages. Eventually the sarcolemma ruptures and debris of the sarcoplasm and degraded myofibrils is phagocytosed by invasive macrophages and digested within heterophagic vacuoles or phagolysomes (Weber, 1964; Watanabe and Sasaki, 1974; Kerr et al., 1974; Fox, 1975).

Some earlier workers believed that enzymes involved in muscle degeneration, influenced by a variety of causes, were derived from macrophages or other phagocytic leucocytes. Weber (1964) concluded that tail muscles of *Xenopus* larvae degenerate without the intervention of lysosomal enzymes, which are mainly concerned with the digestion of degraded muscle products ingested within macrophagic phagosomes. Furthermore, in the involuting tail of *Rana pipiens,* various enzymes were traced in a number of tissues, but not in striated muscle (Kaltenbach, 1971; Fry et al., 1973); in nearby mesenchyme and myosepta (Sasaki and Watanabe, 1983). Others, however, believe that some lysosomal enzymes originate in the muscles themselves and participate in their autolysis (see refs. in Fox, 1975, and Bird, 1975). In muscular degeneration atrophy of mammals, though lyosomes increase in size and number as atrophy progresses (Pellegrino and Franzini, 1963), degenerating myofilaments were not found in their autophagic vacuoles (Schiaffino and Hanzlikova, 1972). In general these workers concluded that extralysosomal enzymes activated myofibrillar degeneration, and those enzymes from lysosomes are involved in final autolysis and phagocytosis. Nevertheless, lysosomal enzymes seem to be involved in sarcoplasmic autolysis in *Rana* tail muscle (Watanabe and Sasaki, 1974; also *vide supra*), though evidence of their activity in the case of the myofibrils is still not convincing (Fox, 1975). The evidence for the presence and activity of muscle lysosomal enzymes has been reviewed by Bird (1975). More recently it was found that in skeletal muscle (and in liver) of hypophysectomized rats, the total lysosomal enzyme activity (cathepsin D, cathepsin B, acid phosphatase, and N-acetylglucosaminidase) increased after administration of physiologic or thyrotoxic dosages of T_3 or T_4, but not in kidney and heart tissues. Skeletal muscle of thyroidectomized rats had these enzyme levels reduced to approximately 50% of normal levels. Thyroid hormone-induced lysosomal enzyme activity and the ensuing protein degradation are probably the result of increased levels of proteases in primary lysosomes and thence autophagocytosis. The severe muscle wastage in human hyperthyroidism is the result of such protein degradation. It is unlikely that such changes in lysosomal enzyme activity are a result of the invasion of muscle (or liver) tissue by phagocytes rich in lysosomes (see De Martino and Goldberg, 1978).

Further elucidation of the origin and function of the muscle lysosomal system, and the possible involvement of extralysosomal proteases in myofibrillar degeneration, is awaited. It seems likely, however, that this system, whether vertebrate or invertebrate, is poorly developed compared with that of other tissues.

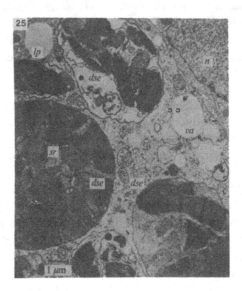

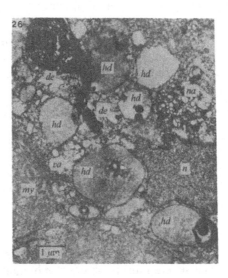

Fig. 25–29. Ultrastructure of degenerating tail muscle of larvae of
Rana temporaria at CM stage 53, near the end of climax (tail a short stub).
Fig. 25. Degenerate myofibrils (sarcolytes) of varied density are present.
Some portions are isolated in membrane-bound areas where they degrade autolytically to a homogeneous substance. Fig. 26. Heavy necrosis
in the sarcoplasm showing profiles of myofibrillar degeneration that includes disorganized and dense myofilaments without striations and
highly degraded homogeneous myofibrillar substance in which vestigial
myofilaments may be present. A nondegenerate nucleus is still recognizable.

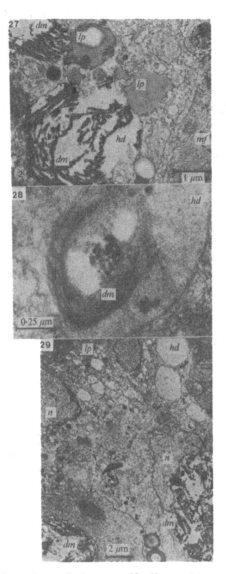

Fig. 27. Highly degraded myofibrillar tissue with vestigial myofilaments. Lipid is present and lesser degraded myofilaments are seen nearby. Fig. 28. Origin of the highly degraded homogeneous substance that is derived from the breakdown of myofibrils. Fig. 29. Varied array of profiles of tail muscle degeneration at climax. They range from striated to unstriated myofibrils, highly electron dense sarcolytes at varying stages of degradation, to fully degraded homogeneous tissue of myofibrillar derivation. Nuclei apparently are the last organelles to degenerate: de, dense myofilamentous tissue; dm, dense myofilaments in final stages of degradation; dse, electron-dense sarcolytes; hd, highly degraded myofilaments; lp, lipid droplet; mf, myofilaments; my, myofibrils; n, nucleus; na, necrotic areas; sr, sarcoplasmic reticulum; va, vacuolation (after Fox, 1975).

19. The Blood

Origin in the Larva

The sites of hematopoiesis in various amphibians vary with age and to some extent with the systematic group. The first blood cells to appear are mesenchymatous in the blood islets of the embryonic blastopore–liver area. Later during larval development erythropoiesis tends to occur where there are blood sinuses and a slow but copious blood supply, with little oxygen and probably a high tension of CO_2. Indeed hypoxia is believed to be a major initiating factor in stimulating erythropoiesis in vertebrates, though lactate is also claimed to be significantly involved in this process (see Frangioni and Borgioli, 1978). This situation can occur in the larval pronephroi and mesonephroi and in the spleen and probably liver of older tadpoles, which together with the marrow of the long bones of adults are generally claimed to be the erythropoietic centers. However, the main erythropoietic sites in larval tadpoles (*R. catesbeiana*) are the mesonephroi and liver (Broyles et al., 1981). Granulocytopoiesis and lymphocytopoiesis occur in the larval kidneys and in the thymus and bone marrow; possibly there are also sites in the wall of the intestine (see Jordan and Speidel, 1923, 1929; Jordan, 1938).

Hematopoiesis in vertebrates including larvae and postmetamorphosed forms of anurans and urodeles was reviewed by Douarin (1966), who also described the presumptive areas of the anuran blastula, where blood islets originate, and their subsequent development.

More recently Turpen et al. (1979) described hepatopoiesis in *R. pipiens* of lymphoid, myeloid and erythroid cell lineages, initially to begin near the end of embryogenesis at Shumway St.24–25, and it is substantial after 30 days at TK St.IV. Granulopoiesis remains stable, but erythropoiesis increases to become the predominant component in hepatopoiesis. Blood cells form within discrete loci in subcapsular or deeper regions of the liver in sub-endothelial sinuses, though the primary origin is from extrinsic stem cells originating in the dorsal embryonic region. Again Carpenter and Turpen (1979) have shown that in the pronephros of *R. pipiens,* osmoregulation apart, granulopoiesis is a major, erythropoiesis a minor (< 10%), and lymphopoiesis a negligible hemopoietic activity. Furthermore, hemopoiesis is here dependent upon colonization of the pronephros by extrinsic stem cells of unknown origin, though they probably do not arise from embryonic blood islets. Turpen and Knudson (1982) suggest that it could be the lateral plate mesoderm in embryos of *Rana pipiens* at Shumway St.14. During embryogeny the pronephros may well be the initial site of granulocyte differentiation.

It is possible that all the blood cells of a developing amphibian may well originate from a primitive stem cell type, a hemocytoblast, which by modification gives rise to the variety of corpuscles ultimately recognized modification gives rise to the variety of corpuscles ultimately recognized in the circulation (Foxon, 1964), whose origin, structure, and differentiation have been described (see Jordan, 1938; Foxon, 1964; Cooper, 1976; Broyles, 1981) and *vide supra*. The amphibian erythrocytes (RBCs) are of particular interest because they are nucleated (except for *Batrachoseps*) (Cohen 1982) and they synthesize different types of hemoglobin (Hb) in larvae and adults, the changeover occurring during metamorphosis (see Herner and Frieden, 1961; Moss and Ingram, 1968a,b; Frieden and Just, 1970; Benbassat, 1974a,b). An exception, however, appears to be *Ambystoma tigrinum nebulosum,* for according to Woody and Justus (1974) the Hb is unchanged at metamorphosis. Another difference is the fact that the content of inorganic phosphate decreases in erythrocytes during metamorphosis, in adaptation to the terrestrial respiration practiced by juveniles and adults (Hazard and Hutchinson, 1978).

Erythrocyte Hemoglobin

Hemoglobin is a conjugated protein composed of four molecules of ferroprotoporphyrin attached to a single globulin molecule. Human hemoglobins have a molecular weight of 68,000, similar to the Hb of the larval and adult *R. catesbeiana* (Riggs, 1951; Trader and Frieden, 1966), and the globin moiety comprises two identical half molecules each of which consists of two different polypeptide chains. In normal human adults there are alpha and beta chains in HbA; alpha and gamma chains occur in normal fetal HbF, and the normal minor component HbA_2 has alpha and delta chains. The synthesis of these four polypeptide chains is controlled by four distinct genes, each one responsible for a separate chain. Hb variants contain altered combinations of the four chains or changed sequences of their amino acids.

Frieden and Just (1970) summarized the main differences between amphibian larval and adult hemoglobins. The biochemical differences are exclusive to the globin portion of the molecule. It seems that most authorities believe that larval and adult anuran Hb do not share a common polypeptide chain (see Sullivan, 1974; Ferigo et al., 1977), though a common alpha chain has been claimed (Hamada and Shukuya, 1966). Larval Hb has no SH groups but they are present in that of the adult (Riggs, 1960; Hamada et al., 1966), a feature used to demonstrate the transition of the larval to the adult Hb. In *R. catesbeiana* there are four major larval Hbs (I–IV) and four major adult Hbs (A–C), some details of which are described by Broyles (1981). The larval Hbs are tetramers of two alpha- and two beta-type subunits. Larval and adult Hb of the bullfrog were shown to have differences in their affinity for O_2, electrophoretic or chromatographic mobility, finger print peptide pattern, and antigenic determinants (see refs. in Benbassat, 1974a, and Ferigo et al., 1977). Again larval erythrocytes of *Pleurodeles waltl* have five molecular types of Hb formed from at least three polypeptide subunits, the major fraction made up of alpha- and beta-like chains the other three fractions of one subunit. There is no identity between the larval and adult globin chains and the switchover at metamorphosis is complete (Flavin et al., 1978).

The switch from larval to adult Hb in the blood of *Pleurodeles waltl* begins before the onset of metamorphosis, which is recognized by the simultaneous synthesis of their respective globin chains in equivalent quantities at this time. Transition is gradual and there is a decline of larval Hb synthesis concomitant with an increase in that of adult Hb (Flavin et al., 1979), a changeover speeded up by treatment with T_4 (Flavin et al., 1982). This ontogenetic pattern is comparable to that which occurs in anuran larvae during metamorphosis. Treatment of *P. waltl* larvae with antithyroid substances to inhibit metamorphosis does not prevent the normal changeover from larval to adult Hb (Flavin, 1973; see Flavin et al., 1979, 1982). This was also found to be the case in *Xenopus* (Maclean and Turner, 1976; Just et al., 1977) (*vide infra*).

In *Xenopus laevis* MacLean and Jurd (1971) using chromatographic and electrophoretic methods, found two types of larval hemoglobins, namely $XHbF_1$ and $XHbF_2$, with different elution peaks that persist throughout larval life. $XHbF_1$ is more plentiful and the proportion of $XHbF_1$ to $XHbF_2$ increases during larval development. Almost all the larval Hb has disappeared after metamorphosis, though traces are still found in adults. Two types of adult Hb also occur and these first appear in late larval stages. $XHbA_1$ comprises the majority of adult *Xenopus* Hb and between 5 and 10% of the total is $XHbA_2$, which is not a polymer of HbA_1. Just et al. (1980) found less than 1% of adult Hb in erythrocytes of larval *Xenopus,* but on completion of metamorphosis 2 wk later there was 90% adult Hb in toadlet red cells and barely 1% larval Hb. During the relatively short transition period these authors suggested that there was a loss of larval red cells and a new adult red cell line originated.

The complete amino acid sequences of the alpha chain (141 residues) and beta chain (117 residues), of component III of Hb of tadpoles of *R. catesbeiana,* were determined by Maruyama et al. (1980) and Watt et al. (1980), respectively. Comparisons of the sequence of the beta chain with that of the adult bullfrog showed that only 50% (59 out of 117) and with that of humans 55% (81 out of 146) are identical. In the case of the alpha chain, 57% (80 residues) of the sequence is identical with that from the human alpha chain, and about the same degree of similarity was also found with the adult bull frog, which so far has only been partially analyzed. On the basis of an estimate that the overall rate of change of Hbs is about 1% every 3.3 million years, then a 50% difference would correspond to about 165 million years, a time scale that can be related to that of the earliest known frogs seen in fossil strata 170–190 million years ago.

The authors speculated that the large differences between the Hb alpha and beta chains in the same organism suggest that gene duplication may have occurred as long ago as the phylogenetic roots of frogs and humans. It is of interest that no specifically embryonic (fetal) Hb distinct from that of adults appears to occur in the viviparous caecilian, *Typhlonectes compressicauda* (see Watt et al., 1980), which may indicate independent multiple origins of the different kinds of Hb in various organisms.

The differences in the physiological properties of amphibian larval and adult hemoglobins (see Sullivan, 1974) are clearly apparent in their relationship to the oxygen dissociation curve (McCutcheon, 1936). The blood of the larval *R. catesbeiana* has an O_2 capacity of 7.58 vol% and that of the adult 11.02 vol%. Tadpole blood has a dissociation curve in the form of a regular hyperbola, a curve suitable when unloading is required at low O_2 tensions, where there may be a scarcity of oxygen. The adult blood dissociation curve is of a sigmoid type, typical of terrestrial air-breathing forms. In young tadpoles the Hb of red blood cells is 50% saturated at 5 mm partial presssure of Hg, but at the same partial pressure adult blood is only 5% O_2 saturated. In amphibians carbon dioxide is eliminated mainly through the skin and even in adults there is little elimination by way of the lungs. It is carried in the blood predominantly as carbonic acid and bicarbonate.

In 1957, de Graf described an apparently normal adult *Xenopus laevis* that was devoid of erythrocytes and Hb. Ewer (1959) found a similar specimen and from its examination concluded that such freaks may not have always lacked Hb, but obviously they can survive for extended periods without it. de Graf (1957) suggested that perhaps oxygen was mainly stored in the blood rather than in transit. Anemic frogs and larvae may thus not suffer to the same extent as in higher vertebrate forms experiencing similar symptoms.

Changeover of Larval to Adult Hb During Metamorphosis

As Frieden and Just (1970) have pointed out, the transition of larval to adult Hb during amphibian metamorphosis is a most promising system to study cellular diversity and hormonal influences. Thyroxine is known to stimulate the production of adult Hb in larval erythrocytes (Moss and Ingram, 1965, 1968a,b). There are several ways such cells could possibly change their biochemical behavior. The same stem cells may continue production of a single population of red cells in various sites, and the changeover to adult Hb would occur within these individual cells. Alternatively a different new population of adult red blood cells may arise from different erythropoietic centers. In this case the larval erythrocytes would be suppressed and disappear and adult ones would be activated. If the first view is correct, then the same erythrocytes should contain both types of Hb during a transition period, with a tendency for an increasing and relatively greater percentage of adult Hb to be present within them as metamorphosis is completed and juvenile froglets are developed. If the second view applies, then individual erythrocytes would only contain one type of Hb, never the two together within the same cell.

The fact that erythropoiesis shifts from the kidney to the spleen and bone marrow during metamorphosis (Jordan and Speidel, 1923) is of no help in clarifying the problem, for migration of some of the original population of stem cells to the new sites may have occurred, or that from their commencement these cells may have included a share of original stem cells. Larval and adult amphibian erythrocytes seem to be generally similar in appearance, though possibly they are slightly smaller in adults of *R. catesbeiana* (McCutcheon, 1936). Those of the larval *R. pipiens* are mainly large and elliptical though some appeared small and round with occasional mitotic figures. Both examples of larval red blood cells had round nuclei and reticular substance in the cytoplasm, which stained with methylene blue. Adult erythrocytes were less pleiomorphic, and had elliptical nuclei and reticular material detectable with methylene blue. Premetamorphic red blood cells contained less Hb and appeared to be more resistant to osmotic lysis than those of adults. However, in media of the same osmolality, erythrocytes of larvae and adults are similar in size, and variation in their volume occurs with changes in the tonicity of the medium. Thus differences in size (Hollyfield, 1967) may really reflect differences in larval and frog osmolarity of the circulation (Benbassat, 1970). Nevertheless the differences noted between larval and adult RBCs in anurans may well be real for there are biochemical differences also in adenyl cyclase activity. In tadpoles of *R. pipiens* and *R. catesbeiana*, it is unresponsive to stimulation by catecholamines, but adenyl cyclase of adult RBCs responds with increased synthesis of cAMP (see Broyles, 1981). Again, Forman and Just (1981) quantified Hb transition by separating larval and adult RBCs (which differ in density) of *R. catesbeiana* on continuous density gradients. The production of adult RBCs began just before TK St.XVIII–XIX, and they had replaced the larval blood cells by the end of climax. Immersion of larvae (at TK St.III–XIV) in thyroxine solution ($2.5 \times 10^{-8}M$ L-T_4) stimulates the production of adult RBCs and within 71 days 90% of circulating RBCs are of the adult type.

Information on the life span of erythrocytes of amphibians reveals values quoted of high variation, and such periods should be treated with some degree of caution. Thus those of the larval *R. catesbeiana* are claimed to survive for about 100 d, but only 24 d in the adult. Adult erythrocytes of *R. pipiens* (200 days) and *Bufo marinus* (700–1400 d) have been reported (see Forman and Just, 1976).

The increase in the synthesis of adult red blood cell Hb in *R. catesbeiana* elicited by thyroxine was associated with smaller, rounded, lighter-staining cells with a relatively larger nucleus (Moss and Ingram, 1968b), similar to the small erythrocytes reported in metamorphosing larvae of *R. pipiens* (Hollyfield, 1966) and *R. catesbeiana* (Benbassat, 1970). However, de Witt (1968), who showed that these cells synthesize adult Hb, considered that they were not normal erythrocytes, but probably appeared as a result of metabolic stress.

It is not clear whether, in *R. catesbeiana*, for example, transformation from larval to adult Hb is fairly rapid during metamorphosis, say within 2 wk (Hamada et al., 1966; Just and Atkinson, 1972) or slower lasting from 4 to 10 wk (Benbassat, 1970). Indeed Theil (1967) reported that the nonhemoglobin protein of the erythrocytes of larval *R. catesbeiana*, present in TK St.XI–XXIII, disappeared together with the larval Hb and was replaced by adult Hb during TK St.XXIII–XXIV, the changeover occurring between 5 and 10 d. Furthermore the entire population of the larval erythrocytes was claimed to be replaced during this period of Hb transformation (Theil, 1970). Nor are we certain whether erythrocytes that synthesize adult Hb are drawn from stem cells (perhaps the smaller erythrocytes, see Moss and Ingram, 1968a,b; Benbassat, 1970), clonally distinct from larval ones. Moss and Ingram (1965, 1968b) concluded that erythrocytes of the larval *R. catesbeiana* may be suppressed and those of the adult activated, possibly when there is a shift in erythropoietic loci during metamorphosis. This view receives some support from the claim that, in metamorphosing bullfrog tadpoles, the erythrocytes contain either tadpole or adult Hb only, never both coexisting within the same cell (Rosenberg, 1970; Maniatis and Ingram, 1971; see also Dorn and Broyles, 1982). In contrast, however, Shukuya (1966) reported the presence of larval and adult Hb within the same erythrocytes, and important results by Jurd and MacLean (1970), using fluorescent immunoblobulin antibodies acting on smears of red blood cells from metamorphosing larvae of *Xenopus laevis*, revealed that up to 25% of them had both larval and adult Hb. The number of cells with mixed hemoglobins decreased as metamorphosis was completed, the cells finally containing only adult Hb. Furthermore, *Xenopus* adults rendered anemic commenced the resynthesis of one of the larval types of Hb in many of their erythrocytes (MacLean and Jurd, 1971). Benbassat (1974b) also concluded that some erythrocytes at climax contained tadpole and adult Hb, and he suggested the possibility that the antisera against bullfrog tadpole or adult blood, used by Maniatis and Ingram (1971), had a higher degree of monospecificity than those used by Benbassat. Alternatively the sandwich method of the latter was more sensitive than the direct double-label method used by Maniatis and Ingram. It is of particular interest that all types of protein may be produced by successive generations of the same cell line in the crystalline lens, and that human embryonic and fetal Hb, and likewise fetal and adult Hb may coexist within the same red blood cell (see Benbassat, 1974a, for references). Nevertheless, whether amphibian larval and adult Hbs are present within the same RBC or are restricted to different RBCs is still not conclusively proven. Sullivan (1974) writes "it seems clear now that the tadpole to frog hemoglobin transition takes place in separate cells in *Rana* and that the control mechanism of interest probably occurs at the tissue level." Broyles (1981) goes on to argue in favor of different larval Hbs elaborated by cells in the liver and mesonephric kidney, respectively, which are separate populations of circulating RBCs differing morphologically. The larval type-2 RBCs (from the liver) are of the same lineage as the adult RBCs. The results of Dorn and Broyles (1982) appear to support the view that a new cell line of adult erythrocytes replaces that of the larva during metamorphosis. However, further evidence is needed to support this interesting hypothesis. The sequential events of erythropoiesis and Hb synthesis in *R. pipiens* and *R. catesbeiana,* based on present knowledge, have been summarized by Broyles (1981), who elegantly emphasised the many as yet unanswered questions.

Thyroid Hormones and the Blood During Metamorphosis

The stimulus for the changeover from larval to adult Hb synthesis is assumed to result from the increase in circulatory levels of thyroid hormones near and at climax. It has been shown the the concentrations of T_4 in serum and pericardial fluid of metamorphosing tadpoles of *Rana catesbeiana* are similar, though T_4 was only detectable from TK St.XIX–XX. Concentration peaked during climax, but by TK St.XXIV it had reduced to TK St.XIX–XX levels. In summer adults the mean T_4 values of these fluids were found by Mondou and Kaltenbach (1979) to be somewhat higher than in larvae at the end of metamorphosis, but measurements were variable, even below the limits of detection in some climactic larvae and adult frogs, so the results should be treated with caution.

In vitro the rate of bullfrog Hb synthesis by normal erythrocytes, from tadpoles immersed in thyroxine ($5 \times 10^{-8}M$ L-thyroxine), is first markedly decreased to be followed by the synthesis of adult Hb (Moss and Ingram, 1965). Forman and Just (1981) likewise prematurely stimulated the production of adult RBCs and their Hb in bullfrog larvae *in vivo,* using T_4 (see also Flavin et al., 1982, *vide supra*). Whether thyroxine acts directly or indirectly on the blood cells is not known. However, *Xenopus* tadpoles at NF St.54, immersed in the goitrogen 0.01% propylthiouracil for up to 18 months, became twice as long and showed almost a sixfold increase in wet weight. The stage number, nevertheless, remained unchanged, though untreated larval controls had metamorphosed 16–17 months earlier. Normal untreated larvae at NF St.54 possess larval Hb in their erythrocytes, but the goitrogenically treated and inhibited 'giants' have erythrocytes with the adult type HbA_1 (MacLean and Jurd, 1971; MacLean and Turner, 1976). Thus either hemoglobin changeover during metamorphosis is independent of thyroid hormones, which would seem to be unlikely, or an infinitesimal production of thyroid hormones that the goitrogen fails to prevent during the extended period results in a stoichiometric effect (see Etkin, 1964, 1968). But if this were the case, then some modest change in stage number (based on external appearance) would also be expected to occur. Again if increased thyroid hormone levels act mainly by suppressing the continued synthesis of larval Hb in erythrocytes, thus stimulating that of adult Hb at climax (Moss and Ingram, 1965), then T_4 inhibition in the goitrogen-treated larvae should result in continued larval synthesis of Hb in the giant *Xenopus* specimens. Perhaps other nonthyroid factors (or possibly thyroid analogs) can influence the synthesis of adult Hb, or that this process occurs as part of a general drift in time, at least in the case of some cells or organs of thyroid-deprived larvae. It is of interest that in these inhibited, but giant larvae, restricted to NF St.54, the pronephroi were absent or could only rarely be distinguished as vestiges (Fox and Turner, personal observations), a feature of organic degeneration that normally is contolled by thyroid hormones during late prometamorphosis and climax. This pronephric degeneration likewise occurs in thiourea-treated larvae of *R. sylvatica* when they are strikingly inhibited in metamorphic development. Hurley (1958) suggested that secondary changes occurring in surrounding tissues, such as the cessation of the blood supply, may be causal factors in this case, though the thyroid does exercise a direct role in pronephric ontogeny.

To summarize briefly: In general it seems that amphibian larval erythrocytes arise mainly from two erythropoietic sites, the mesonephroi and the liver; possibly the erythrocytes in the adult are of the same lineage as those of the larval liver. The pronephros may be the initial site of granulocyte differentiation and lymphocytes originate from the thymus, kidneys, and spleen. In adults the bone marrow is the main locus of erythropoiesis, especially in the spring, together with the spleen. The available (albeit controversial) evidence suggests that Hb synthesis is switched from the larval to adult form within the same cell, via new gene activity. The factors responsible and the role of thyroid hormones in this changeover, indeed the mechanism controlling blood cell development and differentiation among other features of blood biology, are still to be determined.

20. The External and Internal Gill Filaments

The external and subsequently developing internal gill filaments of anuran larvae, and to some extent the entire body surface, provide the machinery for respiratory exchange in an aquatic environment. After the completion of metamorphosis, the terrestrial froglet uses lungs for respiration—the aquatic *Xenopus* also does so—though cutaneous gaseous exchange still occurs in these forms. By this time the pulmonary arterial system is fully developed (see Millard, 1945; Witschi, 1956; Lanot, 1962). However, the gills of anuran tadpoles are probably not essential for the normal uptake of oxygen (Boell et al., 1963); probably they are more important for ionic regulation (Alvarado and Moody, 1970), though generally the skin is permeable to gases, water, and ions (Bentley and Baldwin, 1980).

In the larva of *Xenopus laevis* external gill tufts appear on the 3rd, 4th, and 5th visceral arches at NF St.39 (age 2 d 8½ h) and they reach maximum length by NF St.41 (age 3 d 4 h). They are reduced at NF St.44 when the 5th arch external gill is lost. Remnants of the external gills are present at NF St.48–53, but by NF St.46 (age 4 d 10 h) they are covered by the operculum and not visible externally.

The external gill filaments of *Rana dalmatina* are first recognized at CM St.23 (larva 4 d 4 h old; 4.3 mm long). The external gills on the right side have disappeared at CM St.31 (larva 8 d 12 h old; 9.5 mm long) and those on the left at CM St.32 (larva 9 d 3 h old; 9.7 mm long) (Cambar and Marrot, 1954). In *Bufo bufo* they first appear at St.III$_4$ (larva 4 d 9 h old; 4 mm long). Likewise they disappear first on the right side at St.III$_9$ (larva 8 d 5 h old; 7.5 mm long) and then soon afterwards on the left side at St.III$_{10}$ (larva 8 d 15 h old; 9 mm long) (Cambar and Gipouloux, 1956). In these forms, therefore, the life cycle of the external gills is between 4–5 d.

Michaels et al. (1971) described the ultrastructure of the finger-like external gill filaments, and their degeneration, of *R. pipiens* larvae. They originate on visceral arches 3–5 just before the heartbeat stage, reach a maximum length of 1 mm, and function for 2–3 d. As the external gills are covered by the operculum they finally disappear within 5–7 d, to be replaced by the internal gills. External gills have a two-layered epidermis, some of whose outer cells are ciliated, and below the basement membrane, collagen, mesenchymal cells and 1–2 capillary loops are present. During gill regression lysosomal acid phosphatase is recognized in large autophagic vacuoles of the degenerating epidermal cells. Some of the latter may possibly be phagocytosed by neighboring epidermal cells, though this would seem to be unlikely. Heterophils and macrophages phagocytose most of the necrotic cellular remnants.

In contrast to other anuran larvae, *Xenopus laevis* does not develop internal gills specialized for respiration, though an elaborate vascular network supplies the filter apparatus that thus may have a respiratory as well as feeding function (Millard, 1945).

Internal gill filaments of *Discoglossus pictus,* on visceral arches 3–6, are functional up to the equivalent of TK St.XVIII, when the anterior limbs are clearly visible below the skin (Hourdry, 1974). The internal gills involute from TK St.XIX, at about the beginning of climax, and have disappeared by the end of climax at TK St.XXIV, as is also the case with *R. catesbeiana* (Atkinson and Just, 1975). The filaments enclose capillaries and are covered by a bistratified epithelium that Hourdry (1974) described by electronmicroscopy. Cuboidal cells of type A (see also Beaumont, 1970) have their cytoplasm reduced to a thin perinuclear layer; type B cells have a large infranuclear area with numerous mitochondria and vesicles. Type B cells may possibly subserve a similar function as the chloride-secreting cells of teleosts. In other regions of the internal gill filaments, a bistratified epithelium that is thin and respiratory consists of cuboidal type-A cells. Degeneration of the internal gills is by autolysis and cell debris is extruded into tissue spaces. Lysosomal acid phosphatase is deposited in large cytolysomes and residual bodies, and lipid is present. Possibly nonlysosomal enzymes are also implicated in the cellular necrosis. Phagocytes ingest degraded products and the activity of acid phosphatase increases as more cells regress. The lumina of capillaries within the involuting internal gills eventually shrink and finally the capillaries disappear (Hourdry, 1974). The reduction of the internal gills of *Rana catesbeiana* is characterized by a decrease in the rate of thymidine incorporation into DNA, and of amino acids into protein; there is also an increase in the activity of cathepsin C (see Atkinson and Just, 1975). It is of interest that a highly significant increase (13-fold) in acid phosphatase activity during gill resorption was reported by Benson (1972) in T_3-induced metamorphosis of larval *Ambystoma mexicanum* and *Triturus viridescens.*

In vitro immersion of internal gills and gill arches of prometamorphic larvae of *R. catesbeiana* in Hank's culture medium containing concentrations of $L-T_4$ from 10 to 500 parts/10^9 parts of solution, or concentrations of T_3 from 10 to 100 parts/10^9 of solution, at 22°C, led within 2 wk to the hastened shrinkage of the gills and arch tissue, and hence their necrosis. There was a quantitative relationship between the concentration of T_4 and the rate of tissue regression, and the response by the gill tissue to T_3 was more rapid than the response to T_4. The activity of gill collagenase increased in parallel with the T_4-induced gill resorption, together with a reduction in the levels of tissue collagen. There was also an increase in the activity of acid phosphatase, though not because of a significant increase in the total enzyme activity, but because of the differential loss of total protein of the T_4-induced regressing gills (Derby et al., 1979). It would seem reasonable to assume that lysosomal acid phosphatase was active within the degenerating gill epithelial cells, and that this enzyme, together with collagenase, and presumably other lysosomal enzymes also, participates in the macrophagic ingestion of necrotic tissue. However, the exact causal relationship between lysosomal enzyme activity and gill regression has still to be resolved (see Atkinson, 1981).

21. Balancers

Balancers are transitory organs only found in some urodele larvae. They are lacking in Plethodontidae, Cryptobranchidae, *Ambystoma tigrinum,* and *A. mexicanum,* are rudimentary in salamanders, but occur and are well-developed in some members of the Ambystomidae, Salamandridae, and Hynobiidae. Balancers are slender rod-like appendages, a single one projecting from each side of the head behind the eyes, and they have a club-shaped thickening at the end which secretes mucus. They prop the head up and prevent the larva from falling on its side, hence the name, until the forelimbs have developed (Harrison, 1925; Hamburger, 1950).

In *Ambystoma maculatum* and *A. punctatum* balancers first appear in the larva at Harrison (H) stage 34, before the external gill filaments originate. They attain full size at about H St.40–41, but have disappeared at H St.46, when the forelimbs have three digits. In *A. punctatum* balancers are 1.1–1.3 mm long and 0.14 mm in diameter. Terminal secretion cones that first appear at H St.38 increase in number, but shortly before the balancer disappears their number and the amount of mucous secretion are reduced. The balancer is innervated by a small nerve from the Gasserian ganglion and vascularized by branches from the hymomandibular artery and jugular vein, which form a single loop. Blood is first recognized in the balancer at H St.37 (Kollros, 1940).

In *Pleurodeles waltl* larvae, small conical buds of the balancers first appear at GD St.28 (larva 5.4 mm long, 130 h old) likewise before the origin of the external gill filaments (Gallien and Durocher, 1957). From GD St.36, shortly before hatching, they are about 1 mm long, but they are reduced to half size at GD St.44 (larva 17.5 mm long, 36 d old). Balancers have disappeared at GD St.46 (larva 18.5 mm long, 43 d old) when the forelimbs have four digits. Metamorphosis has been completed at GD St.56 (larva 72 ± 10 mm long, 110 d old).

Examination by light microscopy shows the balancers to be lined by two layers of epithelial cells overlying connective tissue, enclosing an undifferentiated mesenchymatous matrix (Harrison, 1925). The balancer is lost from the larva either by constriction at its base, or autotomy, as in *Diemyctylus* (Bell, 1907) and *Ambystoma punctatum,* or by shriveling, as in *A. opacum* (Kollros, 1940) and *Pleurodeles* (Gallien and Durocher, 1957).

Examination of the ultrastructure of the balancer of *A. opacum* and *A. jeffersonianum* by Anderson and Kollros (1962) revealed a superficial layer of cuboidal and an inner layer of low columnar cells, joined together by desmosomes and overlying a basement membrane (adepidermal membrane) that first appears at H St.34. Later on the epidermal cells become squamous. The balancer was considered to lengthen, probably by change in cellular shape rather than by further cell divisions. By H St.36 basal cells include a well-developed granular endoplasmic reticulum and ribosomes, mitochondria, yolk, and pigment granules. Outer epithelial cells have microvilli and a few cilia at their external surface and in addition include a Golgi complex, multivesicular bodies, and randomly orientated tonofilaments that appear in the basal cells of later stages. Collagenous filaments of the basement lamella, situated below the basement membrane, are first indicated. They are oriented at random; at later stages they show some preferential orientation, but possibly balancers degenerate before any specific orthogonality (Weiss and Ferris, 1954) can be established. The filaments are considered to be mainly of mesenchymal cellular derivation with an ectodermal contribution (Bell, 1907), or ectodermal with a possible mesenchymal cellular contribution (Anderson and Kollros, 1962). Hemidesmosomes are not recognized facing the basement membrane. At H St.38 the basal cells include dense particles, and the outer cells have numerous secretory bodies of varying size and electron density and also secretory globules, which probably contain mucoid substances to be released from the external cell surface. Below the basal cells, elongate mesenchymal cells have a prominent RER, Golgi bodies and yolk.

Degeneration of the balancers at H St.45 and 45+ appears to be autolytic, and the cells develop homogeneous bodies (probably lysosomes), large and smaller vacuoles (probably degenerating mitochondria), and multivesicular bodies. The Golgi complexes and the discrete globules of the outer cortical layer of the cells disappear.

From a series of experiments that involve transplanting balancers between different species and stages of *Ambystoma* larvae, Kollros (1940) concluded that the life span of the balancer is mainly determined by intrinsic factors. Information on this subject is still lacking.

22. Features of the Larval Respiratory System

Among the amphibians the external and internal gills of the larva and the skin throughout life are active in respiratory exchange. Respiration by the lungs is a feature of terrestrial amphibians, mainly after metamorphosis, though this is not as significant or indeed the sole means of respiration as in the case of higher reptiles, birds, and mammals. Indeed, paradoxically the sexually mature *Diemyctylus,* the New England spotted newt, has lungs but is highly aquatic. Its lungless neighbor *Desmognathus* is a persistent terrestrial form. The explanation is that 'wet-skinned' amphibians are mainly skin breathers and the lungs are used as hydrostatic organs (Wald, 1981).

The larynx of *Xenopus laevis* originates in the larva as a gutter-shaped downgrowth of the pharynx, which extends posteriorly into a thin-walled trachea and relatively undifferentiated paired bronchi, and thence the simple sac-like lungs. At NF St.44 the expanded trachea, rounded in cross-section, has a wall one-cell thick. The apical region of the tracheal cells is bounded by compact fibrous bodies and similar individual components are recognized in the bronchi at NF St.47. By NF St.48–49 a pronounced dense fibrous layer, more prominent anteriorly, is present. Such tracheal cells include a smooth and granular endoplasmic reticulum, ribosomes, sparse small mitochondria, a Golgi complex, and lipid droplets. Intercellular junctions are fairly straight though some infoldings may occur, especially amid the fibrous area. The trachea is enveloped externally by collagen and blood capillaries. Cilia were never found to be present at the apical margins of the laryngeal groove, trachea, or bronchi (Fox et al., 1970; Fox et al., 1972). However, ciliated cells, in addition to goblet cells and pneumocytes, were reported by Gonikowska-Witalinska (1982) in the developing lungs of *Salamandra salamandra* larvae. In postmetamorphorphic anurans the trachea is extremely short and the lungs practically lead into the pharynx. According to Tesik (1978), in the short upper respiratory tube of adults of *Salamandra salamandra, Rana temporaria, R. esculenta,* and *Bufo bufo* the microvillous luminal surface includes isolated or small groups of cilia, and the epithelium is generally pseudostratified. Furthermore, granular cells are present and they include similar granular (fibrous) bodies found in the trachea of the larval *Xenopus* (*vide supra*).

The anuran larval lung first begins to function in late stages of metamorphosis (Just et al., 1973), and the rates of incorporation of thymidine and amino acids are much higher than in the lungs of young tadpoles. Early in prometamorphosis (more correctly the end of prementamorphosis), of *R. catesbeiana,* at TK St.X (equivalent to NF St.54–55 of *Xenopus*), the lungs have few septa. Later, numerous septa, many with capillaries, develop and extend into the lumen of the lung: they increase the lung surface area. Some smooth muscle is present in the main lung wall and in many tips of the septa. Injection of triiodothyronine into larvae of *R. catesbeiana* leads to a gain in lung weight and a reduction in that of the internal gills (Atkinson and Just, 1975; see also Atkinson, 1981).

23. The Alimentary Canal

During anuran ontogeny, suspension or microphagous feeding by the larva (Kenny, 1969) is superseded by macrophagous feeding, or eating solid food, in the postmetamorphic froglet and adult. These functional changes are reflected in significant histomorphological alterations and remodeling of the intestine, including the pharynx. At metamorphosis the gut shortens considerably (Janes, 1934; Khan, 1968; Wilczynska, 1981) and there is striking cellular degeneration and regrowth in different regions (Liu and Li, 1930; Bonneville, 1963; Bonneville and Weinstock, 1970; Kopec, 1970; Hourdry, 1971a,b, 1973). Based on earlier investigations by light microscopy (see Janes, 1934; Kaywin, 1936), the alimentary canal of anuran larvae is usually arbitrarily separated into (a) a foregut, which includes the esophagus, stomach region, and a short ciliated region, (b) a larger midgut, and (c) a hindgut (Barrington, 1946; Griffiths, 1961). the larval stomach or "manicotto glandulare" (Lambertini, 1928) was considered by Ueck (1967) to be simply a storage organ. However, examination by electonmicroscopy of various regions of the larval gut of *Xenopus laevis* suggested the use of the terms gastrointestinal region and duodenum, by virtue of their ultrastructure and relationship with the liver and pancreas (Fox et al., 1970; Fox et al., 1972).

The distribution of various enzymes in the esophagus and "stomach" of larval and adult *Rana pipiens* was studied by Lipson and Kaltenbach (1965). Details of the patterns of activity of beta glucuronidase (Varute and More, 1971) and alkaline and acid phosphatase (Jae Chung Hah, 1975; Kaltenbach et al., 1977; Kaltenbach et al., 1981) in the alimentary canals of *R. tigrina* and in *Bombina orientalis* and *R. pipiens,* respectively, throughout metamorphosis have been described. Kujat (1981) found pepsin to appear in the developing stomach of the larval *Xenopus* during climax at NF St.63. Alkaline phosphatase, AMPase, and ATPase are associated with sites of food absorption into cells of the gut during prometamorphosis. At climax when the larva ceases to feed these enzyme activities decrease (Botte and Buanano, 1962; Brown and Millington, 1968). The biochemistry of the intestine has been briefly reviewed by Frieden and Just (1970).

The Ultrastructure of the Alimentary Canal

A description of the origin and development of the mouth and its associated structures in amphibian larvae is more the province of comparative anatomy and embryology and the reader is referred to standard textbooks on the subject (see Nieuwkoop and Faber, 1956; Balinsky, 1970; and others). However, within the framework of the metamorphic cycle, it may be

worthy of mention that the tongue of the bullfrog tadpole is seen as a pro-
tuberance from the floor of the mouth at about TK St.X, near the beginning of prometamorphosis (though its Anlage is recognizable at the early
tail bud stage (see Hammerman, 1969). It increases in size and length
(together with the diameter of the glossopharyngeal nerve) throughout
metamorphosis, to reach maximum size in the adult (Shiba et al., 1980).
Furthermore, in *Xenopus laevis* larvae, premetamorphic chemosensory
gustatory papillae appear on the floor of the mouth in the posterior region
at NF St.47. They disappear at climax to be replaced by anteriorly situated microvillous fungiform papillae that arise at NF St.60 and are retained to function in the adult (Shiba et al., 1979, 1980). A similar replacement seems to be the normal feature of anuran larval development
(Nomura et al., 1979). Whether these morphogenetic developmental
processes are controlled by thyroid hormones is not clear and the subject
could well be investigated.

Electronmicroscopical examination of the pharynx of *Xenopus* at NF
St.43 shows only partial cellular differentiation, and lipid and yolk are
still present in the cells at NF St.44 (Fox and Hamilton 1971; Fox et al.,
1972). The luminal cells have apical granules, probably containing mucus ultimately released into the lumen, and ciliary-microvillous cells are
situated near and also line the pharyngeal dorsolateral grooves. At NF
St.47 the cells include mitochondria, a smooth and granular endoplasmic
reticulum, free ribosomes and polyribosomes, microfibrils, and Golgi
bodies. Lipid may, however, still be present (Fox and Hamilton, 1971).
Ciliary-microvillous cells, numerous in the dorsolateral grooves, still remain there in *Xenopus* and *Rana* larvae just before climax, though by the
end of this period they have disappeared in these regions (Fox et al.,
1972).

At NF St.44 esophageal cells of *Xenopus* have luminal microvilli
and occasional cilia. Most microvilli have disappeared by NF St.47,
when ciliated and vesicular cells alternate to line the lumen. Cilia are still
numerous at NF St.57–59 (at the end of prometamorphosis) and there are
small microvilli, cilia, and forerunners of goblet cells at climactic NF
St.61. Near the end of climax the esophageal lumen is lined entirely by
microvilli, the so-called striated border of light microscopy (Fox et al.,
1970, 1972).

Descriptions of the ultrastructure of the gastrointestinal region and of
the comparable ''manicotto glandulare'' of *Xenopus* during prometamorphosis are broadly in agreement (Ueck, 1967; Fox et al., 1970,
1972). This region has a wall several cells thick. Before climax some luminal cells have cilia that have disappeared by the end (Figs. 30, 31).
Gastrointestinal cells have numerous mitochondria and the large number

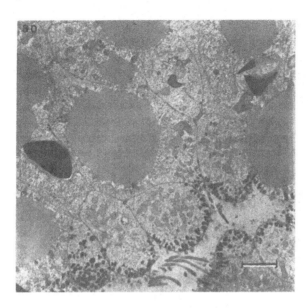

Fig. 30. Gastric region (anterior to the duodenum) of a *Xenopus laevis* larva at NF stage 47. The cells at the lumen surface are ciliated, but without microvilli, and there are dense bodies present in this region that frequently are found further within the cell. Scale mark, 3 μm.

of apical dense granules, more plentiful by the end of climax, are similar in appearance to those in the intestinal cells of the larval *R. catesbeiana* (Bonneville, 1963). By the end of climax the gastrointestinal lumen is bounded by stub-like microvilli with fuzzy coats (Fox et al., 1972). It is still not clear whether the larval gastrointestinal region (or "manicotto glandulare") is a digestive organ, or merely stores food products before digestion and absorption in the duodenum and lower intestine, for the doudenum probably receives digestive products and enzymes directly from the liver and pancreas (see *vide infra,* section on the pancreas) for the extracellular digestion of food. Ueck (1967) believed that the larval foregut of *Xenopus laevis* does not store food, but that that of *Hymenochirus boettgeri,* which has a shorter intestine, is a storage organ. In contrast Dodd (1950) showed the larval foregut of *Xenopus* to store food-laden mucous strings. Both authors deny any digestive function in this region of the alimentary canel. Perhaps the gastrointestinal region of anuran larvae functions partly as a temporary storage organ and also for some modest luminal digestion of food, and the apical granules, seen also in climactic larvae and the postclimactic froglet, are involved. In the larva cilia drive partially digested products toward the duodenum, where nu-

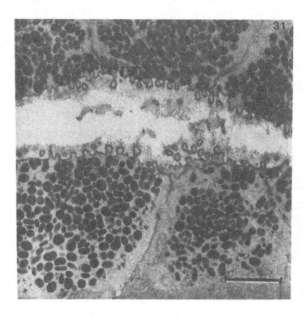

Fig. 31. Gastric region of *Xenopus laevis* (almost a toadlet at NF stage 65 near end of climax). The lumen surface is devoid of cilia, there are short stub-like microvilli and masses of dense granules that are probably digestive in function. Scale mark, 2 μm.

merous luminal microvilli provide a large surface for further digestion and absorption. Sugimoto et al. (1981) suggest that in bullfrog tadpoles protein is partially digested extracellularly in the stomach and then transported to the small intestine, where some at least is absorbed by pinocytosis. Intracellular digestion occurs in large heterophagosomes formed from multivesicular bodies and lysosomes.

The luminal surface of the duodenum, which is more than one-cell thick, of *Xenopus* at NF St.47, includes ciliary, ciliary-microvillous, and large numbers of columnar microvillous cells. Cells of the duodenum, like those of the gastrointestinal region, have numerous mitochondria, but only few apical dense granules (Fig. 32). Cilia have disappeared by the end of climax, when microvilli alone line the lumen (see also Bonneville and Weinstock, 1970).

At NF St.44–47 of *Xenopus* cells of the rectum still contain lipid and yolk. In the hinder region and cloaca, cilia and microvilli line the lumen: only microvilli are present in the more anterior region of the hind gut (Fox, 1970a). Cell profiles of the hinder gut and rectum are generally similar (Fox et al., 1972). By the end of climax the rectal lumen is lined only by cells with profuse short microvilli. The cells contain many small dense bodies, which are fewer in the younger NF St.47, before premetamorphosis.

Thyroid Hormones and Intestinal Changes During Metamorphosis

Thyroxine adminsitration to the larval *Rana pipiens* induced further differentiation of the intestine histologically and enzymatically (Lipson and Kaltenbach, 1965), and injection of triiodothyronine into well-developed tadpoles of *R. catebeiana* at TK St.X–XII induced gut regression, which would typically occur much later in normal development (Carver and Frieden, 1977). Likewise, immersion of *Discoglossus* larvae in thyroxine solution hastened the regression of primary gut epithelium and its replacement by secondary epithelium. Such metamorphic changes are inhibited in *Discoglossus* by treatment with antithyroid substances, and likewise in *Xenopus* after thyroidectomy or hypophysectomy (see Hourdry, 1972; Hourdry and Dauca, 1977).

In *Xenopus* changes in the intestinal epithelium occur at NF St.58, near the beginning of climax. By NF St.61 the degenerating brush border

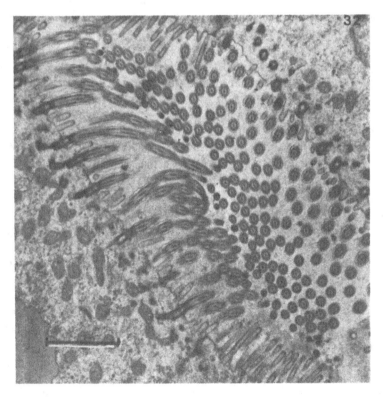

Fig. 32. Duodenal region of a *Xenopus laevis* larva at NF Stage 47 (larva 12–15 mm long; intestine 2 ½–3 ½ revolutions; hindlimb bud just visible). The intestinal lumen surface is highly ciliated and microvillous and there are no dense bodies at the surface. Scale mark, 2 μm.

cells contain numerous secondary lysosomes, and newly differentiating cells proliferate, form microvilli, and line the lumen. The entire process takes 2–3 d (Bonneville and Weinstock, 1970). Changes during metamorphosis, in the intestinal epithelium of *Bufo bufo* (Dournon and Chibon, 1974) and in *Xenopus laevis, Rana catesbeiana* (Bonneville, 1963; Bonneville and Weinstock, 1970), and *Rana temporaria* (Brown and Millington, 1968) have been described using electronmicroscopy. The postmetamorphic intestinal cells originate from small nests of undifferentiated larval basal cells that proliferate and differentiate during climax as the primary surface cells degenerate. The fine structure of the small intestine of the larval *Xenopus,* including its typhlosole, principal microvillous, and gland cells, lymphocytes, and rare endocrine cells, and the proliferation of the nests of cells during metamorphosis to form the future adult intestine, have also been described more recently (McAvoy and Dixon, 1977; Marshall and Dixon, 1978). Furthermore, the small intestine of the adult is about twice the length of the body and it consists mainly of a single-layered microvillous columnar epithelium, goblet, and the uncommon endocrine cells and leukocytes, many of which are probably macrophages (McAvoy and Dixon, 1978).

Hourdry (1971a,b, 1972, 1973, 1977) described intestinal growth and degenerative changes in larvae of *Discoglossus pictus* (see Figs. 33, 34) and *Xenopus laevis* during natural and hormonal-induced metamorphosis, with special reference to the role of lysosomal acid hydrolases. Acid phosphatase increases in amount in the primary degenerating cells, a feature previously noted in the larval intestine of *Rana esculenta* (Botte and Buanano, 1962) and *R. temporaria* (Brown and Millington, 1968).

In gross structure it has been shown that the small intestine of *R. catesbeiana* larvae lengthens from TK St.V to St.XVIII, and thence shortens from TK St.XIX throughout climax to one-eighth of its size by TK St.XXV after metamorphosis. The specific activity of cathepsin D is maximum at TK St.XVIII, thereafter to decrease, but the locus of enzyme activity shifts from the anterior to the posterior intestinal region, a feature similarly recognized with acid phosphatase. Dense granular areas distributed in the intestine in parallel with these enzyme changes, are the sites of lysosomal activity (Sumiya and Horiuchi, 1980).

In organ culture at a low level of T_4 (10^{-8}g/mL), the granular endoplasmic reticulum of the explanted intestinal epitheliocytes of the larval *Discoglossus* increases in amount. At higher levels of T_4 (10^{-7}g/mL), acid phosphatase activity is recognized in lysosomes, the Golgi apparatus, and occasionally in the hyaloplasm. Necrotic cells are extruded into the lumen of the explant, in a manner similar to that which occurs in vivo (Pouyet and Hourdry, 1977).

Shortening in length, the loss of luminal cilia and their extensive replacement by microvilli, glandular changes, and cellular degeneration and reformation are the most obvious histomorphological changes recog-

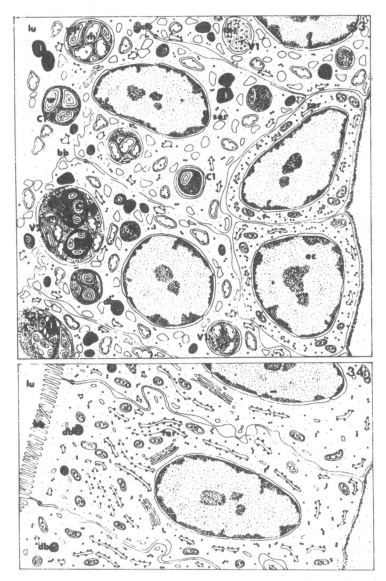

Figs. 33. and 34.   Intestinal epithelium of a *Discoglossus pictus* larva undergoing transformation to the metamorphosed state (diagrammatic). Fig. 33 Epithelium at the beginning of climax. The contents of the primary intestinal cells are evacuated into the lumen: on the right are two secondary embryonic cells, that give rise to the secondary epithelium. Fig. 34 Secondary intestinal epithelium at the end of climax: bb, brush border; CI, primary residual body; C2, secondary residual body; db, small dense body; ec, embryonic cells that give rise to the secondary epithelium; l, lipid droplet; lu, lumen of the intestine; m, mitochondrion; r, ribosome; rer, granular endoplasmic reticulum; ser, smooth endoplasmic reticulum; V1, primary autolytic vacuole; V2, secondary autolytic vacuole; V1 and V2 give rise to the primary and secondary residual bodies respectively (after Hourdry and Dauca, 1977).

nized in the alimentary canal of the anuran larva during metamorphic climax. The mucosal area also greatly increases by surface infoldings (Wilczynska, 1981). These changes reflect the different modes of feeding utilized by larvae and the postmetamorphic froglets and adults, when new neuromuscular mechanisms are required to deal with macrophagous feeding. Microvilli in enormous numbers provide the large surface necessary for the absorption of the digested food.

24. The Pancreas

The structure, degeneration, and thence regeneration at climax, of various functional components of the anuran exocrine pancreas were described by Kaywin (1936) in *Rana catesbeiana* and Janes (1937) in *R. catesbeiana, R. clamitans, R. sylvatica,* and *Hyla versicolor.* Changes in shape and topography during metamorphosis were also reported (see also Beaumont, 1953a,b; Race et al., 1966). Degeneration occurs by autolysis (Liu and Li, 1930; Kaywin, 1936) and remnants of cells slough off into sinusoidal capillaries, though probably some phagocytosis occurs. These results support those earlier reported by Aron (1925) on the exocrine acini and endocrine islets of larval *R. esculenta.*

Race et al. (1966) attributed the profound decrease in pancreatic size, of nearly 70% in weight relative to body weight in the larval *Rana pipiens,* to dehydration, for they found no evidence of widespread cellular necrosis. However, the extensive pancreatic cellular necrosis followed by regeneration recorded during larval development (Kaywin, 1936; Janes, 1937), were subsequently confirmed (Kim and Slickers, 1971; Atkinson and Little, 1972; Bollin et al., 1973; Leone et al., 1976). During natural and T_4-induced metamorphosis, postdegenerate and remaining pancreatic tissue is extensively remodeled by cellular proliferation and differentiation. Pancreatic degeneration involves a massive decrease in cellular DNA and RNAs, and acid phosphatase activity increases up to fivefold. Puromycin, cyclohexamide, and Actinomycin D fail to inhibit the regression and increase in the enzyme activity, and Kim and Slickers (1971) concluded that preexisting enzyme is activated rather than its de novo synthesis during this phase.

In larvae of *Alytes obstetricans* the pancreas is well developed before emergence of the forelimbs. On emergence of the forelimbs at climax the pancreatic volume decreases, mainly because of the reduction of intrapancreatic tissue, partial disappearance of exocrine tissue, and a slight reduction of the endocrine pancreas. The absolute reduction in size is about 80% (Gillois and Beaumont, 1964; Race et al., 1966; Atkinson and Little, 1972). Autolysed cells are phagocytosed *in situ* or the necrotic tissue is lost via the pancreatic ducts and blood capillaries (Beaumont,

1977). After metamorphosis the pancreas increases in size with a relatively larger increase in the amount of endocrine tissue because of the proliferating islets of cells A(× 40) and B(× 12) (Gillois and Beaumont, 1964; Beaumont, 1977). As the pancreas regenerates, new RNA is synthesised and Bollin et al. (1973) concluded that certain predetermined cells need to degenerate in order to permit the remaining rudiment to develop into the adult pancreas.

Investigation of the fine structure of the pancreas of *Xenopus,* at NF St.47, revealed two pancreatic ducts that eventually joined the hepatic duct posteriorly to form a single hepatopancreatic duct opening into the upper margin of the duodenum (Fox et al., 1970). Pyramidal-shaped cells of the exocrine acini, separated from one another by straight intercellular junctions, surround a central lumen. Their cytoplasm includes a well-developed RER, ribosomes, mitochondria, and well-developed Golgi complexes. Remnants of yolk and lipid may still occur. Most acinar cells contain numerous membrane-bound, secretory zymogen granules, up to 1.4 μm in diameter; they are localized apically near the lumen (see Figs. 19–21, Fox et al., 1970). The pancreatic ducts, of a single layer of columnar cells, each with a rounded nucleus, have fairly straight intercellular junctions. The duct margin is bounded by luminal short microvilli and beneath the cell apices are zymogen granules, varying in size and up to 0.8 μm in diameter. There are few mitochondria and a prominent Golgi apparatus is situated alongside the nucleus. The outer surface of the duct is enveloped by collagen and connective tissue cells (see Fig. 22, Fox et al., 1970).

Pancreatic endocrinal or insular tissue is first recognizable at NF St.42 of *Xenopus* (Leone et al., 1976). Insulin is synthesized thereafter throughout larval life and the amount increases at climax (Frye, 1964). In the young anuran larva the endocrinal pancreatic tissue consists of a small number of islets grouped as cells A or B. At climax when exocrine acinar tissue partially regresses, the endocrine tissue eventually proliferates and the islets become mixed. In *Discoglossus* the pancreatic cells B are first recognized when the posterior limb buds appear in the larva, and cells A some days later, when the limbs commence growth. Lipid is reduced but pigment and other granules are present. In contrast acinar cells show only modest differentiation at this time, with only incipient zymogen granule formation (Beaumont, 1970). Acinar cells of *Alytes obstetricians,* cultured at stages equivalent to NF St.50–60 of *Xenopus,* contained typical zymogen granules. Islet cells A and B contained alpha granules probably originating from Golgi saccules. So-called D cells (see Epple, 1966) identified in the larval pancreas were not seen in the cultured pancreatic tissue (Pouyet and Beaumont, 1975).

Using a different cellular terminology, Leone et al. (1976) described acinar A cells of *Xenopus,* at NF St.42, to have a well-developed RER,

Golgi complexes, and numerous large secretory granules. B cells were without an RER and granules, but had well-developed mitochondria. B cells may well be undifferentiated A cells, or because of their location, possibly they are similar to the centro-acinar cells of the adult pancreas. Endocrine C cells located in islets have small membrane-coated granules. The exocrine acini increase in number at NF St.54–56 and necrosis begins at NF St.61 during climax, when the acinar cells become difficult to distinguish. C cells are practically unaffected at this time. The adult pancreas has numerous well-developed acini of A and B cells surrounding lumina and C cells cluster in islets. Lipase activity in the pancreas reaches a peak in *Xenopus* at NF St.54–56, that of amylase at NF St.51. Both enzymes thence decrease their activity to a minimum by the end of climax, corresponding to the period when little feeding occurs. After metamorphosis enzyme activity gradually increased to the adult level. Likewise, Kujat (1981) found that trypsin activity that occurs in the pancreas of larval and adult *Xenopus*, decreases in the larva during metamorphosis.

25. The Liver

The development and topographical changes of the liver in amphibian larvae, typified by that of *Xenopus*, have been described (Nieuwkoop and Faber, 1956). The liver rudiment forms at NF St.35–36 just at the beginning of hatching, and by NF St.40, when the larval mouth is open to the exterior, the liver communicates with the duodenum by hepatic ducts. By NF St.41 hepatic parenchyma cells and a gall bladder are recognized and by NF St.47 the gall bladder is greatly expanded. The liver continues to enlarge and change position before metamorphosis, but during metamorphosis the gall bladder and liver cells do not appear to change significantly, though some cellular necrosis occurs near the gall bladder (however, *vide infra*).

Much of the knowledge about the physiological and pathological action of thyroid hormones is derived from studies on mammals including humans. However, Cohen et al. (1978) have suggested a number of reasons why a study of the ornithine–urea cycle enzymes of the liver of prometamorphic anuran larvae, which are under the control of thyroid hormones, is particularly useful for further elucidation of the thyroidal hormonal mechanism at the molecular level of the cell (Fig. 35). The independently living poikilothermic *Rana catesbeiana*, for example, is a convenient larval form to investigate, particularly for those workers in the USA. *Xenopus* is more widely used in Europe. The rates of molecular and morphological changes, readily determined in experimental or normally metamorphosing forms, are related to specific thyroxine dosages that can be administered either by injection or immersion of tadpoles. During later

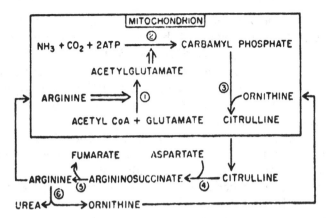

Fig. 35. Urea biosynthesis. The biosynthesis of urea from CO_2 and NH_3 occurs in the liver of the ureotelic animal; carbamyl phosphate synthetase-1 (CPS-1) and ornithine transcarbamylase (OTC) are found in the mitochondria, and argininosuccinate synthetase, argininosuccinase, and arginase are extramitochondrial. Extramitochondrial CPS-II is concerned exclusively with pyrimidine biosynthesis. *N*-Acetylglutamate synthetase is not directly involved in the biosynthesis of urea, but it catalyzes the synthesis of *N*-acetylglutamate, an activator of CPS-1. The enzyme steps in urea biosynthesis are: 1, *N*-acetylglutamate synthetase; 2, carbamyl phosphate synthetase-1; 3, ornithine transcarbamylase; 4, argininosuccinate synthetase; 5, argininosuccinase; 6, arginase. Double arrows indicate activation of the reaction 1 by arginine and reaction 2 by *N*-acetylglutamate (from Cohen et al., 1978).

stages of metamorphosis, larvae cease to feed, so the effects of diet are abolished. The hepatocytes of the larval liver remain as a stable population of cells (Cohen, 1970), for these workers recognize little proliferation of liver cells during prometamorphosis, and various enzymes synthesized particularly mitochondrial carbamyl phosphate synthetase, can be adequately measured. Such hormonally activated enzyme biosynthesis involves regulatory factors beyond that of specific protein biosynthesis and these changes may be investigated in liver cells in vivo and in vitro.

The Fine Structure of the Liver

Fox et al. (1970) described aspects of the ultrastructure of hepatocytes, the gall bladder, and the bile duct of *Xenopus laevis* at NF St.47 (approximately equivalent to TK St.II of *R. pipiens*). Liver parenchyma cells form small lobules and microvilli line the small lumen of the bile canaliculi. Hepatocytes have a well-developed RER and there are numerous ribosomes; glyocogen granules and lipid droplets are present.

The gall bladder has a wall of columnar cells about 7 μm thick. Numerous short microvilli with fuzzy coats line the cavity. Ribosomes and vesicles with smooth walls are plentiful and many mitochondria aggregate towards the apical region of the cells. Large lipid droplets are still recognizable at this stage (see Figs. 13–15, Fox et al., 1970).

Ciliated and microvillius cells, often adjacent to one another, line the lumen of the bile duct, whose wall is composed of a single layer of cells 6–10 μm thick. Some ciliary-microvillous cells also occur. Dense rounded bodies are frequently found at the apical region of some of the microvillous cells, mitochondria are common and ribosomes abound. Cilia are present along the entire luminal surface of the bile duct, extending from the short nonciliated neck adjacent to the gall bladder, to the region near where the hepatopancreatic duct opens into the duodenum (see Figs. 16–18, Fox et al., 1970).

The fine structure of the liver of the normal *Xenopus laevis* larva, from its origin and through metamorphosis, has been described by Spornitz (1978) and in *Bufo bufo* larvae also by Villani (1980). It is clear that in addition to the profound ultrastructural changes in the hepatocyte, including increases in the number and amount of various organelles such as the RER related to new biosynthetic activity *(vide infra)*, hepatocytes also have an important function in the synthesis, storage, and utilization of glycogen. In *Bufo*, Villani (1980) showed that during metamorphosis glycogen granules are metabolized and resynthesis occurs at the end of metamorphosis when feeding recommences. In this regard it is of interest that Farrar and Frye (1979) showed epinephrine and glucagon to stimulate a hyperglycemic response in larval *R. pipiens*, which also elicited the release of glucose from liver slices of tadpoles and frogs. In contrast Spornitz (1978) showed that in *Xenopus* embryonic glycogen has disappeared by NF St.42. However, at NF St.46, when larval feeding has begun, newly formed smaller alpha and beta glycogen particles are present in the hepatocytes situated independently of the SER. From NF St.54–66 (to the end of climax) the glycogen content increases from 0.2 to 10% of liver weight. Adult *Xenopus* may have a liver with 20% or more glycogen. The results contrast with the glycogen depletion in the liver of T_4-induced metamorphosis of *Xenopus* (Kistler and Weber, 1975) and likewise the reduction in the liver of normally developing *Bufo bufo* until the end of metamorphosis (Villani, 1980). Perhaps one explanation of these conflicting results is that it may be caused by larval feeding (or fasting) behavior under the different conditions of the larvae investigated.

Hepatocytes are mainly unchanged in appearance during climax, but Spornitz (1978) described the involution of *Xenopus* liver parenchyma in some regions during NF St.63–66 and invading macrophages are found in the sinusoids. Likewise, Liu and Li (1930) mentioned some liver autolysis in *Rana nigromaculata* during metamorphosis, and features of hepatocyte necrosis were briefly reported in *Bufo bufo* larvae during prometamorphosis by Villani (1980). The main features of liver development in larval *Xenopus laevis* are shown in Table 2.

Liver Cellular Changes During Metamorphosis

During natural or T_4-induced metamorphosis of anuran larvae, the liver shows a striking increase in the production of the urea-cycle enzymes, albumin, hydrolases, and various other enzymes and ribonucleic acids. The changeover from the excretion of ammonia by larval anurans to urea in adults is well-known (Paik and Cohen, 1960; Cohen, 1970; Frieden and Just, 1970), though the aquatic adult *Xenopus* retains the larval method of excretion (see Dodd and Dodd, 1976). Again the bile salts of larval anurans are less complex than those of adults. In larvae of *R. catesbeiana*, for example, after TK St.XVIII (near the end of prometamorphosis) 26-deoxy-5α-ranol is hydroxylated to 5α-ranol and at least two other bile alcohols are synthesized. Anderson et al. (1979) regarded tadpole bile alcohols as biochemical precursors of those in adults. The biochemistry of the liver of anuran larvae has been extensively reviewed by Cohen et al. (1978) and Smith-Gill and Carver (1981). These varied biochemical changes occurring in the larva during metamorphosis

Table 2

Significant Features of Liver Cells of *Xenopus laevis* During Development, With Particular Emphasis on Ultrastructural Characteristics and Glycogen Content[a]

NF stage	First appearance or enlargement of hepatocyte component
35	Bile canaliculi
35–36	Larval hatching
37–40	Golgi apparatus
38–39	Prominent development of RER
41	Formation of sinusoids and endothelia, perihepatic layer, bile ducts
42	Embryonic glycogen disappears
42–43	Dilation of bile canaliculi
43–45	Absence of detectable glycogen
44–46	Larva commences feeding
45–46	Adult alpha- and beta-glycogen
46–47	Bile secretion and pronounced development of the SER
47	Glucose-6-phosphatase activity and disappearance of yolk platelets
49	Inosindiphosphatase activity in bile canaliculi
51	Macrophages with melanin granules first seen
60	Proliferation of macrophages
63–66	Some liver parenchyma involution
66	Glycogen content 10% of liver

[a]Larval stages NF are of Nieuwkoop and Faber (1956) and climax ends at NF St.66 (details after Spornitz, 1978).

would appear to be expressed *pari passu* with ultrastructural changes, which are visualized by the electronmicroscope.

Two-dimensional aggregates of primary cultures of liver cells of the larval *Xenopus*, examined by electronmicroscopy, appear to retain and develop characteristics of typical liver parenchyma cells (up to 80–90% of the culture population), including bile canaliculi and microvilli and the excessive development of microfilaments. Synthesis of liver-specific protein albumin occurs in vitro, even in cultures from thyrostatic tadpoles. Thus, hepatic cellular albumin synthesis may not be dependent upon thyroid hormones (Wahli et al., 1978). It is known, however, that such morphological differentiation in liver cells appears fairly early during normal larval development, before premetamorphosis (Spornitz, 1978; see Table 2), when thyroid hormones would not be expected to have any significant role. Nevertheless, biochemical differentiation of larval liver cells is stimulated in vivo by T_3 and T_4 treatment, including the increase in production of albumin (Herner and Frieden, 1960). If therefore cultured liver cells can synthesize some proteins independently of thyroid hormones, then perhaps in some cases thyroid hormones may act by derepressing (or inactivating) hepatic inhibitory factors present in the larva, but not in the in vitro cultures. The hormonal influence in the case of liver albumin synthesis, for example, could thus perhaps be indirect, in contrast, for example, to the direct stimulation of vitellogenin synthesis in the liver elicited by T_4-stimulated ovarian estrogen or by exogenous estradial-17β.

Changes in the liver cell RER of anuran larvae during T_4-induced metamorphosis were first shown by Tata (1967, 1968a), who suggested a relationship with the biochemical changes occurring at this time (Figs. 36, 37). Ultrastructural changes in the hepatocytes of tadpoles of *Rana pipiens* during metamorphosis were also described by Spiegel and Spiegel (1970). These changes include pleomorphism of the mitochondria, the presence of glycogen in nuclei of late stages, and its disappearance in the cytoplasm, probably to form glucose. Pinocytotic vesicles occurred between the hepatocytes, and these are found only in the larva. The Golgi complexes appear to decrease in size by the end of metamorphosis. The RER occurs in rows of cisternae throughout metamorphosis, though the number of cisternae decreases near the end of climax, which coincides with an increase in the amount of SER. The adult hepatocyte ressembles that of earlier larval stages. Likewise, Bennett and Glenn (1970) and Bennett et al. (1970), investigating larvae of *R. catesbeiana* during natural and T_4-induced metamorphosis, found changes in nuclear appearance from euchromatic to heterochromatic, and increases in the size and altered appearance of mitochondria and the Golgi complexes of hepatocytes. The RER proliferated and the cisternae dilated. A list of the cytological changes that occur in hepatocytes of *R. catesbeiana* larvae after in vivo administration of thyroid hormones, together with references

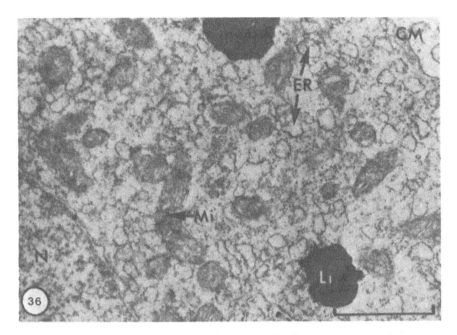

Fig. 36.   Electronmicrograph of a liver cell of a premetamorphic *Rana catesbeiana* larva showing little differentiation of a granular endoplasmic reticulum. Scale mark, 1 μm.

to the relevant authors, has been prepared (Cohen et al., 1978; see also Smith-Gill and Carver, 1981). Such ultrastructural changes are now generally presumed to be related to the increased biosynthetic activity and secretion by the larval liver during the metamorphic cycle.

Liver Cellular Proliferation During Metamorphosis: Do the Same Cells Change Their Biosynthetic Activity?

There is still some disagreement among workers whether hepatocytes proliferate or not during metamorphosis, especially at climax, coincident with the new biosynthesis at this time. Champy (1922), Kaywin (1936), and Cohen (1970) reported little or no increase in the number of anuran larval hepatocytes during metamorphosis, though Elias and Sherrick (1969) reported that the number does increase during larval development, which must indeed be true at least during early stages. In *Xenopus* "the liver and gall bladder are hardly influenced during metamorphosis but near the gall bladder there is some cellular degeneration" (Nieuwkoop and Faber, 1956). Spornitz (1978) also reported some hepatocyte involution at NF St.63–66, as did Villani (1980) in the liver of prometamorphic *Bufo* larvae. Campbell et al. (1969), however, wrote that "there is no

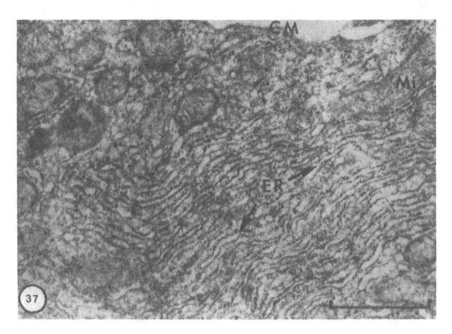

Fig. 37. Electronmicrograph of a liver cell of a metamorphosed *R. catesbeiana*, induced from a larva 6 d previously with triiodothyronine, showing a profuse lamellated granular endoplasmic reticulum. Scale mark, I μm. CM, cell membrane; ER, granular endoplasmic reticulum; L, lipid droplet; Mi, mitochondrion; N, nucleus (After Tata, 1967).

evidence of changes in cell number or of mitotic index in the tadpole liver during metamorphosis.''

Paik and Baserga (1971) showed the hepatocyte DNA-labeling index to increase during T_4-induced metamorphosis, but that the total amount of liver DNA did not change significantly. There is little DNA synthesis or repair turnover according to Pearson and Paik (1972); nor does tadpole liver weight change during metamorphosis (Paik et al., 1961). Nevertheless, Atkinson et al. (1972) reported an increase in the concentration and biosynthesis of nuclear and mitochondrial DNA in hepatocytes, and found no evidence of polyploidy, after T_3-induced metamorphosis of bullfrog larvae. Smith-Gill (1979) and Smith-Gill et al. (1979) showed that the rates of DNA synthesis in liver cells of *R. pipiens* larvae, during spontaneous and T_3-induced prometamorphosis and climax, fluctuate preceding corresponding fluctuations in the DNA content per cell. DNA synthesis and its cell content increase to a maximum at TK St.XVI–XVIII, towards the end of prometamorphosis, but they decrease thereafter to rise again during late climax stages (St.XX–XXIV). The DNA cellular changes precede corresponding expressions of biochemical differentiation represented by the cellular arginase content. Atkinson et al. (1972) and these latter authors concluded that such changes may not have occurred in a "fixed population" (Cohen, 1970) of hepatocytes. Mainly hypertrophy of hepatocytes was claimed to occur by Finamore and Frieden (1960) (see also Kistler et al., 1975b), together with modest (albeit not significant) hyperplasia (Kistler and Weber, 1975), during hormone-induced anuran metamorphosis. The liver of *Xenopus* was shown by Oates (1977) to undergo a biphasic pattern of growth, changing from cellular hyperplasia to hypertrophy after NF St.51. The hepatocytes show a large reduction in their mitotic index, which is extremely low or nil during prometamorphosis and climax. After NF St.51 the hepatocyte mean diameter increases from 13.7 μm to about 17 μm and the calculated cell volume doubles from about 300 to 600 μm^3. Cell volume thence decreases after NF St.57, possibly because of the decrease in glycogen content (but see Spornitz, 1978, *vide supra*), and/or changes in the number and dimensions of some intracellular components.

After an investigation of the histone metabolism of larval liver cells of *Rana catesbeiana* during T_4-induced metamorphosis, Morris and Cole (1978, 1980) likewise concluded that the extensive metabolic changes occur in the absence of any significant cellular division. In vivo liver cell histone phosphorylation of larval *R. catesbeiana* showed phosphoryl H_1, H_{2a}, and H_4 at low levels, peaking after 2–8 d of T_4 treatment to over $\sim$ 2–5-fold for histones H_1 and H_{2a}, which correlated with increased levels of various liver enzymes. The low basal rates of synthesis of histones and DNA were unchanged, a result that did not equate with any increase in hepatocyte proliferation. Perhaps any assessment of hepatocyte proliferation during metamorphosis, in terms of DNA content per cell (presumably DNA/tissue volume or weight), may not sufficiently take into account changes in hepatocyte volume during late prometamorphosis and climax (see Oates, 1977).

Kistler and Weber (1975) described biochemical and morphometrical changes in hepatocytes of *Xenopus* larvae during T_4-induced metamorphosis. Initially, individual hepatocytes increase in volume together with their cytoplasmic organelles, especially the RER and mitochondria. Likewise the extrahepatic space enlarges. During late metamorphosis the liver decreases in size because of the reduction of the extrahepatic space to its original volume and a relatively greater reduction in the size of the individual hepatocytes, mainly involving the cytoplasmic ground substance. These results are broadly in agreement with those later reported by Oates (1977), though Kistler and Weber (1975) suggested that some cellular necrosis probably occurs [also mentioned by Nieuwkoop and Faber (1956) and Spornitz (1978) in the *Xenopus* liver at climax and by Villani (1980) in the liver of *Bufo bufo* during prometamorphosis], but that there was little evidence of any liver dehydration. Hepatocyte nuclear volume may well increase during the growth period and there appeared to be a greater number of smaller nuclei near the end of metamorphosis, though the nuclear numerical changes were not found to be significant. It appears that adult hepatocyte mitochondria are larger and of higher electron density than those of larval hepatocytes, and the changeover is stimulated by thyroid hormones (Smith-Gill and Carver, 1981).

Thus as yet there is no unequivocal evidence that any significant hyperplasia occurs in the anuran larval liver during the period of extensive new biosynthesis that proceeds in hepatocytes during metamorphosis. As Smith-Gill and Carver (1981) have concluded, the relationship of DNA synthesis and possible hepatocyte turnover (cell necrosis and proliferation) at late prometamorphosis and climax, and the concurrent developmental and biochemical changes, still require further elucidation. Yet because there seems to be only limited cellular necrosis, it is more likely that the metabolic changeover occurs within the same, albeit hypertrophied, cells. Subsequently as the liver enlarges in the postmetamorphic froglet and juvenile, presumably there is proliferation of the original clonal population of hepatocytes, and these cells continue to function biochemically in the adult manner.

Finally, evidence from morphological or biochemical changes in hepatocytes during normal or T_3 and T_4-induced metamorphosis, suggests that it is as well to exercise some degree of caution in equating such results. Experimentally induced changes may not exactly reflect normal developmental changes, which usually proceed more slowly and in orderly sequence, are of less intensive nature, and presumably are less likely to suffer induced artefacts, which could well arise under the prevailing experimental conditions.

26. The Pronephros

Since the classic review of the amphibian larval pronephros by Field (1891), there have been several detailed accounts of this paired organ system (Goodrich, 1930; Fox, 1963; Oates, 1977). Using light microscopy, aspects of the structure and degeneration of the anuran pronephros were described by, among others, Jaffee (1954), Hurley (1958), Fox 1962a,b), Michael and Yacob (1974), and Oates (1977). Pronephros (and duct) ultrastructure has been described in urodeles *Ambystoma opacum* (Christensen, 1964), *A. mexicanum* (Poole and Steinberg, 1981), and *Triturus alpestris* (Lehmann, 1967), and in anurans *Rana pipiens* (Gibley and Chang, 1966), *Rana temporaria* (Fox, 1970b, 1971), and *Xenopus laevis* (Fox and Hamilton, 1971; Oates, 1977). Cell profiles of pronephroi of *Rana* and *Xenopus* appear extremely similar (Fox, 1970b; Oates, 1977).

Origin in the Larva

The first indication of the pronephroi, in anurans and urodeles, is revealed on fate maps of their blastulae, by the use of agar impregnated with dyes such as Nile blue sulfate, neutral red, or Bismarck brown. The future pronephric tissue is situated on each side 90° to the sagittal plane and somewhat concentrated in the future 3rd and 4th somitic regions (see Pasteels, 1940). The presumptive pronephric tissue can thence be followed to its ultimate location as the future pronephrotomes on each side of the larva behind the otic region.

In *Xenopus* the pronephric Anlagen are first recognized as small thickenings of the intermediate mesoderm (between the upper somite and the lateral plate) of post-otic somites 3–5 in the young embryo at NF St.21. Nephrotomes develop at NF St.23. The pronephric duct tissue gradually extends caudalwards (see Poole and Steinberg, 1981), to reach the 5th trunk somite by NF St.27 and the cloaca by about NF St.33–34. The ducts join the rectal diverticulae by about NF St.37–38 and functional continuity between the pronephros, via a luminated duct, with the cloaca is achieved by about NF St.43 (Nieuwkoop and Faber, 1956).

Typically in anurans the future pronephros and duct tissue are located below visible somites 3–5; in urodeles the pronephros is situated below somites 3 and 4 and the duct below somites 5–7. The ontogeny and phylogeny of the vertebrate kidney, including the pronephros (and mesonephros) of amphibians, have been considered by Fox (1963, 1977c), to which reference should be made for further details.

Functional Activity of the Pronephros

The paired pronephroi initially are the sole ammonotelic excretory (and probably osmoregulatory) system. During anuran metamorphosis an important metabolic change occurs from ammonotelism to ureotelism, activated via the urea-cycle enzymes of the liver, the changeover in the bullfrog larva mainly proceeding at about TK St.XIX–XX (see Ashley et al., 1968), at the beginning of climax (Etkin, 1968). The blood urea concentration in *Xenopus* is higher during metamorphic climax and in juveniles than in earlier larval forms, and the main nitrogenous product excreted is ammonia. Urea loss is prevented so as to maintain a higher osmolality of the body fluids. Whether in *Xenopus* there is a reduced excretion of urea by the mesonephroi themselves, or a reduced excretion rate of urea owing to reduced permeability through the skin, however, is not clear (Schultheis and Hanke, 1978a,b). Also, in *Xenopus* the blood ammonia level remains fairly constant during metamorphosis.

Among anurans, however, as with *R. pipiens,* the pronephroi and mesonephroi function simultaneously during prometamorphosis (Jaffee, 1954), until the pronephroi disappear during climax leaving the ureotelic mesonephroi as the sole adult kidneys. Presumably until climax the anuran mesonephroi excrete only small quantities of urea. The pronephros functions early in larval development, for example, at Shumway St.19 in *R. pipiens* (equivalent to about NF St.33–36 of *Xenopus*) (Rappaport, 1955), NF St.37–38 in *Xenopus* (Nieuwkoop and Faber, 1956) and at about Harrison St.36 in *Ambystoma* (Fales, 1935). Biphronephrectomized young larvae soon become edematous (Cambar, 1947; Rappaport, 1955). Early in larval life water and other substances, which diffuse from the surrounding tissues into the coelom, are the main source of fluid traversing the pronephros to the exterior. Later, filtration from the dorsal aorta through the paired glomi into the coelom is probably more important (Rappaport, 1955). The pronephric (and mesonephric) tubules of *Xenopus,* from NF St.55 during prometamorphosis, appear to be capable of phagocytic activity, for they ingest carbon particles and mamalian erythrocytes drawn from the body cavity (Turner, 1969). The contribution by the pronephros in hematopoiesis (see Carpenter and Turpen, 1979) has been described in the section on blood (*vide supra*).

Relationship of the Pronephros to the Thyroid Hormones

Present available evidence supports the view that pronephric growth, differentiation, and ultimate degeneration are mainly controlled by the circulatory thyroid hormones. Certainly its degeneration does not depend upon the occlusion of the pronephric duct and/or nephrostomes (Hurley, 1958; Fox, 1962a). Nor does a remaining phonephros, after the larva has been unilaterally pronephrectomized, degenerate histologically or temporally in any other manner than normally (Fox, 1962b). Antithyroid goitrogens (Fox and Turner, 1967; MacClean and Turner, 1976) and surgical hypophysectomy or thyroidectomy inhibit larval development, including that of the pronephros and mesonephros, and thyroxine treatment stimulates their development (Hurley, 1958; see Fox, 1963 for refs.). In *Eleutherodactylus martinicensis* (which lacks an aquatic larval stage in its development), treatment of the embryonic stages in the egg with phenylthiourea, thiourea, or thiouracil inhibits pronephric degeneration; cessation of the treatment permitted thyroid recovery and thence ultimately degeneration of the pronephros. Precocious development and then rapid degeneration of the pronephros occurred in these specimens treated with thyroxine (Lynn and Peadon, 1955). Similar results were obtained by Hurley (1958) with larvae of *Rana sylvaticus*. It was concluded that the thyroxine circulatory concentration, temperature, and the larval stage when thyroid hormones are active, are of supreme importance for metamorphic change, including that of pronephros (Fox, 1971). Histological changes of the thyroid would seem to be related to the degree of pronephric degeneration (Michael and Yacob, 1974), doubtless a reflection of the level of thyroid hormones secreted by the thyroid near and at climax. However, whether thyroid hormones primarily influence pronephric cells exclusively at the level of the genome, and/or indirectly, is not clear. Hyperfunction of the pronephros can maintain or increase the size of various pronephric tubular components; hypofunction, by reduction of intratubular fluid tension, leads to a reduction of lumen volume of the pronephric tubules and duct (Fox, 1956, 1957, 1960). Chopra and Simnett (1969a,b, 1970, 1971) concluded that inhibitory growth factors, or chalones, are involved in the growth of the pronephros and mesonephros.

Pronephros: Gross Structure and Size

Examination by light microscopy of the amphibian larval functional pronephros reveals a complex, tightly arranged mass of tubules, somewhat oval-shaped in form, with three (anurans) or usually two (urodeles) ciliated nephrostomial tubules opening by nephrostomes into the coelom. The nephrostomial tubules join a common proximal convoluted tubule, which leads into a short ciliated intermediate segment followed by a nonciliated distal tubule, and then the pronephric duct (Figs. 38A–E). The pronephros is situated within the postcardinal sinus (Fig. 39), receives oxygenated blood from the dorsal aorta, and individual spinal nerves lead over the nephrostomial tubules, reflecting an ancestral segmental relationship. Alongside the inner margin of the pronephros in the vicinity of the nephrostomes, a glomus comprises a small mass of capillaries originating from the dorsal aorta. The basic arrangement of the mesonephric nephron is similar to that of the pronephros (Fox, 1963).

RNA is detectable in the pronephric proximal tubule of *R. pipiens* at Shumway St.23 (equivalent to NF St.42 of *Xenopus*), as the tubular regions becomes differentiated. RNA levels stay high until pronephric differentiation is completed. As would be expected histochemical activities accompany pronephric differentiation; PAS staining of brush borders of the tubules and resorption of protein droplets are similar to those reactions found in mesonephric and metanephric kidneys (Jaffee, 1963). Alkaline phosphatase activity has been demonstrated in the Anlage and developed pronephric tubules (especially the proximal tubules) of *R. pipiens* larvae, though its function is not clear (Piatka and Gibley, 1967). The activity likewise increases in the region of the brush border of pronephric tubules of *Bombina orientalis* during larval development, but it decreases and eventually disappears, simultaneously with pronephric degeneration at climax. Alkaline phosphatase activity of the mesonephric tubules, however, remains high throughout matamorphosis (Jae Chung Hah, 1974).

Oates (1977) found the mitotic index of pronephric tubule cells of *Xenopus* to decrease at NF St.51, though the tubule lumen volume and

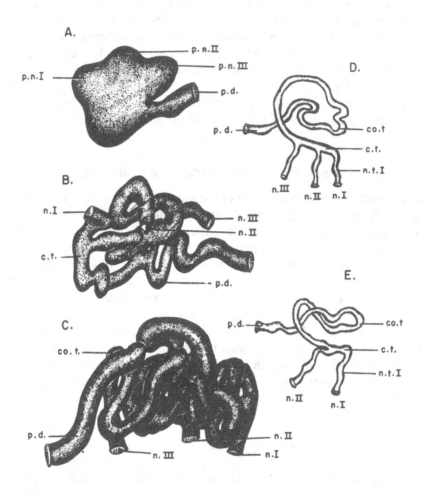

Figs. 38.   Right pronephros of larvae of *Rana sylvatica*. A, Larva 3.5 mm long, 4 d after fertilization, median view; B, median view of a functional pronephros when the duct is open into the cloaca; this larva is 3–5 mm long, 30 h older than in A; C, ventral view of a pronephros of a larva 8 mm long, 2-3 d older than in B; D. pronephros somewhat diagrammatically drawn in a larva at stage between B and C; E, pronephros of a urodele larva, e.g. *Ambystoma* or *Triturus*. Anuran pronephroi usually have 3 nephrostomes and nephrostomial tubules and urodeles usually have pronephroi with two nephrostomes and nephrostomial tubules: ct, collecting tubules of pronephros; cot, common tubule of pronephros; n, I–III, nephrostomes; pd, pronephric duct; pn, I–III, presumptive nephrostomes; n, nephrostomes (After Field, 1891).

their length increase between NF St.45–55. There is a decrease in pronephric size by NF St.61. An earlier, more detailed quantitative analysis of pronephric growth and degeneration in *R. temporaria* by Fox (1962a), showed that from CM St.29–47 (Cambar and Marrot, 1954) the nuclear population of the pronephros and its length, tissue, lumen and thus overall volume, and the tubule internal surface area increase. Measurements of these components at CM St.49 and thereafter showed them to be substantially reduced until the final disappearance of the pronephros by the end of climax at CM St.54. The pronephric tubular individual cell volume ranged between 4000 and 7000 μm^3, but as the tubules regressed this component lessened to between 2000 and 5000 μm^3. The maximum number of cells forming a pronephros (judged from its nuclear population) is about 10,000 at CM St.47. During climactic CM St.51–52 the number has reduced to about 2000 and at CM St.53 to about 400. The glomus reaches a maximum length of 0.3 mm at CM St.47 and then is reduced to a vestige by the end of climax. The mesonephros lengthens threefold from CM St.29–50. At climax it appears to shorten slightly, when the body shortens overall at this time. Afterwards the mesonephros increases in size to attain adult form.

The origin of the mesonephros of *Bufo* and *Xenopus* embryos has been reconsidered by Gipouloux and Hakim (1976) using light microscopy. The mesonephric blastema of each side arises from the upper external area of the lateral plate mesoderm near the pronephric duct. Furthermore, after more than a decade when amphibian embryos apparently were of little overt interest in considering causal mechanisms of induction, in terms of the pronephric duct and the formation of tubules from the mesonephric blastema (Fox, 1963), the subject has again been investigated by Gipouloux and Delbos (1977) using the electron microscope. They described closely apposed pseudopodial protrusions of duct and blastema cells, and extracellular mucopolysaccharide (glycosaminoglycans) material and collagen fibers occurred at their interfaces during the induction process. The future differentiation of the mesonephric tubules may therefore depend upon this extracellular material; indeed mesonephric cells may well reciprocally influence the further differentiation of the mesonephric duct.

Fine Structure of the Pronephros

Electronmicroscopical examination of the fully developed prometamorphic pronephros of *Rana temporaria* (Fox, 1970b), showed the cells of the functional proximal tubules to be bounded externally by a basement membrane and some free collagen fibrils. There is substantial infolding of the plasma membrane and lateral interdigitations occur between the cells (Fig. 42). Presumably this arrangement provides a large surface area to facilitate the transport of substances between cells and into (or from?) the postcardinal sinus. These cells possess a large rounded or oval-shaped nucleus, often with a recognizable nucleolus, and there are numerous mitochondria in the basal region. A Golgi complex near the nucleus is not commonly seen in *Rana,* but it is prominent in *Xenopus* (Fox and Hamilton, 1971) and in the pronephric cells of *Ambystoma* (Christensen, 1964). The proximal tubule lumen is lined by microvilli 2 μm long, and there are about 6000 of them per millimeter length of the tubule surface. Pinocytotic vesicles occur between the microvillous bases and within the cells. Doubtless they originate by pinching off of the lateral tubule walls. The rest of the cell includes a granular and smooth endoplasmic reticulum, ribosomes, and polyribosomes and small lipid droplets. The distal tubule is similar in appearance, except for the absence of the cellular microvilli and the intercellular junctions are less infolded. The short intermediate tubular segment is ciliated at the lumen surface. The cells have a high ribosomal content and lipid and pigment may be present.

The cells of the nephrostomial tubules of *Xenopus* at NF St.47 (Fox and Hamilton, 1971) have straight intercellular junctions, the mitochondria are irregular in shape and variable in number, and cilia line the lumen margin. The pronephric duct has a particularly high degree of infolding of the plasma membrane, similar to that in *Triturus* (Lehmann, 1967). The duct lumen surface is fairly smooth, with only a few short projections. Generally the ultrastructure of the cells is similar to that of the distal tubule.

Pronephric Degeneration

At first sight different anuran species seem to vary in the time when pronephric degeneration commences. In *Xenopus* it begins at NF St.53 with the first sign of atrophy at NF St.54 (Nieuwkoop and Faber, 1956). Indeed acid phosphatase activity occurs in the pronephric tubules of *Xenopus* at NF St.51 (Goldin and Fabian, 1971). However, Oates (1977) found well-formed tubule lumina at NF St.55, though incipient connective tissue bordered the coelom near the nephrostomes. At NF St.57 the luminal border of the tubule cells was ragged. Pronephric degeneration in *Rana sylvatica* begins simultaneously with the rapid growth of the hindlimbs (Hurley, 1958). In *R. temporaria* maximum pronephric size occurs at CM St.47, when the hindlimbs are well-developed (Fox, 1962a) and its degeneration is first apparent between CM St.47 and 49 (equivalent to NF St.55–56 and 57–58 of *Xenopus* and TK St.XII–XV or *R. pipiens*). In general, however, in anurans incipient pronephric involution seems to commence around the middle of prometamorphosis, with massive breakdown at climax.

Pronephric degeneration in *R. temporaria* is uneven and variable in rate (Fox, 1970b). Adjacent tubule cells frequently differ in their rates of necrosis and even within the same cell different (or similar) organelles vary in their degree of degeneracy: some are almost wholly so, others are hardly degenerate. Nevertheless, through climax pronephric involution is continuous and the organ gradually disappears amid surrounding extrarenal cells of the postcardinal sinus (Figs. 39–41). Large autophagic vacuoles, secondary lysosomes, cytolysomes (Novikoff, 1963), termed degeneration bodies (Fox, 1970b), and recognized at all levels of the degenerating pronephric cell, are often up to 8 μm in diameter (Figs. 43–46). They are centers of intense necrosis and include mitochondria at varied degrees of degeneration, vesicles, myelin figures, dense osmiophilic bodies, lipid, and pigment. The background matrix is of a disorganized granular composition. These areas are delimited, usually by one or several delicate membranes (Figs. 44, 45). Acid phosphatase activity is registered within them. Pronephric nuclei ultimately become dense, shrunken, and necrotic. DNA synthesis ceases in the pronephros of *Xenopus* by NF St.51, well before it starts to degenerate, though RNA synthesis is proceeding at NF St.50, perhaps involved in the production of lysosomal enzymes (Goldin and Fabian, 1972). Microvilli and cilia are still recognizable and thus they are slow to degenerate in climactic degenerating tubules of *Rana*. Eventually the basement membrane and overlying collagen disappear, plasma infoldings are lost, and the tubules change to solid strands of necrotic cellular tissue whose debris ultimately becomes unrecognizable amid the surrounding postcardinal cells (Fig. 41).

It is of interest that pronephric degeneration of prolactin-treated *Rana temporaria* larvae (CM St.47–48, just before the onset of climax) is delayed when compared with untreated controls, an example of thyroid hormone antagonism (Vietti et al., 1973). The mechanism is not clearly understood; perhaps prolactin acts by stabilizing the membranes of lysosomes (Campantico et al., 1972; Giunta et al., 1972), which normally develop in the degenerating pronephric tubules (Fox, 1970b).

Within the postcardinal sinus the pronephric tubules are surrounded by blood cells that increase in number during prometamorphosis (Fig. 39). There is a high preponderance of erythrocytes to leukocytes. The stem cells of erythrocytes, in the intertubular spaces of the pronephros and mesonephros of *R. pipiens* larvae, are claimed by Hollyfield (1966) to develop *in situ*. Horton (1971) described lymphoid histogenesis in the thymus and pronephros of *R. pipiens* at TK St.I (equivalent to NF St.46 of *Xenopus*). He referred to the pronephros and mesonephros as the main centers of hemopoiesis during larval life, though Carpenter and Turpen (1979) found that lymphopoiesis is a negligible activity in the pronephros, whose main hemopoietic activity is granulopoiesis (see section on

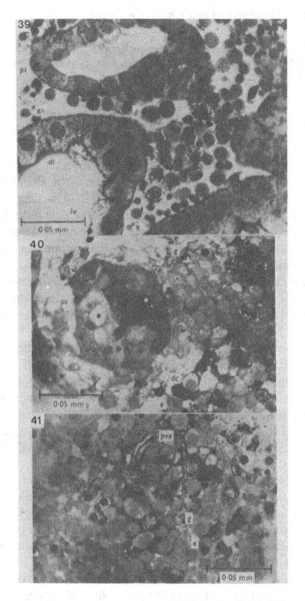

Figs. 39–41. Micrographs of pronephric tubules of *Rana temporaria* larvae. Fig. 39. At CM stage 45–46 before climax, the tubules are functional and nondegenerate with blood cells in the intertubular regions of the postcardinal sinus. Fig. 40. At CM stage 51–53 during metamorphic climax, degenerate strands are relics of the tubules and almost digested pronephric tissue is recognizable amid extrarenal cells and phagocytes. Fig. 41. At CM stage 51–53 as the pronephros disappears by autolysis and phagocytosis, highly degenerate vestigial pronephric tissue is submerged amid the extrarenal phagocytic cells. Granulocytes are particularly abundant. dc, dark cell; dt, distal tubule of pronephros; e, erythrocyte; en, endothelial cell nucleus of blood vessel; g, granulocyte; lu, lumen of pronephric tubule; ps, pronephric strand; pt, pronephric tubule; pve, pronephric vestige (after Fox, 1970b).

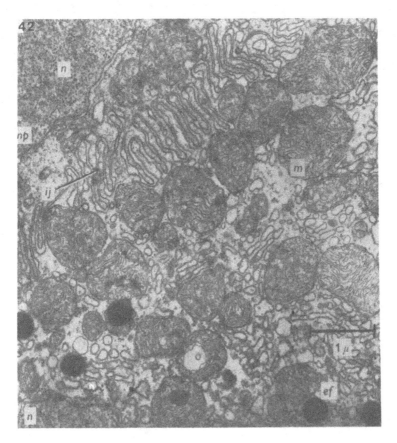

Fig. 42. CM stage 48–49 larva of *Rana temporaria* just before climax. Electronmicrograph of region of pronephric tubule before degeneration showing a high content of mitochondria, smooth endoplasmic reticulum, and a wavy intercellular junction between adjacent pronephric cells (arrowed): ef, emulsified fat droplet; m, mitochondrion; n, nucleus; np,nucleopore (after Fox, 1970b).

blood). Whether such renal Anlagen include lymphocytic stem cells is not certain. It is probable, however, that erythrocyte, granulocyte, and lymphocyte stem cells originate elsewhere and later in larval development concentrate between and around the pronephric and mesonephric tubules (Foxon, 1964; Carpenter and Turpen, 1979). Oates, (1977) reported the first appearance of intertubular lymphocytes in *Xenopus* at NF St.53 and erythrocytes were numerous at NF St.55. Agranular lymphocytes, granulocytes, and also erythrocytes, recognized around the pronephric tubules of prometamorphic *R. temporaria* at CM St.45–46 (equivalent to NF St.54 of *Xenopus*), increase in number around the climactic degenerating tubules (see also Hurley, 1958 in *R. ˙sylvatica* larvae). In *R.*

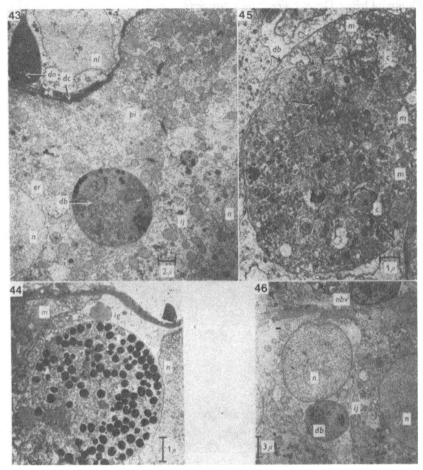

Figs. 43–46. Electronmicrographs showing pronephric degeneration in *Rana temporaria*. Fig. 43. CM stage 48–49 near onset of climax. There is a large 'degeneration body' (cytolysome) in the peripheral region of the tubule, which was probably still microvillus. A dark cell (phagocyte) and a lymphocyte-type cell are present outside and against the tubule surface. Fig. 44. CM stage 51–53 at climax. Large cytolysome in a degenerating tubule. These areas are positive for acid phosphatase and are lysosomal. Fig. 45. CM stage 51–53. The large 'degeneration body' shows aggregation of smaller cytolysomes (arrowed). The entire area is limited by cell membranes and peripherally mitochondria are still recognizable. The area gradually becomes more extensive as autolysis proceeds. Fig. 46. CM stage 51–53. Topographic relationship of a pronephric degeneration area and a still nondegenerate nucleus. There is much variability in the rate and appearance of degeneration in different regions of the pronephric tubules during climax: c, cytolysome; db, degeneration body; dc, dark phagocytic cell; dn, dark cell nucleus; er, endoplasmic reticulum; ij, intercellular junction; lg, lipid granule; m, mitochondrion; n, nucleus; nl, nucleus of lymphocyte; nbv, nucleus of blood vessel; pi, infolding of basement membrane of pronephric cell (after Fox, 1970b).

temporaria characteristic 'dark' cellular extensions, often nucleated, appear to invade degenerating pronephric tubule cells; they are probably profiles of phagocytic cells (Fig. 43). The role of the granulocytes and lymphocytes associated with the involuting pronephros is not clear. Probably pronephric necrotic debris, derived from autolytic degeneration, is ingested by the 'dark' phagocytes and granulocytes.

27. The Lymph Glands

The anuran larval lymph glands, located on each side of the body in the branchial cavity, situated ventroposteriorly to the thymus (Cooper, 1967), like the pronephros, mesonephros, and thymus, have an important role in hematopoiesis (Horton, 1971), and they also synthesize serum antibodies (Cooper et al., 1971). Indeed the lymph glands, thymus, and spleen are the major if not the only organs of the larval immune system. In *Xenopus laevis* larvae, in addition to the lymphoid tissue of the thymus, spleen, kidneys, and liver (Manning and Horton, 1969), there are also pairs of four ventral lymphoid cavity bodies (VCBs) and two dorsal lymphoid cavity bodies (DCBs) in the branchial region (first detected at NF St.50), and a variable number (30–50) of scattered lympho-epithelial tissues (LET) along the entire length of the alimentary tract (first detected at NF St.48). These branchial and intestinal LET are of maximum size at NF St.55–58. Such bodies of relatively large size are found in the intestine of metamorphosed toadlets (Tochinai, 1975). Pharyngeal lymph glands have disappeared after metamorphosis (Horton, 1971), probably because of the influence of the raised circulatory level of thyroxine at climax (Riviere and Cooper, 1973). It is of interest that in the larva, lymph glands and the pronephroi are lost while the thymus glands and the mesonephroi are retained after metamorphosis to deal with the specific requirements of adult life.

28. The Skin of Amphibian Larvae

Earlier descriptions of the structure of amphibian skin using light microscopy subsequently were reported in more detail by the use of electronmicroscopy (see refs. in Fox, 1974, 1977a, 1981; Whitear, 1974, 1975). These investigations demonstrated that the larval epidermis and dermis comprise a complex arrangement of many different cellular components, each of which differing from the other in structure, function, and topographical relationships, and frequently in their time of origin and longevity (Fox, 1978). The skin of the anuran (and urodelan) larval tail and body progressively becomes more elaborate and cosmomolitan in its cellular composition as development proceeds. However, some kinds of epidermal cells, such as ciliary, hatching, and cement gland cells originate early in larval life and disappear before the onset of metamorphosis. These cells subserve specific functions during larval life; other individual cell types are recognizable throughout ontogeny, including that of the adult. Presumably their function is the same in an aquatic larva or a terrestrial frog. Nevertheless, though the ultimate fate of some specialized cell types is not clearly known, it is likely that, apart from the germinative cell layer, most if not all epidermal cells (and much of the dermis too) ultimately degenerate and disappear, some to be regularly replaced throughout life, though obviously tail skin is not renewed after tail involution at climax. The skin of an amphibian is thus a dynamic cellular system, highly adapted to a complex and changing mode of life.

Early Epidermal Cellular Specialization

Ciliated and nonciliated mucous-containing cells are probably the first differentiated body epidermal cells to become distinguishable within the epithelium in young amphibian larvae (Edds and Sweeny, 1961; Billet and Gould, 1971; Kusa et al., 1976), though mitochondria-rich cells also appear early in larval development. About one-third of the surface cells of the embryonic *Ambystoma mexicanum* are ciliated (Landstrom, 1977). Ciliated cells are first recognized at NF St.19 of *Xenopus* (Billet and Gould, 1971) and at the surface of the tail bud of *Rana temporaria* at CM St.22–23 (equivalent to NF St.29–33 of *Xenopus*) (Fox and Whitear, personal observations). A truly demarcated bilaminar epidermis is formed slightly later (*vide infra*). The ciliary cells eventually disappear, though there are vestiges of them at NF St.38–39 of *Xenopus* (Steinman, 1968), or even later at NF St.43 (Fox and Hamilton, 1971). Ciliary cells are still recognizable in body epidermis of *R. temporaria* at CM St.34 (equivalent to NF St.44 of *Xenopus*) (Fox and Whitear, personal observations) (Fig. 47). Cilia are still present in cells at the surface of the external gills of Ambystoma mexicanum (150 mm long) before metamorphosis.

In order to assist hatching specialized, elongate, bottle shaped cells of the hatching glands (HGCs) differentiate in the epidermis of the frontal region of the head, and to a lesser extent along the dorsal midline to the level of the ear vesicle in prehatching stages of *Xenopus laevis* and *Rana chensinensis* (Fig. 48). They are first detected in *Xenopus* at NF St.22 and in *R. chensinensis* at Shumway St.17 (equivalent to about NF St.24 of *Xenopus*) (Yoshizaki, 1973, 1975; Yoshizaki and Katagiri, 1975). During hatching the two outer jelly layers surrounding the larval *Xenopus* rupture, mainly because of water inhibition. The HGCs secrete a hatching enzyme(s), probably a protease, that partially degrades the fertilization envelope (Urch and Hedrick, 1981). Subsequently, movements of the

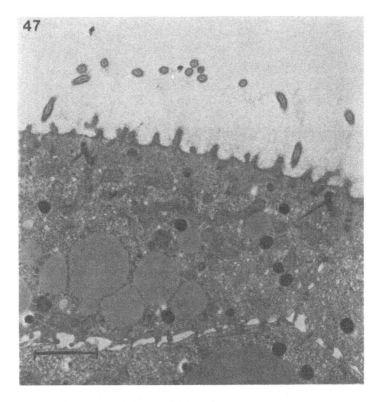

Fig. 47. Ciliated epidermal cell from the tail of *Rana temporaria* at CM stages 32–33 (the external gills of the larva are almost covered by the operculum). Note there is still much lipid in the cell. Ciliated cells soon disappear during development. Scale mark, 2 um.

larva rupture the weakened membrane to permit hatching (Carroll and Hedrick, 1974; Yoshizaki and Katagiri, 1975; Katagiri, 1975).

HGCs, about 45 μm high in *R. chensinensis,* have prominent apical microvilli, a cytoplasmic granular endoplasmic reticulum, and mitochondria. Apical granules formed in the Golgi complex of *Xenopus* at NF St.24 subsequently increase in size and number and are secreted during NF St.24–38 (Fig. 48). In pre-hatching stages the apical granules and Golgi bodies contain polysaccharides other than glycogen. In hatching stages, the hatching of *Xenopus* begins at about NF St.35–36—acid phosphatase-rich granules fuse with membrane-bound bodies within the cell, to ultimately form an elaborate phagolysome, an expression of cellular degeneration (Fig. 49). Probably a functional change occurs in the HGC soon after hatching, from the secretion of carbohydrate-rich granules to those containing acid phosphatase or lysosomal bodies (Yoshizaki and Katagiri, 1975). HGCs slowly diminish in size and in *Xenopus* begin to degenerate by autolysis at NF St.39 after hatching (Yoshizaki, 1974).

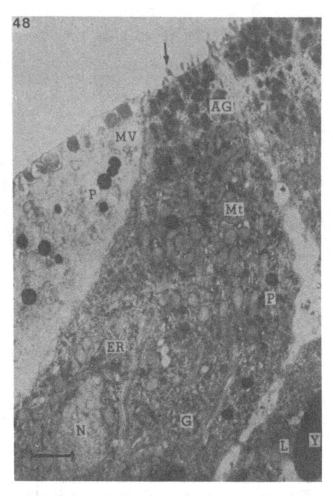

Fig. 48.   Well-developed hatching gland cell from a *Xenopus laevis*
larva at NF stage 34 (early tailbud stage just before hatching).There are
apical microvilli (arrow) and granules (AG) in the pigment-free apical cytoplasm. A common epidermal cell with mucous vesicles (MV) is on the
left. Scale mark, 2 μm. ER, granular endoplasmic reticulum; G, Golgi
complex; L, lipid droplet; Mt, mitochondrion; N, nucleus; P, pigment
granule; Y, yolk body (after Yoshizaki, 1973).

Explants of cells of the presumptive HGC region treated with
Actinomycin D fail to differentiate secretory granules, and ciliogenesis of
adjacent ciliary cells is also inhibited. Cessation of the treatment leads to
the resumption of the production of secretory granules in the HGC; presumably DNA-dependent RNA synthesis influences granular development in the HGCs and also the ciliary cells in early larval stages
(Yoshizaki, 1976).

It is of much interest to find that the differentiation of HGCs and other skin cells of embryonic ectoderm is influenced in vitro by factors present in the culture medium. Thus explanted presumptive ectoderm of early gastrulae of *R. japonica*, cultured for 4–7 d in standard salt solution, differentiated cement gland cells (CGCs), ciliary cells (CCs), and common epidermal cells (CECs). Explants treated with LiCl and thence transferred to Barth's solution differentiated HGCs and pigment cells: the

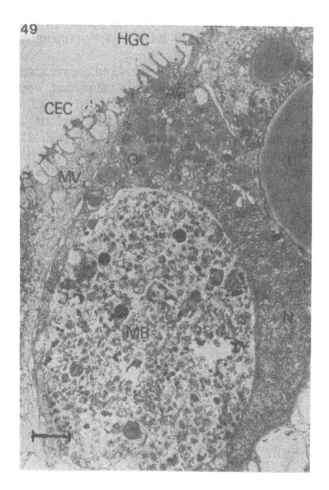

Fig. 49. Hatching gland cell from a *Xenopus laevis* larva at NF stage 46 (larva 9–12 mm long before there are incipient hindlimb buds). There is acid phosphatase activity (lysosome activity) within a large autophagic vacuole or membrane-bound body (MB) in the cell, which is degenerating. Scale mark, 4 μm. AG, apical granule; CEC, common epidermal cell; G, Golgi complex; HGL, hatching gland cell; L, lipid droplet; MV, mucous vesicle; N, nucleus (after Yoshizaki, 1974).

superficial outer cell layers of dorsal presumptive epidermis of St.10 (Tahara, 1959) embryos showed the best ability to differentiate HGCs. LiCl-stimulated explants of slightly younger embryos (St.8,9) differentiated nerve and pigment cells, but slightly older embryos (St.11) ciliary and common epidermal cells. Under optimal conditions up to 70% of explanted tissue formed HGCs. The culture medium of the LiCl-induced tissue of HGCs showed a jelly digesting activity, which suggests that the cells that were induced functioned similarly to HGCs normally formed *in situ* (Yoshizaki, 1979). Immersion of superficial layers of ectoderm cells of embryos, at St.10–11, in specific concentrations of NaCl in Barth's solution for up to 7 d, resulted in about an 85% proportion of cement gland cells (Yoshizaki, 1981) *(vide infra)*.

The Anlage of the future paired pear-shaped cement glands (CGCs), or oral suckers, of *Xenopus* first appears in the preoral, ventro-anterior region of the early neurula stage, at about NF St.15 (Perry and Waddington, 1966; Weets and Picard, 1979). The maturity and activity of the CGCs of *Xenopus* are at their peak by NF St.35–36 (Picard, 1976; Weets and Picard, 1979) (Fig. 50); they degenerate after NF St.40 and have disappeared before NF St.50, the entire life cycle lasting less than 15 days (Nieuwkoop and Faber, 1956). In *Hyla regilla* the suckers are

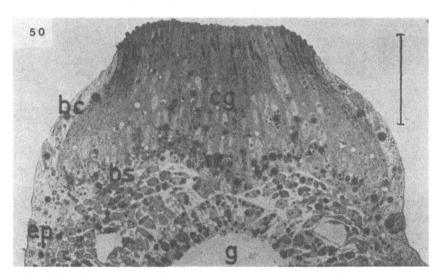

Fig. 50.   Section through the cement gland of a *Xenopus laevis* larva at NF stage 28 (prehatched larva 3.8–4.0 mm long)showing the elongated cement gland cells (Cg), the bordering cells (bc), the basal cells (bs) that form a sheet-like structure, the surrounding epiblast (ep), and the gut (g). The cement gland is well-differentiated at this stage below the incipient stomodeal invagination. Scale mark, 0.05 mm (after Picard and Gilloteaux, 1976).

first adhesive soon after the late tailbud stage, when the secretory granules are recognizable (Eakin, 1963). The cuboidal cells of the cement gland possibly are derived directly from epithelial cells (Billet and Gould, 1971). Fine unmyelinated nerve endings, 1 μm or less in diameter, originating from the trigeminal ganglion, lead to the cement gland (Roberts and Blight, 1975). The CGCs subsequently develop an extensive membrane system producing a mucous-like secretion that is elaborated in the Golgi apparatus and packaged in secretory granules. The secretory adhesive mucin (Eakin, 1963) is probably a glycoprotein (Ling and Lyerla, 1976). The cytoplasm also includes microtubules and microfilaments.

Picard (1976) described five zones in the elongated CGC of larval *Xenopus,* differing in structure and function (Fig. 50). These are: (a) an apicical zone (5% of cell zonation) with alpha and beta vesicles (Fig. 51), (b) a transit zone (30%) including many microtubules with secretion vesicles migrating towards the cell apex, (c) a zone of biosynthesis (40%) with concentric regions of RER and an extensive Golgi apparatus, (d) a zone containing an elongated nucleus (15%) and little lipid and yolk, and (e) a storage zone (10%) characteristically with lipid and yolk, which may occupy up to 50% of the cell volume (see Figs. in Picard, 1976). Two types of cells are recognizable that differ in their staining intensity with methylene blue: 20% stain weakly (type B) and 80% stain more strongly (type A) and they have fewer organelles and more clear vesicles. At NF St.35 there is a greater preponderance of type A cells, suggesting that cellular involution has commenced. Degenerating CGCs develop large autophagic vacuoles and the necrotic cells are ultimately ingested by phagocytes (Perry and Waddington, 1966). Probably lysosomal enzymes are involved in the regression of the cement glands (Hsu and Lyerla, 1977). Cement glands of *Hyla* are poorly differentiated when neurulae are treated with Actinomycin D, probably because of interference with DNA-dependent RNA synthesis (Eakin, 1964).

Among anuran tadpoles some epidermal cells are modified early on to fashion a paired horny beak (Fig. 52) that functions throughout larval life and then disappears at climax (Luckenbill, 1965; Kaung, 1975). The horny jaws of *R. pipiens* larvae are clearly apparent early before premetamorphosis at Shumway St.24 to TK St.III, and though smaller and having fewer keratinized cells they are representative of jaws of older TK Stages from V to XV (Luckenbill, 1965) (equivalent to NF St.45/46–50 and 52–58 of *Xenopus,* respectively). Nevertheless, epidermal cellular differentiation of the jaws may begin as early as Shumway St.21 (equivalent to NF St.40 of *Xenopus*), when the mouth has first broken through to the exterior (Kaung, 1975).

Below each serrated edge of the horny beak there is a column of cells; these are flatter near the base. Apically the cone-shaped cells nestle into each other. At the tip of the column the sharply angled cone cells are

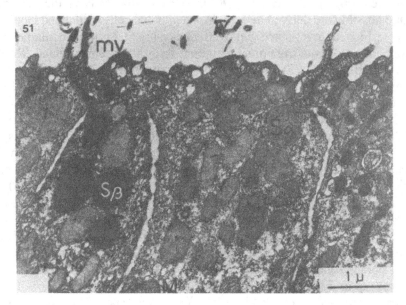

Fig. 51.   Apical zone (5% of total length) of cement gland cells of *Xenopus laevis* larva at NF stage 28, showing moderately (Sα) or heavily (Sβ) electron dense secretion vesicles. The central cell profile contains only Sα vesicles: mv, microvilli; M, mitochondrion (after Picard, 1976).

keratinized. Together the cells of the columns make up a pallisade and adjacent cells interdigitate and join one another by desmosomes (Fig. 53). Apical cell loss, which continues throughout the existence of the beak (Fig. 54), is accelerated by thyroid hormones (Kaung and Kollros, 1977).

General Structure of Larval Skin and a Consideration of Its Cellular Diversity

The epidermis proper is formed when the outer epithelial layers of the skin are delimited by a basement membrane. This membrane and the first whisps of collagen of the future basement lamella first line the inner margin of the second epithelial layer of cells, of the early developing tail of the larval *R. temporaria,* at CM St.25 (equivalent to about NF St.39 of *Xenopus,* Fox and Whitear, personal observations). The epidermis of the body and tail of larval anurans consists of 2–3 layers of cells, increasing to 5–6 layers in the body at climax, when outermost cells may slough (Fox, 1977a). In general epithelial cells of the tail and body epidermis are similar ultrastructurally. Adjacent outermost cells join laterally by tight junctions, and in all layers desmosomes are present between them. Inner cells are usually cuboidal or sometimes irregular in shape, with some interdigitation between adjacent cells. Nuclear profiles are irregular and more flattened in surface cells. The cytoplasm includes a granular endoplasmic reticulum, often quite extensive within some inner tail epidermal cells, ribosomes and polyribosomes, mitochondria, and a Golgi apparatus, which is frequently seen to be well-developed and associated with numerous smooth-walled vesicles. Mucous vesicles line the external epidermal surface (they open into a surface mucoid cuticle), and they also occur within cells of the inner layers. Membrane-bound pigment bodies, lipid, and lysosomes are present, the latter becoming larger and more numerous in outer cells of the epidermis at late prometamorphosis and climax (Fox, 1974). Tonofilaments are more profuse in epidermal cells of older larvae. In the bilayered epidermal cells of *Xenopus* larvae, the thickish bundles of the tonofilaments, associated with basal desmosomes, twist around each other, which explains the elastic filamentous curled meshwork in the apical region of the cells. Actin filaments are also arranged as a meshwork, sometimes in bundles, intermingling with the tonofilaments, and surface membranes and desmosomes are delineated by actin and contain α-actinin (Kunzenbacher et al., 1982).

The inner margin of the basal cells is lined by hemidesmosomes (Weiss and Ferris, 1954), which are associated with the masses of tonofilaments that form the Figures of Eberth (Eberth, 1866), described by electronmicroscopy in the tail of *R. clamitans* by Chapman and Dawson (1961) and in the larval belly and back epidermis of *R. clamitans, R. gryllio, R. catesbeiana,* and *Triturus viridescens* by Singer and Salpeter (1961). The plaques (bobbins or hemidesmosomes) of the

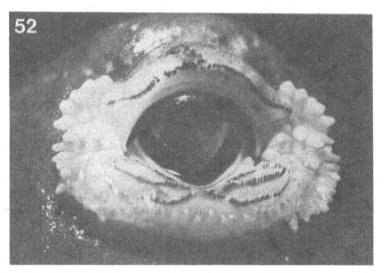

Fig. 52. Mouth of *Rana pipiens* larva at stage X (Taylor and Kollros, 1946) just before prometamorphosis showing the horny jaws. ×19.

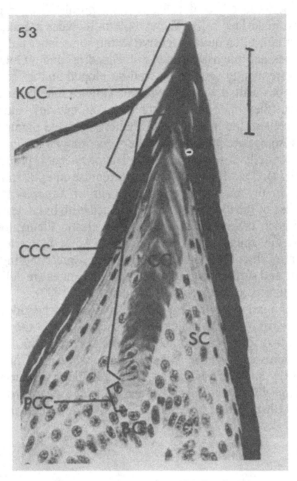

Fig. 53.   Sagittal section of the upper beak of a TK stage X larva of $R.$
pipiens; external surface on the left and internal (oral) surface on the
right. Scale mark, 0.05 mm. BC, basal cells; CC, column cells; CCC, cone
cells; KCC, keratinized cone cells; PCC, precone cells; SC, sheath cells.

basal cell plasma membrane are considered to be hubs on stays, from
which the skeins of fibrils of the Eberth bodies extend intracellularly. The
function of such a system is not clear. It may serve for mechanical support
of the epidermis as a whole (see Lane and Whitear, 1980), or perhaps
could be concerned with the structuring or the fibrogenesis of the underlying basement lamella (Singer and Salpeter, 1961).

Underlying the epidermis are the adepidermal space with its constituent lamellated bodies (Nakao, 1974) and adepidermal membrane
(Salpeter and Singer, 1959), and then the orthogonally oriented collagen
fibrils of the basement lamella in the viscous-like ground substance

(Weiss and Ferris, 1954; Fox, 1972a, 1974). In the tail, anchoring fibrils and larger anchoring fibers lead from the adepidermal membrane and traverse the basement lamella; the anchoring fibers ultimately join subdermal collagen surrounding muscle tissue (see Nakao, 1974; Fox, 1976).

In addition to the epithelial cells, larval tail and body epidermis may include surface mucous goblet cells (Kelly, 1966b) and intra-epithelial Leydig cells (Hay, 1961b; Rosenberg et al., 1982), which are more common in urodeles. The so-called "Kugelzellen" (ball cells) in larval *Xenopus laevis* epidermis (Fröhlich et al., 1977) may be comparable structures. In *Taricha torosa* the large Leydig cells, numerous in midlarval stages, disappear at metamorphosis (Kelly, 1966a), as they do in other urodeles such as the axolotl (Fährmann, 1971). They have a clear cytoplasm, scattered vesicles, dense granules and a network of tonofilaments (Langerhans network) at the periphery. Leydig cells probably secrete mucus into subsurface extracellular compartments of the epidermis to prevent desiccation. Merkel cells are found in the epidermis of larval and adult amphibians (Nafstad and Baker, 1973; Fox and Whitear, 1978; Tachibana, 1978, 1979), and "Stiftchenzellen" or chemosensory cells, in several species of *Rana*, but only in their larvae (Whitear, 1976). All amphibian larvae and adult perennibranchiate urodeles possess epi-

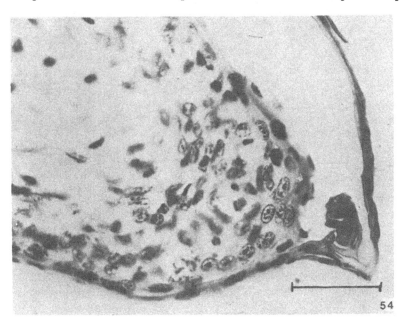

Fig. 54. Tip of upper beak of TK stage *XX* larva of *R. pipiens* (beginning of climax) showing the shedding beak and internal keratinized band of beak cells. Only a few modified column cells remain. Scale mark, 0.05 mm. (Figs. 52–54 after Kaung, 1975).

dermal neuromast organs that disappear at climax in anurans, but are retained in the aquatic *Xenopus* (Dijkgraaf, 1963; Shelton, 1970; see review by Russell, 1976). Specialized electroreceptor ampullary organs have also been reported in the skin of the caecilian *Icthyophis*. (Hetherington and Wake, 1979), and in the lateral line organs of urodeles (Fritzsch and Wahnschaffe, 1983). Lateral line neuromasts differentiate normally in aneurogenic amphibian larvae of *Ambystoma* (Tweedle, 1977). Mitochondria-rich cells and mitochondria-rich flask cells of climax and adult stages also occur (*vide infra*; see Whitear, 1977; Brown et al., 1981).

Merkel cells of *R. temporaria* are first clearly recognizable in the epidermis of the tail at CM ST.33–34, and the body at CM St.35 (equivalent to NF ST.44–45 of *Xenopus*). In *Xenopus* actually they were first found in relatively slightly older larvae at NF St.49–50 (tail), St.50 (hindlimb digits) and St.52 (body) (Fig. 55). They appear to originate and thence differentiate from rounded undifferentiated, interstitial, precursor cells, located between the epidermal epithelial cells (Fox and Whitear, unpublished; see also Fox et al., 1980). Tachibana (1978) and Tachibana

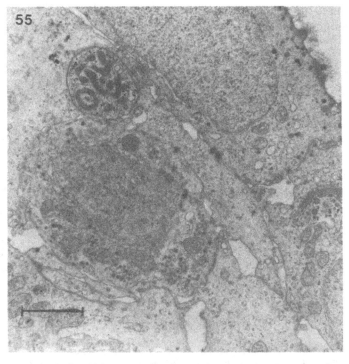

Fig. 55.   Merkel cell from the back epidermis of a *Xenopus laevis* larva at NF stage 55 (beginning of prometamorphosis). Merkel cells are also found in adult epidermis. Scale mark, 2 μm.

et al. (1980) reported their presence in the epidermis of the labial ridge of the larval *R. japonica,* among a group at TK St.I–X (equivalent to NF St.46–54 of *Xenopus*), and in the tentacles of *Xenopus* at NF St.46 respectively. The first Merkel cells were recognized as elliptical cells with cored granules, situated near the base of the epithelium of the labial ridge at Tahara's St.23 (Tahara, 1959), when the larva has a fused operculum (Tachibana, 1979). They appear to originate from undifferentiated epithelial cells in the epidermis, in the tentacles of *Xenopus laevis* larvae (Tachibana et al., 1980). These cells receive synaptic contact from nonmyelinated nerves and are thought to serve a mechanoreceptive function in the tentacles (Ovalle, 1979).

Merkel cells were claimed to derive from the neural crest and migrate to the skin (Winkelmann and Breathnach, 1973; Winkelmann, 1977), but as yet there is no convincing evidence for this view. Indeed, since Merkel cells still occur in the epidermis of *Ambystoma maculatum* larvae after the previous removal of the neural crest, such a derivation would seem to be unlikely, at least in amphibians (Tweedle, 1978).

The young roundish-oval shaped and differentiated Merkel cells of *Rana* and *Xenopus* larvae, about 7 μm long and 6 μm wide, have desmosomes joining adjacent epithelial cells. Several finger-like processes of maximum length of about 1 and 0.1 μm wide, may appear in a single cell profile (Fig. 56). Up to 50 membrane-bound granules may be seen in a single cell profile; they probably originate from Golgi saccules (Fox and Whitear, 1978). The fate of Merkel cells is unknown. Since they occur in amphibian larvae and adults seemingly of similar rarity (about 0.3% of the epidermal cell population, Nafstad and Baker, 1973), then either the same cells survive throughout life, which seems unlikely, or old Merkel cells ultimately involute and are replaced by new ones from existing pre-senile and dividing Merkel cells, or from epidermal basal germinal cells.

Tweedle (1978) showed that Merkel cells can differentiate in the epidermis of larvae of *Ambystoma maculatum* before nerves appear, though the content of cored granules appears to be reduced. The independent development of Merkel cells was confirmed for regenerated epidermal skin of adult *A. tigrinum* by Scott et al. (1981), who also showed that these cells can survive apparently perfectly normally in density and appearance for at least 6 months. After experimental denervation, reinnervated skin of *A. tigrinum* regained its mechanosensory function and the neurite–Merkel cell complexes reformed. A normal neurite–Merkel cell association and function develop in regenerated and thence reinnervated skin. Scott et al. (1981) concluded that Merkel cells appear to act as a target for mechanosensory axons, either by some sort of attraction or because they have specific markers so that ingrowing nerves can recognize them. After innervation Merkel cells lose this target quality. Merkel cells have well-

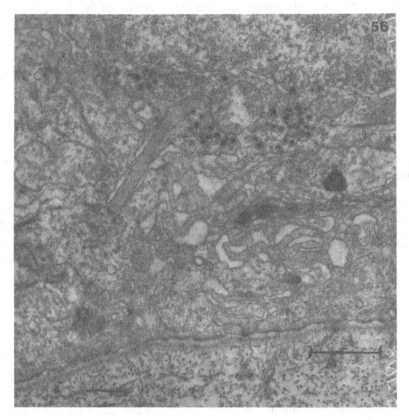

Fig. 56.   Portion of a Merkel cell with a process (finger), from the epidermis of the hindlimb of a *Xenopus laevis* larva at NF stage 57 (during prometamorphosis). Scale mark, 1 μm.

pronounced reciprocal synapses with nerve terminals, and they may function either as slowly adapting sensory touch or mechanoreceptors (Winkelmann and Breathnach, 1973; Andres, 1974), or rapidly adapting mechanoreceptors (Parducz et al., 1977; Roberts and Hayes, 1977; Scott et al., 1981), or mediate a trophic influence (Munger, 1977), or act in some way as transducers (see Fox and Whitear, 1978). However Gottschaldt and Vahle-Hinz (1981) concluded that the nerve endings and not the Merkel cells are the mechano-electric transducer elements, at least for those components in cat hair follicles.

The youngest larva of *R. temporaria* found to possess a clearly recognizable differentiated "Stiftchenzelle" (in the dorsal tail fin epidermis) was at CM St.28–29 (equivalent to about NF St.41 of *Xenopus*). Typically the cell had apical microvilli, a characteristic pear-shape form, an apical core of microfilaments and desmosomes joined it to adjacent epidermal cells (Fig. 57). Large lipid droplets and yolk granules were present in the cytoplasm, but they reduce as the cells fully differentiate.

"Stiftchenzellen" appear to have synaptic associations with nerve fibers (Whitear, 1976). In late prometamorphic larvae, enlarged vacuoles in the cells may be a feature of degeneration. The fact that "Stiftchenzellen" are not found in the skin of adult frogs led Whitear to suggest that they degenerate near (or at) climax, which would account for the high proportion of degenerate looking examples found at this time. The cells may function as chemoreceptors.

The primordium of the amphibian lateral line organs (Singh-Jande, 1966), i.e., as for example in the larval *Triturus pyrrhogaster,* typically originates and thence migrates from pre- and post-auditory placodes from which sensory and supporting cells differentiate in the epidermis. They associate closely with nerve endings, though Schwann cells are located below the epidermis. Lateral line organs are almost wholly differentiated at the time of larval hatching. Each receptor cell has a surface kinocilium

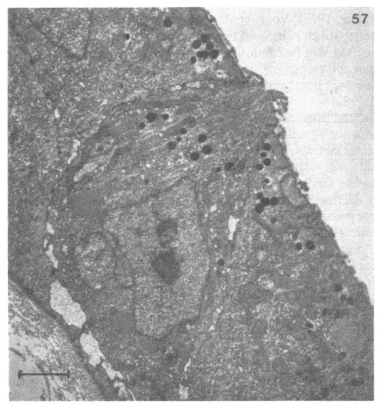

Fig. 57.   'Stiftchenzelle' (chemosensory cell) from the back epidermis of *Rana temporaria* at CM stage 34 (the external gills are completely covered by the operculum). These cells are not found in the adult; indeed so far they have only been reported in *Rana.* They presumably disappear during climax. Scale mark, 2 μm.

and a number of stereocilia. Two types of supporting cells have a granular endoplasmic reticulum, mitochondria, and a Golgi apparatus. Type I cells are distinguished by an abundance of mucous granules; type II cells may be engaged in cupola formation (see Sato, 1976; Sato and Kawakami, 1976).

Mitochondria-rich cells are occasionally found in the surface layer of the epidermis of the larval tail and body; they were recognized in the body epidermis of *R. temporaria* larvae at CM St.33–34 (equivalent to about NF St.44 of *Xenopus*) (Fox and Whitear, personal observations). Mitochondria-rich cells are also found in the epithelium of the bladders of *Bufo marinus* (Choi, 1963), *Rana catesbeiana* (Strum and Danon, 1974) and *Salamandra salamandra* (de Piceis Polver et al., 1981) and of the frog palate (Whitear, 1975). In the adult the flask cells of the epidermis are mitochondria-rich (Masoni and Garcia-Romeu, 1979); they react strongly for carbonic anhydrase (Rosen and Friedley, 1973) and possibly are involved in osmoregulation or ion transport (see refs. and discussion by Whitear, 1977). Voûte et al. (1975) claim that they produce carbonic anhydrase, which is induced by aldosterone, to regulate sodium permeability in frog skin by controlling the pH in the extracellular space next to the outer cell membrane. Indeed, MRCs of larval *Salamandra* react for carbonic anhydrase, suggesting a relationship with flask cells (Lewinson et al., 1982).

Ilic and Brown (1980) have suggested that the flask cells may play a role in assisting adaptation of adult amphibians (*Xenopus*) to different ionic environments. They found that flask cell number, shape, and structure and the glycogen content vary according to osmotic differences in the surrounding fluid medium. Flask cells, which are first recognizable in the body epidermis at climax, make up about 10% of adult epidermal cells and they probably originate *in situ,* after cell division from the basal layer (Whitear, 1975). They differentiate from cells of the stratum intermedium, near those that they replace according to Masoni and Garcia-Romeau (1979). The epidermis and gill epithelium of intrauterine larvae of *Salamandra salamandra* includes occasional pea-shaped slightly protruding cells that stain intensely with toluidine blue–borax. They correspond to the so-called Langerhans cells (see Greven, 1980) of unknown homology and function; perhaps they relate to the mitochondria-rich dark cells of Warburg and Lewinson (1977).

The epidermis also includes immigrant melanocytes (Gartz, 1970; Bagnara, 1976), polymorphonuclear neutrophiles, lymphocytes, granulocytes, and mesenchymal macrophages, all without desmosomes, and there are distributed nonmyelinated nerve fibers (Fox et al., 1980).

Below the basement lamellar collagen the rest of the dermis of an anuran climactic larva also includes melanophores (Fig. 58), xanthophores, and iridophores (see Gartz, 1970; Berns and Narayan, 1970; Taylor and Bagnara, 1972; Bagnara, 1976), granulocytes and other

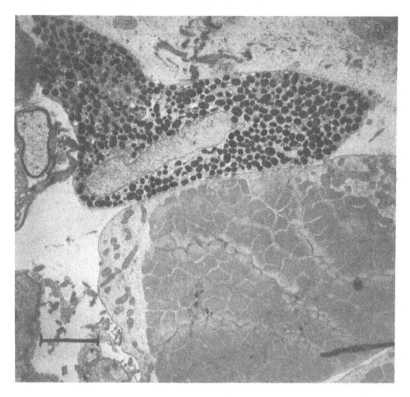

Fig. 58. Melanophore from the tail dermis of a *Xenopus laevis* larva at NF stage 49 (larva with distinct hindlimb buds). Note the nearby nerve and the underlying tail muscle. Scale mark, 4 μm.

leukocytes usually in capillaries, mesenchymal fibroblasts and macrophages, and multicellular mucous and granular (serous) glands (McGarry and Vanable, 1969; Kollros, 1972; Whitear, 1974; Dodd and Dodd, 1976), which open by ducts at the skin surface at climax (Bovbjerg, 1963; see also recent account of amphibian dermal glands by Fox, 1983). There are also muscle and nerve components, including Schwann cells (see Webster and Billings, 1972; Fox et al., 1980). There is extensive vascularization (Saint-Aubain, 1982).

Most of these components, with the exception of the multicellular skin glands, are present in the dermis of the larval tail. Iridophores and xanthophores, however, are rarely if ever found in tails of *R. temporaria* and *Xenopus laevis*. Nevertheless, what appeared to be an incipient or pre-iridophore was recognized in a tail section of *R. temporaria* at CM St.33–34, and pre-iridophores and pre-xanthophores occurred in body epidermis of larval *R. temporaria* at CM St.33 (Fox and Whitear, personal observations). Iridophores, however, occur in the larval tail of *R. catesbeiana* (Ide, 1973).

Table 3
Various Cellular Components Found in Amphibian Skin

	Larva		Adult body	Epidermis	Dermis
	Tail	Body			
Epithelial cells	+	+	+	+	−
Surface keratinocytes (keratinized)	+	+	+	+	−
Keratinized beak cells	−	+	−	+	−
Hatching gland cells	−	+	−	+	−
Cement gland (oral sucker) cells	−	+	−	+	−
Ciliary cells	+	+	−	+	−
Mucous-containing cells	+	+	+	+	−
"Stiftchenzellen" (anurans) (*Rana*)	+	+	−	+	−
Merkel cells	+	+	+	+	−
Goblet cells (few in *Rana*)	+	+	−	+	−
Leydig cells (usually larval urodeles)	+	+	−	+	−
Flask cells	−	+	+	+	−
Mitochondria-rich cells	+	+	+	+	−
			Palate, *Rana* bladder, *Rana*, *Bufo, Salamandra*		
Melanophores	+	+	+	+	+
Xanthophores	−	+	+	−	+
Iridophores	+	+	+	−	+
	(*R. catesbeiana*)				
Mesenchymal macrophages	+	+	+	+	+
Mesenchymal fibroblasts	+	+	+	−	+
Granulocytes	+	+	+	+	+
Lymphocytes	+	+	+	+	+
Polymorphonuclear leukocytes	+	+	+	+	+
Nerve fibers	+	+	+	+	+
Schwann cells	+	+	+	−	+
Neuromast cells (with kino- and stereocilia)	+	+	+	+	−
			Xenopus perennibranchiate urodeles		
Muscle tissue	+	+	+	−	+
Mucous and granular gland cells	−	+	+	+ (neck)	+ (rest of gland)
Myoepithelial cells (glands)	−	+	+	−	+
	(prometamorphosis, climax in anurans)				

Skin Color and Color Change

There are three basic types of amphibian pigment cells or chromatophores in the skin that have their origin from the neural crest. The melanophores are black, brown, or red; iridophores contain reflecting platelets (refractosomes) arranged in stacks, whose orientation determines their pigmentary function; xanthophores or erythrophores are yellow-orange or red.

In adult amphibians, and frequently larval forms as well, the epidermal melanocytes, the only chromatophores found in this layer of the skin [except for xanthophores in *Ambystoma macrodactylum* (Oliphant, 1973); and erythrophores in *Notophthalmus viridescens* (Forbes et al., 1973)], are usually long, slender, and spindle-shaped with dendritic processes: the dermal melanophores are typically stellate in form. Their average length may be about 300 μm in some ranid larval forms (Bagnara, personal communication). In passing it may be mentioned that some workers choose to call them melanocytes for those cells in the epidermis, reserving the term melanophore for those in the dermis. Melanophores contain the pigment melanin (located on a subcellular melanosome about 0.5–1.0 μm in diameter, Taylor and Bagnara, 1972), which is a complex polymer of tyrosine derivatives and protein. Its formation is catalyzed by tyrosinase that is present in premelanosomes (see Bagnara, 1976). Melanophores of *Xenopus* embryos, from cultured neural crest, are retarded in the formation of pigment by Actinomycin D. Pigment formation is thus either controlled via DNA-dependent RNA synthesis, or the chemical influences some other cytoplasmic biosynthetic activities (Kramer, 1972, 1978). Such inhibition of melanophores also results in poor development of cytoplasmic components and the fibrillar and granular areas of the nucleolus separate. The changes in the size and arrangement of the granular nucleolar region (probably the source of rRNA precursors) may be related to the reduced number of free ribosomes in treated cells (Kramer, 1980). Treatment of embryos of the urodele *Cynops pyrrhogaster* with Actinomycin D led Matsuda and Kajishima (1978) to conclude that presumptive melanoblasts are determined during the middle and late gastula stages.

Iridophores have the purines guanine, adenine, and hypoxanthine in crystalline form deposited in their reflecting platelets. In reflected light iridophores exhibit a metallic gold or silver effect; in transmitted light they reflect blues, greens, and reds.

Xanthophores contain carotenoids, derived from the animal's diet, that are fat-soluble and stored in small or large vesicles to appear as oily droplets. Pteridines are synthesized and contained in pterinosomes. The color of xanthophores is determined by the pattern of the contained pteridines and carotenoids.

Melanosomes of melanophores were believed to be derived from premelanosomes pinched off from the Golgi complex. More likely, however, they are formed from vesicles of the endoplasmic reticulum that fuse with vesicles containing tyrosinase of the Golgi cisternae. Their fusion results in the fully formed eumelanin-containing melanosomes of the vertebrate melanophore. Reflecting platelets of iridophores and pterinosome vesicles of xanthophores also appear to be derived from the endoplasmic reticulum (Bagnara et al., 1979).

Dermal chromatophores of postmetamorphic froglets and adults are usually organized in a "chromatophore unit," the xanthophore situated outermost, then the intermediate iridophore and innermost is the melanophore. Together they integrate the total chromatophoral pigmentary response to hormones and light in these animals.

Bagnara et al. (1979) have suggested that melanophores, iridophores, and xanthophores are all derived from basic neural crest cells containing primordial organelles that originate from the endoplasmic reticulum. These primordia differentiate into the pigmentary organelles: mosaic cells containing more than one type of pigment (melanins, pteridines, and purines) have been found. The controlling mechanisms switching on (or indeed restricting) the various forms of pigmentary biosynthetic activity have yet to be discovered.

It has long been known that hypophysectomized amphibian larvae are silvery in color (Smith, 1916; Allen, 1916). The pars intermedia of the pituitary was subsequently found to be the source of a chromatophore-stimulating hormone (Allen, 1930) and called intermedin, but now known as melanophore-stimulating hormone or MSH. It is probaby a tridecapeptide. When larvae are hypophysectomized they have contracted melanosomes within their melanophores (see Pehlemann, 1972). In normal larvae iridophores have their reflecting platelets situated near the center of the cell; in hypophysectomized larvae the platelets are dispersed. In contrast MSH stimulates the dispersion of melanosomes and contracts the iridophores and aggregates the reflecting platelets (Bagnara, 1964; Bagnara and Hadley, 1973), such organelle movements possibly regulated through cAMP (Hadley and Goldman, 1972). In some amphibians at least, the xanthophores are expanded in the presence of MSH and contracted in its absence. Furthermore, the pteridine content of xanthophores is lower in hypophysectomized than in normal larvae, but when hypophysioprivic larvae are injected with MSH the pteridine levels return to normal. Hypophysectomized larvae also show xanthophores with a diminished content of carotenoid.

During larval development there is a primary phase when larvae cannot adapt their color, and they remain dark whether situated on a light or dark-colored background. A secondary phase develops when larvae can adapt and they are able to blanch in darkness. *Xenopus* larvae respond to background changes from very young stages onwards, larvae of *Rana* and *Hyla* are slower to respond, and *Bufo* larvae probably do not acquire a secondary phase. Eyeless larvae have the same paling reaction as normal larvae (Hanke, 1974) and though the view is not unequivocal, on the basis of pinealectomy, it is generally believed that the secondary larval phase is controlled by melatonin from the pineal, which substance directly contracts dermal melanophores and also inhibits the release of MSH. The enzyme hydroxy-indole-O-methyl transferase (HIOMT) is essential in the biosynthesis of melatonin produced in the pinealocytes. The melatonin mechanism is restricted to larvae.

Thyroid hormones may possibly influence larval chromatophores either directly (Bagnara, 1964; Bagnara et al., 1978) affecting their pigment, or their numbers (Collins, 1961; see reference in Bagnara, 1976), or again by antagonizing the MSH darkening of frog skins (Bagnara and Hadley, 1969, 1973). The body blanching reaction of larvae is controlled by melatonin from the pineal, and in adults epinephrine and norepinephrine affect chromatophoral form and activity. It is of interest that Nielsen and Bereiter-Hahn (1982) showed that in the adult *Hyla arborea,* in vitro αMSH, βMSH and ACTH rapidly disperse melanosomes in the melanophores, and contract entire xanthophores and iridophores (and also change the orientation of the platelets). These effects on the chromatophores are antagonzed by melatonin. Furthermore, MSH and epinephrine stimulated melanosome dispersal in all mitosing phases of proliferating melanophores in vitro (Ide, 1982).

In summary: In the larva MSH influences the division, dispersion, and differentiation of melanophores, the contraction and the numbers of iridophores, and stimulates the biosynthesis of pteridines (and possibly carotenoids) in xanthophores. Thyroid hormones, the catecholamines, epinephrine, norepinephrine, and pineal melatonin, and the steroid hormones estrogen, progesterone, and testosterone, also influence amphibian pigmentation in a variety of ways (see Bagnara, 1976; Richards, 1982). During and soon after metamorphic climax, the adult pigmentary patterns develop, and these involve migration and rearrangements of the chromatophores, differentiation of new ones, and mitoses of existing ones. There are also changes in chromatophore morphology: new associations and new classes of pigment are synthesized. Nevertheless, apart from T_4-implantation experiments, there is little direct evidence of thyroid hormones influencing chromatophoral morphological changes during metamorphosis. Perhaps steroids are more closely involved in color patterning than heretofore believed (Richards, 1982).

The subject of the anuran skin pigmentary system has been reviewed by Smith-Gill and Carver (1981), who should be referred to for further details.

Degeneration of Larval Tail and Body Skin

Throughout metamorphic climax the anuran tail progressively degenerates in the distal to proximal direction. Tissue necrosis becomes more widespread as the tail involutes to a small stump, which finally disappears. In general epidermal cells of the larval tail and body degenerate in a similar manner, except that the outer layer of the body shed at climax, is replaced by new cells derived from the basal germinative layer.

At climax outermost epidermal cells of the distal tail region, and likewise those of the body, include fibrous or other dense rounded bodies, which are probably lysosomes. Such cells become electron-dense, flattened, dehydrated, and cornified as keratin is deposited within them. Cell components autolyse and eventually are unrecognizable (Fox, 1977a,b). Ultimately only tonofilaments, fused together in a homogenous mass, remain within the cell. An outer thickened dense membrane below the cell surface (slightly thicker in the adult epidermis), envelops larval keratinized epidermal cells (Farquhar and Palade, 1965; Fox, 1974; Whitear, 1975), which are finally shed as desmosomes are degraded, probably by the action of lysosomal enzymes (Douglas et al., 1970). Inner epidermal cells likewise show features of autolysis, and many develop large cytolysomes before they cornify and are shed when they reach the surface. It is unlikely that there is any significant macrophagic activity during tail epidermal degeneration, as reported by Kerr et al. (1974). During later stages of prometamorphosis epidermal cells show acid phosphatase in small lysosomes, autophagic vacuoles, and cytolysomes. The primary lysosomes probably originate from Golgi cisternae of GERL (Novikoff et al., 1964). At the height of climax there is heavier deposition of acid phosphatase in larger cytolysomes, though the outermost highly cornified cells show little of the enzyme since most of the organelles, apart from the tonofilaments, have disappeared by this time (Fox, 1974). Epidermal cellular degeneration of the tail, body, and external gill filaments of larval *R. temporaria* and *Xenopus laevis,* and of their adults, is similar except that larval external gill filaments do not keratinize. Presumably lysosomal enzymes initiate the autolysis of cellular organelles of degenerating epidermal cells (Michaels et al., 1971; Fox, 1972a, 1974, 1977a; Lavker, 1974). Probably all kinds of amphibian postclimactic epidermal cells, apart from the basal germinal cells, whatever their derivation, eventually degenerate and are shed and replacement occurs. The details of these processes in the case of some specialized cells are not known.

The fine structure of the nondegenerate prometamorphic skins of the operculum, which encloses the chamber within which the forelimbs develop, is similar to that of the rest of the larva. During metamorphic climax it degenerates and disappears (Reichel, 1976), in a manner similar to that of tail skin (Fox, 1974, 1977a), except that opercular epidermal cells apparently do not keratinize.

Near the beginning of climax opercular epidermal cells of *R. pipiens* initially proliferate on extensive granular endoplasmic reticulum, and the Golgi complex is large and prominent. Thereafter autolysis occurs, probably elicited by lysosomal enzymes, and abnormal mitochondria, various large spherical vesicles and residual bodies develop. Finally, the nucleus becomes pycnotic, but the Figures of Eberth within cells of the inner epidermal layer are still recognizable until near the end of skin shedding. Mesenchymal macrophages invade and phagocytose the basement lamellar collagen. The skin becomes transparent against the forelimb, probably because of the destruction of its cellular and extracellular components and some mechanical stretching. Ultimately the remains of the opercular skin are shed (Reichel, 1976).

A combination of factors may well be responsible for the degeneration of opercular skin including, among others, the direct effect of thyroid hormones and mechanical pressure of the emerging forelimb. The introduction of thyroxine-cholesterol pellets into the operculum elicits degeneration of opercular skin and the external gills, and the transformation of adjacent body skin into the postclimactic adult-type. Kaltenbach (1953b, 1968) therefore concluded that changes in opercular skin at climax are the result of the direct action of thyroid hormones.

During metamorphic climax the dermal basement lamellar collagen of the anuran larval tail is invaded by macrophages that engulf the collagenous tissue within heterophagic vacuoles, where they are digested utilizing lysosomal enzymes (Usuku and Gross, 1965; Gona, 1969; Fox, 1972a). The fate of other dermal components at this time, such as melanophores, is not clear. They may well survive and reach the body of the newly metamorphosed froglet. In *R. pipiens* larvae treated with exogenous thyroxine, macrophages are stimulated to similarly invade the basement lamella of back skin and presumably phagocytose collagen (Kemp, 1963). The same process seems to occur in *Xenopus* larvae (Pflugfelder and Schubert, 1965). A new stratum spongiosum develops from mesenchymal cells and the stratum compactum, or adult lamella, forms from collagen fibrils and mesenchymal cells. Polymerization of new dermal collagen continues through the frog's life, and it would appear that the stratum compactum is continuously remodeled, for phagocytosis of its collagen by macrophages occurs in the back skin of the adult *R. temporaria* (Fox, 1977a).

It would seem that in all cases where larval epidermal cellular degeneration has been investigated, autolysis occurs, usually demonstrated by autophagy and the presence of cytolysomes. Some necrotic cells, such as those of degenerating hatching glands and cement glands, are probably phagocytosed by neighboring macrophages. Likewise, most of the remnants of autolysed epithelial cells of the external gill filaments probably suffer phagocytosis. Epidermal epithelial cells, however, autolyse and keratinize, and most of them appear to be shed from the body and tail surface. Perhaps a small number of them are phagocytosed, but this feature is likely of minor importance in both the tail and body. Flask cells of the postclimactic froglet and adult do not keratinize, and presumably they are shed with the slough (Whitear, 1975).

Acid phosphatase and other lysosomal enzymes are present in degenerating larval epidermal cells near and at climax (also in degenerating hatching gland cells, *vide supra*), and in surface presloughing and sloughing epidermal cells of adults. These enzymes are also known to occur in a variety of degenerating larval tissues (Kaltenbach, 1971; see also refs. in Fox, 1974, 1977a, and other sections of the present work). The levels of the lysosomal enzymes increase in the larval tail towards climax (Atkinson, 1981) and it is known that protein synthesis is necesary for tail regression to occur (Tata, 1966; Weber, 1969a; Eeckhout, 1969; see Fox, 1974, for other refs.). The deposition of acid phosphatase is heavier in necrotic tail tissues and in macrophages near and at climax (Salzmann and Weber, 1963; Weber, 1964; Robinson, 1972). Collagenase production is stimulated by thyroxine (Lapiere and Gross, 1963) and T_3 (Smith and Tata, 1976), and the enzyme continues to be synthesized in the tail during its regression (Gross, 1966; Davis et al., 1975). Thyroxine also induces increased hyaluronidase activity in back skin of *R. catesbeiana* larvae during metamorphosis (Polansky and Toole, 1976).

However, the exact nature of cellular degeneration and the role of the lysosomal enzymes during amphibian metamorphosis is still not clear. They may indeed have a variable function in different tissues, as for example, in tail muscle (Fox, 1972b, 1975, 1977b). Smith and Tata (1976) concluded that T_3-induced tail regression could well result from the activation of regulatory "proteolytic cascades," and they found no significant increased synthesis of new protein during the first 3–4 d of treatment of tails in culture.

It would therefore seem reasonable to suppose that autophagy of degenerating larval tissues, including those cells of the epidermis, and the heterophagy exercised by macrophages, function by the utilization of lysosomal enzymes, though often nonlysosomal proteolytic enzymes could well be involved.

Thyroid Hormones and Larval Skin

The influence of thyroid hormones on the origin, development and, in many cases, subsequent degeneration of specific larval epidermal (and dermal) cells would seem to be variable, though doubtless all tail skin degenerates at climax when the circulatory thyroid hormonal level is maximum. These hormones doubtless act at this time through the agency of lysosomal enzymes and the synthesis of keratin, though whether the same endocrinal influences operate in amphibian larvae and adults to activate epidermal cell death is not clearly understood.

In young larval stages of *R. pipiens*, immersion in T_4 (up to 300 μg/L) at TK St.III and TK St.IX–XI (equivalent to stages at premetamorphic and early prometamorphic NF St.49–50 and 54–55 of *Xenopus*, respectively), elicited changes in the mitotic index and growth fraction of different layers of the epidermal cells of the hindlimb bud and limb (Wright, 1977). Implantation of T_4-impregnated pellets into prometamorphic larvae of *R. pipiens* causes localized molting of body skin (Kaltenbach, 1953a).

Keratin synthesis in larval skin of *Xenopus laevis* was induced precociously in vivo and in vitro by triiodothyronine, at a developmental age corresponding to NF St.50–52 (NF St.50 is about 15 d old). Keratin normally appears in the skin about 20–24 d later at NF St.57 when 40 d old, near the beginning of climax (Reeves, 1977). The physiology and biochemistry of keratinization in amphibian larvae and adults are still little understood. Tonofilaments are unlikely to be specifically related to keratinization, nor is it likely that the small granules released into the intercellular spaces of the epidermis of *R. pipiens* (Lavker, 1974) preceding the process are directly involved. However, there is an increase in the amount of endoplasmic reticulum and of the Golgi apparatus during keratinization, and this may be directly related to the process in terms of enzyme synthesis and/or granule production. Extirpation of the pars distalis of adult *Bufo* disturbs the normal synchrony of keratinization, but the factors involved are unknown (Budtz, 1979). Molting is elicited after a specific time lapse of about 17–24 h by ACTH or adrenocorticosteroids, and doubtless molting is hormonally controlled in some manner via the pituitary–thyroid–adrenal endocrine axis. The specific roles of different hormones in the mechanism of epidermal cellular proliferation and keratinization (and in lysosomal enzyme synthesis also), and the processes of separation and ultimate shedding of the cells during the anuran (and urodele) molting cycle are as yet unknown. The pituitary-adrenal axis may well be more concerned with cell separation and shedding rather than with keratinization, which may well be controlled through the pituitary–thryoid axis. The subject has been reviewed by Budtz (1979).

It is of interest that some adult anurans of diverse taxa that live in extremely arid habitats, avoid desiccation by burrowing underground for much of the year and forming cocoons on their body surface. These forms include some Australian leptodactylid frogs, and *Neobatrachus pictus* develops a cocoon of a single layer of flattened epidermal cells 0.5 μm thick (Lee and Mercer, 1967). *Lepidobatrachus llanensis*, a ceratophryd frog from Argentina, has an epidermal cocoon 0.02 mm thick, of more than 50 keratinized cell layers; each flattened cell is about 0.2 μm thick with recognizable remnants of nuclei, organelles, and desmosomes (McClanahan et al., 1976). An underground-living adult ranid *Pyxicephalus adspersus* of South Africa (living up to 8 months of the year underground), likewise has a cocoon of up to 70 keratinized epidermal cells, including their nuclear remnants that cover the entire body except for the nostrils (Parry and Cavill, 1978, and personal communication). *Leptopelis bocagei* from South Africa also forms a cocoon that reduces water loss to about 20% of the rate for with noncocooned frogs (Loveridge and Crayé, 1979). The endocrinological mechanisms responsible for cocoon formation are unknown. Perhaps a reduction in pituitary activity in some way influences the accumulation of layers of keratinized cells that are not shed. In this regard the presence of desmosomes between these cells may well be of significance. Thus it is known that molting is inhibited in both anuran and urodele adults after hypophysectomy. Thyroidectomy, however, does not inhibit molting in adult anurans, in contrast to urodeles when there is a piling up of unshed keratinized layers of cells (Larsen, 1976).

Epidermal cellular proliferation in the adult newt *Notophthalmus viridescens* seems mainly to be influenced by prolactin (Hoffman and Dent, 1977), a result supported by Iwasawa et al. (1978) for *Cynops pyrrohogaster*, who also reported the prolactin-treated newts to have larger and better differentiated (and possibly more numerous) neuromasts than in controls.

Anuran larval premetamorphosis is a period of low thyroid activity and rapid body growth; indeed the thyroid shows only incipient differentiation at this time (Kaye, 1961; Etkin, 1964; Turner, 1973; also see section on the thyroid). The circulatory thyroid hormonal level is extremely low or infinitesimal (Etkin, 1964). Turner (1973) concluded that in *Xenopus* NF St.46–53 are premetamorphic, NF St.54–57 are prometamorphic, and climax then continues to end at NF St.66. However, premetamorphosis of *R. pipiens* larvae (TK St.I–XI) seems to be equivalent to NF St.46–55 of *Xenopus* (see Etkin, 1968; Nieuwkoop and Faber, 1956). Premetamorphosis extends through TK St.XI–XX (equivalent to NF St.55–60). Premetamorphosis is probably a period of larval development practically independent of thyroid function for hypophysectomized larvae of *Scaphiophus* and *Hyla* (Eakin and Bush, 1957) and of *R. pipiens* (Kollros and McMurray, 1956; Decker, 1976) never proceed beyond about TK St. VII (equivalent to NF St.53 of *Xenopus*). Likewise, *Xenopus* larvae hypophysectomized at about NF St.33–36 (tailbud stages) and at NF St.49–50 (first trace of hindlimb bud), ultimately develop only to NF St.52 and NF St.54, respectively (Streb, 1967). They barely, if at all, commence prometamorphosis.

A consideration of different skin cells during anuran larval development reveals that epidermal ciliary cells of *Xenopus* differentiate early and thence have practically disappeared by NF St.43. Hatching gland cells disappear by NF St.39 and cement gland cells by NF St.50. Indeed the life cycle of cement glands, cultured in vitro in Barth solution, was found to be similar to that in vivo, probably independent of hormonal controls, though T_3 did increase the rate of regression of the cement glands in vivo and in vitro, a process whose rate increases with concentration (Weets and Picard, 1979).

Differentiated "Stiftchenzellen" are first apparent in *R. temporaria* larvae at CM St.28–29 (equivalent to NF St.41 of *Xenopus*), and Merkel cells at CM St.33–35 (equivalent to NF St.44–45 of *Xenopus*). Though Merkel cells are first recognizable in the *Xenopus* tail at NF St.49–50 (Fox and Whitear, unpublished), they appear in the tentacles at NF St.46 (Tachibana et al., 1980). It is of interest that they are well-developed with a substantial content of microfilaments (possibly a feature reflecting a long-lived Merkel cell), in the tail of a giant *Xenopus* larva (Fox and Whitear, personal observations) maintained at NF St.54 for 18 months by immersion in the goitrogen propylthiouracil (MacClean and Turner, 1976). Thus presumably the presence of thyroid hormones in a larva is not essential for the origin, maintenance, and perhaps longevity of Merkel cells. Mesenchymal fibroblasts, which probably synthesise the procollagenous fibrils of the dermal basement lamella, are recognized in the tail of *R. temporaria* at CM St.25 (equivalent to NF St.39 in *Xenopus*). In *Xenopus* melanophores first appear dorsally in the head of the larva at NF St.33–34 and xanthophores in the peritoneum and outer surface of the eyecup at NF St.46. Lateral line organs have differentiated into individual sensory components at NF St.41 and they are conical at the skin surface at NF St.51 (Nieuwkoop and Faber, 1956). They are highly differentiated at NF St.54 (Shelton, 1970), a feature recognizable earlier in larval development by Fox and Whitear (unpubl.). Epithelial cells of the external gill filaments of *Xenopus* also differentiate early on in larval development and are merely remnants before the end of premetamorphosis.

Hypophysectomized larvae of *R. pipiens* at the tailbud stage do not subsequently develop skin glands (Bovbjerg, 1963), and goitrogens such as potassium perchlorate inhibit glandular development. Skin glands develop prematurely near implanted T_4-cholesterol pellets (Kollros and Kaltenbach, 1952). McGarry and Vanable (1969) concluded that thyroxine treatment of cultured forelimb skin of *Xenopus* stimulated glandular cell division, and in some way assists their development. At later larval stages thyroxine only seems to influence the rate of maturation of the epidermal glandular rudiments. These authors found the older larval skin glands to have an increased sensitivity to thyroid hormones, a conclusion that confirmed that of Heady and Kollros (1964) and Vanable and Mortensen (1966), in tadpoles of *R. pipiens* and *Xenopus laevis,* respectively (see also Kollros, 1972).

Thus the available evidence so far suggest that the life cycles of ciliary, hatching gland, cement gland, and external gill epithelial cells are probably practically independent of thyroid hormones. So at least are the primary origin and differentiation of "Stiftchenzellen," Merkel cells, beak cells, goblet cells, and Leydig cells, melanophores, pre-iridophores, pre-xanthophores, and mitochondria-rich cells, cells of the neuromasts, and the mesenchymal fibroblasts. Whether the degeneration of "Stiftchenzellen" is influenced by the high level of circulatory thyroid hormones at climax is not known.

The origin and differentiation of the epidermal flask cells of the larval body and of the skin glands, and in addition the loss of the beak cells, appears to depend upon a threshold level of thyroid hormones.

Skin Biochemistry

The biochemistry of the skin of amphibian larvae has been reviewed (see Frieden and Just, 1970; Dodd and Dodd, 1976). Among the variety of differences that occur between larval and adult skin, the ability of adult skin to deiodinate thyroxine begins during climax (Dowling and Razevska, 1966). Again the glycosaminoglycans and hyaluronidase of prometamorphic and climactic larval skin of *R. catesbeiana* seem to be absent in the adult (Lipson et al., 1971). Tadpole skin contains hyaluronic acid and the skin of the frog has a mixture of sulfated and nonsulfated glycosaminoglycans. The changeover from the larval to the adult type occurs in *Rana catesbeiana* at TK St.XX to be completed by TK St.XXII at climax, probably activated by collagenase and hyaluronidase (Lipson et al., 1971). The larval and adult skin also differ in their permeability to Na^+ and Cl^- ions (Alvarado and Moody, 1970). Likewise the levels of Na-K-ATPase and the response of ATPase of the skin of larval *R. catesbeiana* to thyroxine differ from that of the adult (Kawada et al., 1969, 1972). Perhaps these features are related to the fact that, when treated with T_4, the metabolic rate (measured in terms of O_2 consumption) of larval *R. pipiens* skin increases to a maximum at climax, and later decreases to the adult level (Barch, 1953). It is of interest that thyrotropin releasing factor (TRF) was found to be present in the skin of the adult *R. pipiens* in up to four times the concentration found in the hypothalamus (Jackson and Reichlin, 1977, 1979). They suggest that the skin acts as a huge endocrine organ that synthesizes and secretes the hormone. It is possible, though, that the skin merely serves as a storage organ for this substance. More recently, TRF (and 5-HT) were located in the granules of the granular glands of adults of *Rana pipiens* and *Xenopus laevis,* and they are discharged after stimulation by adrenergic agents (Bennett et al., 1981). The amount of TRF is low during metamorphosis, but it increases to high levels in adults. Clothier et al. (1983) suggested that TRF is synthesized in the glands and thereafter is released into the blood in small amounts.

It is not clear whether any specific cellular components, such as flask cells, are directly involved in any of the listed biochemical processes. The epithelial cells of larval and adult epidermal skin of amphibians show similarities in morphology and behavior (Fox, 1977a), and the functional differences that presumably occur are doubtless in adaptation to different modes of life.

Section IV

Cellular Differentiation, Ontogeny and Molecular Biology

29. A Consideration of Cellular Diversity

There is a vast array of different cell types within the animal kingdom. There are probably more than 200 discinct cell types, and many of these include a number of different varieties (Alberts et al., 1983). In gross visual terms they are recognized in different species, either of Protozoa or Metazoa, at different levels of evolutionary structural and functional complexity. In animals throughout ontogeny, cellular diversity is clearly recognizable at the level of light microscopy. Visualized under the electron microscope, cells contain what would appear to be a multitude of subcellular organelles of infinite complexity, which by virtue of their number, arrangement, and disposition confer upon the cell its distinctive ultrastructural specificity.

In essence, however, the morphological diversity of cells is a result of differences in their size, shape, and organelle content, ultimately expressed in diversity of function. These features arise as a result of the individual cell's ontogenetic experience, broadly explained in terms of inductive phenomena, and thence chemical interactions originating at the level of the target cell's nuclear genes, which have evolved to the requisite degree of molecular complexity.

With the general qualification that synthesis of the great variety of specific subunit molecules and their integrated three-dimensional arrangement within different subcellular organelle macrocomponents are extremely complex phenomena, nevertheless the number of basically different types of macrocomponent that make up an organelle is relatively small. As with the 20 or so amino acids whose varied combinations of individual type, arrangement, and number give rise to the enormous variety of proteins, so can relatively few organelle components, by increasing the structural complexity within cells of differing size and shape, create the clearly recognizable wide range of diverse cells.

Molecular Structure

The basic subunits of subcellular organelles, the proteins, range from those whose molecular weight is about 10,000 to others of over 1 million: proteins may contain from 100 to 10,000 different amino acid residues. The linear polypeptide chains, the protomers, exist either singly or as dimers or tetramers. Proteins are either elongated fibrous molecules that usually play a mechanical role within or outside the cell, or more usually are folded spontaneously in a complex manner into tertiary three-dimensional structures, as β-sheets or α-helix form—the globular proteins. In this form they are active either as enzymes or comprise subunits of cell organelle components or their surface membranes. Other proteins in the quarternary state are multiple aggregate systems that form multienzyme complexes. Hemoglobin is a tetramer whose configuration depends upon the assembly of the 2 alpha and 2 beta chains into a globular aggregate in specific relationship with heme, Hb. It is this integrated molecular system that effectively binds and transfers oxygen. Denaturation by urea disassociates the alpha and beta subunits and the capacity to bind oxygen disappears. Removal of urea allows the automatic reassembly of the Hb and consequently oxygen binding again.

The range of proteins synthesized by ribosomes via mRNA is enormous; the number of transcribed mRNAs likewise is presumably very large. There are far fewer tRNAs with their anticodons; definite assignments of specific amino acids are known for 61 out of the possible 64 codons, and thus tRNAs, the other three being chain-terminating codons. However, other anticodon bases, such as inosine (arising by enzyme modification of a completed tRNA base), may be responsible for a larger possible number of tRNAs (Watson, 1976).

Allosteric proteins are a special class of enzymes that assume different functional states depending upon their three-dimensional configuration. In combination with small effector molecules such as inducers, an allosteric repressor protein can change its function from an active to an inactive state: countless examples are now known of such allosteric interactions. An allosteric repressor protein has combining sites for an operator gene and for an effector. Recognition by the repressor of a nucleotide sequence of an operator gene is governed by its three-dimensional structure and its relationship with the effector. In the final analysis a protein's 'teleonomic performance' is determined by its stereoscopic properties, that is by virtue of their shape and molecular configuration they have the 'ability to recognize' other molecules including protein. The simplest type of organism such as the bacterium *Escherichia coli* (wt: 5×10^{-13}g; 2 μm long) may contain 2500 ± 500 different proteins, and for higher mammals including humans the figure could well be of the order of one million (Monod, 1972), among the 5×10^{12} (approx.) cells that make up a human being (Watson, 1975).

The broad sequence of events, proceeding from the beginning of development of an organism through embryogenesis, starts with the folding of polypeptide chains to form globular proteins with stereoscopic properties. Associative interactions between proteins and other cellular constituents produce cellular organelles, such differentiating cells mutually interacting to form tissues and organs. The entire process embraces coordination and differentiation of a large variety of chemical activities, through allosteric-type interactions determined initially by genetic information acting via the synthesis of polypeptide sequences (see Monod, 1972).

Cell Size and Shape

A host of examples of significant differences in size and shape of cell types within and between different groups of vertebrates are well-known. It seems that small cellular dimensions are characteristic of tissues that are continuously replaced during life: the largest cells, including those of striated muscle and neurones, usually multiply only during embryogenesis. Neuroblasts divide when they are without processes. Within their relative locations in the CNS, cellular processes develop and interconnect, creating complex neuronal morphological patterns. At first glial cells are few, neuroblasts small, and axonal processes short, in order to reach their termination. During embryogenesis glial cells proliferate, interconnections between nerve cells are newly acquired, and gradually restricted as the neurones enlarge with further elongation of their axons (Szarski, 1968).

Among amphibians, comparable cells of urodeles are larger than those of anurans. In general within this class of animals, hepatocytes, renal cells, and erythrocytes show a descending order of magnitude, though kidney cells may often be larger than hepatocytes. Furthermore, in different individuals of the same species apart from distinct type cells, comparable cells may also differ in size, seen for example, in the case of hepatocytes. The largest liver cells were found in the urodele *Rhyacotriton olympicus* (Szarski and Czopek, 1965).

Typically larval (and adult) epidermal Merkel cells are larger in urodeles than in anurans; those of caecilians seem to be intermediate in size compared with the other two orders (Fox and Whitear, 1978). Again in *R. temporaria* larvae, individual cell volume of the pronephros (about 5000 μm^3) is significantly superior to that of the thyroid (800 μm^3) (Fox, 1962a, 1966). Liver cell volume in the larval *Xenopus* is about 600 μm^3, though the cells hypertrophy at climax (Oates, 1977). In general amphibian metamorphosis is accompanied by a decrease in cell size, though during the final stages of metamorphosis and afterwards cell size may increase (Szarski, 1968; Szarski and Czopek, 1965).

Differences in the shape of various cell types occur between and within different tissues. Again similar cells can, and frequently do, change their shape at different stages of the metamorphic cycle. Thyroid columnar follicle cells elongate at climax and thereafter reduce their height to become squamous cells in the adult; throughout life, epidermal cells flatten as they keratinize and slough at the body surface; larval melanophores expand and iridophores and xanthophores contract and change in appearance under the influence of MSH (Bagnara, 1976); the irregular and constantly changing shape of mesenchymal fibroblasts, macrophages, and blood cells presumably is a regular feature of their life histories. Furthermore, urodele intestinal epithelial microvilli are suppressed by high hydrostatic pressure, to quickly reform, within 1.5 h, when this is reduced (Tilney and Cardell, 1970).

Polarity, often expressed in terms of shape but better considered with reference to the specific location of organelles, or in the uniform or gradient distribution of organelle and metabolic systems, is a fundamental property of cells. Multiple gradients likewise may be present within or among cells, especially during embryogenesis, with high points in specific regions of a cell diminishing progressively along a recognizable axis. Gradients of concentration or of size of yolk and pigment particles in cells of amphibian cleavage stages, and gradient concentrations of 'inducer substances' along the dorsal axis of the gastrula, are typical examples. Cell polarity may be determined initially through external stimulation by a variety of stimuli evoking differential permeabilities of the plasma membrane. Such stimuli could initiate ion movements or electric field phenomena, which would thence influence the arrangement and distribution of organelles or other cell constituents. In this manner cell shape and form could at least in part be determined by microtubules and microfilaments acting as subcellular cytoskeletal components whose orientation and location within the cell would strongly influence its asymmetry.

Yet notwithstanding variations in cellular size, shape, and behavior, the main diagnostic criterion of cellular specificity is expressed in the diversity of its subcellular structure.

Cell Organelles

The spontaneous assembly of a variety of subunit molecules and macromolecules results in the creation of macrounits, which in turn integrate to form components of organelles and enzyme reaction sites of lesser or greater complexity. Most cellular organelles are polytypic, consisting of lipid and/or nucleic acid subunits, with proteins associating in complex arrangements, as with the multienzyme systems wthin lipoprotein membranes of mitochondria or cell surface membranes. Other enzyme systems may not be membrane-bound, but occur freely in the cytoplasm. Organelle macrounits can be broadly listed into a limited number of categories that include the following:

Aggregate Solid Inclusions

Polytypic ribosomes are complex conglomerates of stereospecific heterogeneous subunits that include rRNAs and various proteins. Their assembly is temperature-dependent and does not occur at 0°C. Ribosomes of *E.*

coli have been examined in great detail. There are about 15,000–30,000 per cell, each spherical structure is about 20 nm across and of molecular weight of about 3 million, and they are composed of 50S and 30S subunits (the larger subunit containing 34 different proteins and the smaller one 21) and RNA. Eukaryote ribosomes each have 60S and 40S subunits held together by ionic linkages. Different proteins operate in amino acid incorporation and possibly rRNA assists in binding tRNA and mRNA to ribosomes during the polypeptide synthesis. Functional groups of ribosomes or polyribosomes may be linked by a fine strand of mRNA, probably spatially associated with the smaller ribosomal subunit; the newly synthesized proteins are more closely associated with the larger subunit (Novikoff and Holtzman, 1970). However, the exact chemical details of ribosomes may well be only realized decades ahead (Watson, 1976), a view that may yet prove to be overoptimistic.

Microfilaments

These structures are of varied length and diameter, either of a single linear molecular assemblage or as entwined chains in helical configuration, seen with alpha and beta chains of the extracellular tropocollagen (again each chain composed of three polypeptide chains). Microfilaments are ubiquitous structures in eukaryotic cells. For example, the cytoskeleton of cells includes among other components three classes of filaments, of diameter 9–11 nm, 6–8 nm, and 2–3 nm. A protein-rich microtrabecular lattice (proabaly not artifactual, though this is still controversial), extends throughout the cell and an important part includes the 2–3 nm microfilaments and the actin filaments (Schliwa et al., 1982). The intermediate unbranched cytoplasmic filaments, of diameter 7–11 nm, are of a size between that of the myosin filaments (15–16 nm) and the microtubles (22 nm), and the actin microfilaments (~6 nm) and the 2–3 nm microfilaments. Microfilaments are described and discussed, among others, by Franke et al. (1982), Holtzer et al. (1982) and Liem et al. (1982). Actin and myosin contractile proteins forming the filaments of striated muscle cells are well-known examples, though they occur widely in a variety of other tissues. In muscle thin F (fibrous) actin myofilaments of globular subunits together with other tropomyosin and troponin molecules are organized in two helically arranged strands, with about 13–14 molecules per twist. Subjected to different ionic strength solutions, the strands either dissociate into subunits of G (globular) actin molecules (each 5.5 nm in diameter; MW 43,000), or are polymerized into F actin filaments of variable length, a feature controlled by factors yet unknown. One molecule of ATP binds firmly to one of the G actin molecules and phosphate is split from the ATP during the polymerization of G to F actin. The structure, arrangement and packaging of actin microfilaments, and their binding to one another by macromolecular bridges, have recently been described by Rosier and Tilney (1982). Again Lind et al. (1982) showed that the protein gelsolin (MW 91,000) of human platelets, binds reversibly to actin in the presence of Ca^{2+}, and in some manner appears to regulate microfilament length. Likewise vinculin, a globular 130K protein of diameter 85 Å, from chicken gizzard, binds to muscle actin and induces filaments to form bundles, of a configuration to interact with the plasma membrane (Isenberg et al., 1982). In amphibian cells a feature of interest is the discovery by Colombo (1982) that actin microfilaments form an actin shell around yolk platelets, in developing *Xenopus* embryos, though its function is not known. Myosin molecules are of characteristic shape; seen under the electron-microscope they comprise a 'head' and 'tail' extension. The proteolytic enzyme papain separates the 'tail' (molecular weight 210,000) from the 'head,' itself composed of two similar components each called SI (MW, 120,000). The tail can be further separated by trypsin into two unequal portions, light meromyosin (LMM of MW 150,000) and S2 (MW 60,000); in essence it consists of two alpha helices wrapped round each other like a double-stranded rope. Myosin crossbridges have a regular arrangement. X-ray diffraction has shown that they probably appear as spokes from the thick myofilaments in sets of three, spaced 120° apart. The next three arise 14.3 nm along the thick myofilament, but rotated through 40°, so that the original pattern repeats after 42.9 nm. Tropomyosin (MW 68,000) comprises about 0.8% of muscle and is composed of rods (40 nm long), each formed from a double alpha helix ressembling the tail of a myosin molecule. Troponin (MW 85,000), making up about 0.2% of muscle, is composed of at least three subunits. In solutions of low ionic strength, individual myosin molecules aggregate to form rods that are visible under the light microscope. Experiments show, however, that such orderliness can occur quite naturally and requires no more elaborate control than needed for the growth of a crystal (see Wilkie, 1976).

Using immunoflourescence with antibodies specific for various structural components of the Z disk, it has now been shown that in avian embryonic and adult smooth, cardiac, and skeletal muscle cells, a wide variety of protein components make up the muscle filaments. These include myosin (MW 200,000), tropomyosin (MW 68,000), troponin (MW 85,000), actin (MW 43,000), α-actinin (MW 100,000), filamin (MW 250,000), thermin A and thermin B (MW 68,000–70,000 daltons), and the components of the intermediate filaments, desmin (MW 50,000), vimentin (MW 52,000), synemin (MW 230,000), and paranemin (MW 280,000). The desmin and vimentin intermediate filaments and their associated proteins are integrated into the Z disks. Their structure and composition are regulated during development in response to specific functions of the filaments in the differentiating cells (Lazarides et al. 1982).

Epidermal keratin is formed from a large number of different cytoplasmic structural proteins bonded together by disulfide linkages rich in cystine. Constituent keratin fibrils are of alpha helical form situated within a matrix of globular proteins of various kinds. Indeed keratins have a wide range of chemical structure in different classes of vertebrates (Spearman, 1977). Comparable though chemically dissimilar epidermal tonofilaments are likewise common; they are seen in profusion forming the Figures of Eberth in larval amphibian basal epidermal cells.

Recently Lazarides (1980) categorized five chemically distinct classes of intermediate filaments (about 100 Å in diameter), distinct from actin, myosin, and the microtubules. These are (1) keratin (tono) filaments in epithelial cells; (2) desmin filaments predominantly in smooth, skeletal, and cardiac muscle cells; (3) vimentin filaments in mesenchymal cells [so-called because immunofluorescence reveals a wavy (Latin *vimentus*) filamentous system]; (4) neurofilaments in neurones; and (5) glial filaments in glial cells. Commonly two types of these filaments can exist in the same cell (See also Osborn et al., 1982). Intermediate filaments likewise occur with other microfilaments and microtubules in the same cell. For example, the cytoplasm of the neurone has been arbitrarily separated by Lasek and Brady (1982) into a nucleated "transitional region," where polysomes (corresponding with the basophilic Nissl substance) are located, and an extension or "expressional region," the enlongated axon, whose axoplasm is distinguishable by, among other things, the linear intermediate neurofilaments, actin filaments, and microtubules (the cytoskeletal complex), tubulovesicular elements, mitochondria, and the proteinaceous matrix.

Microtubules

These organelles are present in most eukaryotic cells and are especially prominent in cilia, flagella, and the mitotic spindle. They are unbranched cytoskeletal structures, probably concerned with cell shape and the transfer of intracellular substances, and can arise and disappear in cells at different stages of their life cycles. Elongated microtubules, of variable length and about 25 nm in diameter, have a wall of globular subunits (each of MW about 60,000), of the protein tubulin, which can be made to recombine in vitro after dissociation to form structures resembling microtubules (Stephens, 1968). The microtubule wall is formed from a left-handed helix with 13 polypeptide subunits per turn. Tubulin is a 6S dimer of one alpha and one beta component, each of different amino acid sequence. Polymerization of microtubules by self-assembly of subunits occurs at 37°C and dissociation at 0°C when the concentration of Mg^{2+} ions is low. Colchicine inhibits polymerization of the subunits and causes dissociation of microtubules by binding to the tubulin subunits (see

Grant, 1978). Present evidence now suggests that a pool of soluble tubulin, accumulated during oogenesis, supplies the subunits of the microtubules formed during cellular development. Constant tubulin pools are maintained by further synthesis and in vivo microtubule subunits are assembled at specific nucleation sites, or microtubule organizing centers. Nontubulin microtubule-associated proteins, which appear to stabilize microtubules, may be necessary for their assembly (Raff and Raff, 1978).

Microtubles exhibit a steady state of opposite end-to-end assembly and disassembly (treadmilling) in vitro that could well occur in vivo. Such an activity may be utilized to influence the flow of cell substances in the tubules (Wilson and Margolis, 1982). Microtubule self-assembly and the genetic control of α- and β-tubulin synthesis, have been described by Cleveland and Kirschner (1982).

Membranes

Membranous structures of varied composition are ubiquitous features of all cell systems. They surround the cell (plasma membrane) and occur intracellularly to surround the nucleus and forming organelles such as the endoplasmic reticulum, Golgi apparatus, mitochondria, and vesicles of various kinds. The molecular structure of cell membranes is mentioned here only briefly and detailed descriptions can be obtained from an extensive literature on the subject (see Chapman and Wallach, 1968,1973,1976; Weissman and Claiborne, 1975; Houslay and Stanley, 1982). The classical concept of the cell membrane by Davson and Danielli (1943; see also Danielli, 1975) proposed a lipoprotein structure usually containing up to 1–4 times by weight of protein to lipid (fatty acids, phospholipids, and steroids). As lipids have molecular weights of the order of 100 and proteins over 10,000, there are thus usually far more lipid than protein molecules in membranes. Seen under the electronmicroscope, a trilaminar unit membrane comprises a central lighter staining layer 3–4 nm thick bounded by dense inner and outer layers each about 2 nm thick. The middle layer is envisaged as a bilayered core of lipid whose hydrophilic and hydrophobic molecules face outwards and inwards respectively. The dense outer and inner layers are composed of protein molecules, the whole forming the lipoprotein sandwich. This idea, however, is doubtless too simplistic (Gomperts, 1977) and subsequent ideas by others suggest far greater complexity in the arrangement, distribution, and content of cell unit membranes. A current idea proposes a more fluid dynamic system, where protein and probably other molecules also, are less stable and move around within the membrane (Ferber, 1973; Chapman, 1975; Cherry, 1976; Houslay and Stanley, 1982). For example the membrane 'iceberg' theory (Singer, 1975) proposes a fluid mosaic model where proteins (like icebergs) float in a 'sea' of lipids. Proteins that are large enough span the entire thickness of the membrane; possibly they are amphipathic, and protruding hydrophylic and interior hydrophobic terminals are similar to lipid molecules in this regard, though the protein molecules are much larger.

Membrane proteins are termed peripheral (40% of the total) when they are located at the cytoplasmic face, and they include those such as spectrin of mammalian erythrocytes, ankyrin or syndein, and actin. The integral membrane proteins, of five different types, include fibrous and globular proteins. Other ectoproteins and endoproteins are simply anchored to the surface of the membrane (Hously and Stanley, 1982).

The nuclear envelope (NE) is a porous, double-membrane system whose nucleopores are about 70 nm in diameter. There is a morphological continuity between the NE and the endoplasmic reticulum (ER), and during NE regeneration, at the end of mitosis, it incorporates membranous material from the ER. Furthermore, the outer nuclear membrane is also in continuity with the Golgi apparatus. Indeed, it is widely postulated that the NE is merely an ER cisterna located around the nucleus, specialized for nucleocytoplasmic transport of ribonucleoproteins via the nucleopore complexes and possibly other well defined sites near their peripheries. This view is strongly supported by the fact that there is great similarity in the biochemical composition of the NE and ER (see Wunderlich et al., 1976). A more recent review by Franke et al. (1981b) has added further confirmation. Nuclear membranes are chemically similar to ER membranes in their lipid pattern, the large number of proteins and enzymes, the carbohydrate pattern of glycoproteins, in their lectin-binding properties, and perhaps also in their patterns of hormone receptors and several components defined as antigens.

In some cases at least, membrane self-assembly probably requires a pre-existing template to which new components of different molecules, synthesized outside the membrane (See Houslay and Stanley, 1982), may be added. Membranes from different organelles appear to originate from different sources. Those of a mitochondrion form within it: endoplasmic reticulum may well be an extension from the nuclear membrane: vesicles, lysosomal membranes, and some sections of the plasma membrane arise from the Golgi apparatus, though the derivation of the latter is not clear. Furthermore, most membranes are dynamic structures, manifesting molecular translocation and flip-flop, continuous degradation of old and incorporation of new molecules of protein and lipid during life. Presumably, various enzymes synthesized on ribosomes have some structural affinity with the 'homing sites' on membranes to which they migrate. In this manner sequentially in space and time a variety of subcellular enzyme sites form that confer upon the membrane its biological specificity.

It is clear that notwithstanding any configuration or other possible biophysical or biochemical factors limiting the spatial arrangement or association of constituent molecules, the molecular make up of similar and of different macrounits of organelles may show wide variation. This results from the enormous number of different proteins, apart from any other type of chemical component, whose presence allows the possible macrounit composition of molecules to be practically limitless in its variation. Nevertheless, one can broadly generalize that at a relatively gross level of molecular organization, though still one at the level of the electron microscope, from a limited set of integral macrounit components the array of different organelles formed creates a wide range of cellular structural diversity.

Indeed, a belief that "the features which distinguish one cell type from another represent the amplification or exaggeration of a set of basic components, is one of the unifying concepts of cell biology" (Lasek and Brady, 1982).

Within an individual cell, membranous structures for example located around the cell or nucleus, or forming the different kinds of vesicles (including the clathrin-coated vesicles; the coat consisting of a basket-like network of three-legged molecules or triskelions), enclosing a multitude of different secretory chemical products, an endoplasmic reticulum, Golgi complex, sarcoplasmic reticulum, or T-tubular system, etc., by virtue of their molecular composition, size, shape, and arrangement have assumed a variety of biological functions. Extensive diversity of functional biosynthetic activity occurs when membranes associate spatially with other cellular constituents, as with ribosomes of the RER and when mitochondrial membranes integrate with specific molecular aggregations involved in ATP generation during oxidative phosphorylation. The extent of the biosynthetic activity of the Golgi apparatus is now known to be far greater than heretofore believed (See Rothman et al., 1982), and the list of its activities is growing continually as more is learned about it. Golgi complexes consist of sets of flattened disk-like stacks of smooth-surfaced membranous cisternae—like a pile of plates; they are usually located juxtanuclearly. A complex has two distince faces; the *cis* (forming face) and the *trans* (the maturing face), nearest to the plasma membrane. Proteins enter from the ER on the *cis* side and leave at the *trans* side after being processed. Smaller complexes are also recognized near the cell surface and these are considered to participate, by their vesicles, in the synthesis of material for the plasma membranes of adjacent embryonic cells of *Xenopus* (Sanders, 1973). In amphibians at least, Golgi vesicles, among other things, are also implicated in the reformation of the cytoarchitecture of metamorphosing *Xenopus* intestinal brush border cells (Bonneville and Weinstock, 1970). Again it has been shown by Lemanski and Aldoroty (1977) that in the young developing larva of the axolotl, yolk platelets of differentiating cells degrade by membranous unraveling or delamination, associated at their peripheries with acid phosphatase activity. This enzyme is first detectable in the Golgi complex and thence packaged in lysosomes that, at this very early stage, migrate to the yolk platelets where their enzymes are released. A similar mechanism in *Xenopus* eggs and larvae was suggested by Decroly et al. (1979).

We can summarize the foregoing on cell diversity as follows: Within developing cells self-assembly of a variety of protein molecules, in association with lipid compounds, results in the formation of specific macrounits basically limited structurally in their three-dimensional form, which are recognizable to make up the architecture of the cellular organelles. The variation in arrangement, quantity, disposition (including the spatial association with other macrounits), and the chemicophysical activities of these resulting organelle components confer upon the cell its individual structural and functional specificity. Grouped together, the like cells form tissues, but more often diverse cells associate to form complex compound tissues and organs that interrelate with each other in the living animal. For a full account of the cell, the reader is referred to the recent extensive work on the subject by Alberts et al. (1983).

30. Early Cleavage Cells and Their Nucleocytoplasmic Relationship

In order to test the developmental potential of nuclei of cells during their early cleavage after fertilization of the oocyte, Spemann (1901) demonstrated in a classical experiment that each separated blastomere with its nucleus of a *Triturus* egg ligatured at the two-celled stage, subsequently developed into a complete larva. The presence in each cell of part of the grey crescent is essential, though this fact was unknown at the time (Deuchar, 1966, and *vide infra*). This result by Spemann confirmed earlier work by Driesch (1891) on cleaved sea urchin eggs, for before this time Roux (1888) had killed one of the two cleaved blastomeres of a *Rana* egg by pricking it with a hot needle, and later obtained an incomplete embryo, probably as a result of the associated dead cell impeding the progress of the remaining living one. One could surmise that later cleavage nuclei may eventually lose their ability to behave like zygote or first cleavage nuclei, possibly because subsequent cellular divisions allow insufficient cytoplasm for these blastomeres to develop and initiate their full potential. Spemann (1928) devised an experiment to settle the question. A fertilized *Triturus* egg, partially divided by a fine hair into a dumbbell shape and leaving the nucleus at one end, subsequently cleaved, while the uncleaved other half of the cytoplasm did not. After about four cleavages (16 nuclei), the loosened knot permitted a random nucleus (one that happened to be nearest to the cytoplasmic bridge) to pass into the uncleaved cytoplasm. Subsequently the two separated halves developed into complete embryos. The earlier Weismann hypothesis that nuclear substance is divided unequally during cleavage, leading to the segregation of specifically different determinants in different cells, in other words that cells possess differing constituents of the original zygote nucleus, was thus convincingly disproved. This totipotentiality of early cleavage blastomeres also occurs among invertebrates (Horstadius and Wolsky, 1936) and mammals (Tarkowsky and Wroblewska, 1967).

In order to investigate whether nuclei remain stable and unchanged, in terms of functional potential, after a large number of cleavages, more difficult and sophisticated techniques of nuclear transplantation were devised by Briggs and King (1952) and Moore (1960, 1962) in the USA and later by Gurdon and his colleagues at Oxford (see *vide infra* and Deuchar, 1975). It appears that a blastula endodermal nucleus of *Xenopus*, transferred to one of its previously enucleated eggs, can directly control normal embryonic and larval development in at least up to 30% of cases investigated (Gurdon, 1962). Ectodermal and mesodermal nuclei generally behave in a similar manner. In these experiments failures of normal development could be a result of irreversible changes in the transplanted nuclei during their previous divisions, or probably in many cases occur because of their mechanical damage during manipulation (see Briggs and King, 1952, 1953; King and Briggs, 1954). It is significant that a higher percentage of normally developing eggs was obtained as techniques improved. In some fewer cases nuclei from neural plates of frog embryos (di Berardino and King, 1967), from the prosencephalon of advanced neurulae (NF St.22) and epidermis of hatching larvae (NF St. 36–42) of *Xenopus* (Brun and Kobel, 1972), or from cultured cells of highly differentiated tissues of *Xenopus*, such as intestinal epithelium of feeding larvae (Gurdon and Uelinger, 1966) and adult skin and liver, when transplanted to enucleated eggs can produce normal metamorphosed animals (Gurdon, 1962; 1968; Gurdon and Lasky, 1970). Such nuclei presumably have thus retained their totipotency, that is, an ability to again switch on previous functional activity, apparently switched off and latent after an earlier functional phase. Brain nuclei that eventually cease to synthesize DNA and RNA commence synthesis when transplanted back into egg cytoplasm (Gurdon, 1968). One cannot but conclude that the cytoplasm contains or progressively acquires factors that directly or indirectly influence genomal function, and at specific stages of embryonic or larval development in some manner they are active or inactive. Other evidence available also reveals that the normal functional potential of transplanted nuclei (or at least of some of their components), can be maintained unchanged or recreated, for mRNA of rabbit hemoglobin injected into a *Xenopus* egg 'instructs' the synthesis of rabbit hemoglobin (Gurdon et al., 1971). Thus protein synthesis normally determined through the agency of nuclei of species X, in the cytoplasm of species Y is still of the species X type.

Nevertheless, full nuclear potential, at least in some cases, may well reduce gradually relative to the number of cell divisions, perhaps owing to the accumulation of random mutations and/or chromosomal abnormalities and to the elimination of variable quantities of chromosomal substance (see di Berardino and Hoffner, 1970). Again aging cytoplasm may in some manner eventually elicit irreversible nuclear changes, so that such nuclei transferred to younger cytoplasm have lost their ability to switch on functions previously only latent, but now permanently inactive. The indefinite retention of complete nuclear totipotency from highly differentiated tissue cells may thus be the exception rather than the rule, though the registered successes obtained so far imply that such cells do retain the same complement of genes of the zygote, and that there is a constancy in the amount, composition, and hybridizing capacity of the nuclear DNA (see Truman, 1974).

More recently, some research on amphibian metamorphosis has tended to concentrate upon the biochemical mechanisms of causality of cellular differentiation at the level of molecular biology; for example, the causal mechanisms involved in effecting embryogenesis and the way in which the pre- and prometamorphic cells respond to the influence of thyroid hormones. The problem can be rephrased into one that is concerned with causality in the creation of cellular diversity, and hence the emergence in embryos and larvae (or fetuses of mammals for that matter) of specific organs and tissues that make up an integrated complex and fully developed animal. It is now known that cellular differentiation, with its resulting structural and biochemical specificity, involves intimate complex feedback relationships between the nuclear genes and, what to the investigator is likely to be, a depressingly large number of organic and inorganic components in the ambient cytoplasm. The synthesis of specific proteins that are integrated uniquely in terms of ultrastructure distinguishes one type of cell from another; muscle cells, for example, containing myofilaments of myosin and actin are quite different from erythrocytes, which manufacture hemoglobin. Fairly accurate measurable concentrations of thyroidal triiodothyronine and of thyroxine, acting on anuran larvae at specific threshold levels (obtained by immersion or injection of larvae), can stimulate surgically treated (hypophysectomized or thyroidectomized) or normal specimens to respond and reach certain stages of organobiochemical development, depending upon the concentration of hormone. The larval developmental stage achieved (classified from its external appearance), usually reflects the level of its organic cellular differentiation. Implants into a larva, at chosen locations, of agar pellets impregnated with T_3 or T_4 also influence adjacent circumscribed regions. The close relationship, so described, between the degree of larval cellular development and the thyroid hormonal concentration, and the exploitation of modern highly sophisticated microbiochemical techniques, provide an invaluable opportunity to investigate causal mechanisms eliciting cellular diversity.

The pioneer discoveries on pneumococcal bacteria by Griffiths in 1928, followed by the work of others on DNA, such as Avery, Hershey and Chase, Watson and Crick, Wilkins, and Nierenberg and Matthei (see Moore, 1957), inevitably led to the subsequent concepts on the control of cell mechanisms enunciated by Jacob and Monod (1961), from their work on *Escherichia coli*. Their ideas embracing the operon system and genomal repression and depression, provide a broad basis for envisaging somewhat similar mechanisms between the cytoplasmic constituents and genes of vertebrate cells, in terms of RNA templates complementary to the base sequences of nuclear DNA. The mRNA triplet coded 'read out' is then translated into cellular protein synthesis at the level of the cytoplasmic ribosomes (see Bonner, 1965; Watson, 1976; Alberts et al., 1983). In terms of hormonal activity such nuclear-cytoplasmic interactions doubtless operating in variable ways, presumably appertain in multicellular organisms (see Gorski et al., 1969; and also discussions by Tata, 1975; Hourdry and Dauca, 1977).

The complexity of such mechanisms should not, however, be underrated. Notwithstanding those operating at the level of transcription, the biochemical regulatory mechanisms effecting conversions after enzyme synthesis at the ribosomes, initiating cellular differentiation, are likewise extremely complicated. At least nine possible levels effecting conversion of a cell metabolite to one of another kind have been postulated by Ashworth (1973). These include protein modification and degradation, alterations in the concentration of prosthetic groups and of effector metabolites, and there are also metabolite transfer phenomena and alterations in the concentrations of substrates and cell products.

31. Cellular Differentiation During Embryogenesis to Premetamorphosis

In order to evaluate causal mechanisms that are responsible for early cellular differentiation during embryogenesis, it is necessary to accept as truisms certain basic assumptions. First, we may state the obvious, that groups of cells of a living organism, derived by successive cleavages from a single fertilized egg, ultimately differ from one another in structural and functional specificity. Second, the genes of nuclei from cells of different groups of cells, organs, or tissues are generally equivalent, that is, the nuclei remain potentially totipotent. Third, it may be accepted with some degree of confidence that there are major factors influencing cellular differentiation which, at least in embryonic and early larval stages, originate within or can enter from outside (probably from adjacent cells) the cell target cytoplasm, to influence gene expression and thus initiate structural and functional changes.

The Amphibian Zygote, Cleavage, and the Blastula

In amphibians the 'natural' position of the primary oocyte reveals a darkly pigmented uppermost animal pole and a lighter colored, more heavily yolked, ventrally situated vegetal hemisphere. The nucleus is located near the animal pole. The egg has a distinct cortical granular cytoplasmic layer that is pigmented and generally thicker in the region of the animal pole. During oogenesis the oocyte of *Xenopus* synthesizes significant quantities of various classes of RNA (Mairy and Denis, 1971). Staining by methyl green-pyronine reveals an animal–vegetal gradient of RNA (corresponding to the cytoplasm between the yolk platelets), which is richer in the animal half of the egg, though RNA-rich islands also occur in the vegetal half probably belonging to the germinal plasm. Yolk platelets of varying size fill the more central regions of the egg and many of the larger ones are situated in the vegetal hemisphere (Czolowska, 1969). They contain phosphorylated proteins that are a source of amino acids and inorganic phosphate for the embryo (Denis, 1974). Probably the yolk platelets of eggs of different species of amphibians differ in their protein and lipid content and in the amount of activity of such enzymes as phosphoprotein phosphatase and cathepsins. Primary vesicular yolk of the oocyte of *R. pipiens* arises from a Golgi complex that includes numerous Golgi vesicles, multivesicular bodies, small granular–vesicular bodies and particulate vesicles; subsequently they are distributed throughout the ooplasm where stages in growth of yolk precursor complexes are seen (Kessel and Ganion, 1980). The animal region of the egg contains more mitochondria, Golgi vesicles, and ribosomes and smaller quantities of endoplasmic reticulum. Mitochondria, for example, accumulate and are stored during oogenesis for future use by the embryo. Indeed in a mature oocyte of *Xenopus* there has been a 10^5-fold enrichment of mitochondria compared with a somitic cell and 300–500 times as much mitochondrial DNA as chromosomal DNA. The rates of synthesis of mitochondrial DNA of such oocytes have been measured by Webb and Camp (1979). As oocytes mature the DNA polymerase activity associated with mitochondria decreases, and the replication of mitochondrial DNA molecules terminates midway through oogenesis. The mitochondria probably function similarly in oocytes and embryos as in somitic cells, and as they multiply and are distributed among the developing organs, they supply them with the energy they require by oxidative phosphorylation (Denis, 1974). Oxidation rates, however, were reported to increase during larval development in *Bufo arenarum* mitochondria (Legname et al., 1979).

In the vegetal regions there is a germ plasm, that ultimately becomes included in cells of the blastula floor, destined to be the future germ cells that migrate to the presumptive gonads (Blackler, 1958; Czolowska, 1969).

In amphibian eggs, cleavage and the ensuing cell cycle, the formation of the grey crescent and of bilateral symmetry, and the establishment of a dorso-ventral axis, are initiated by sperm entry at fertilization. During early cleavage an asymmetric distribution of vegetal materials establishes the dorsoventral axis, and more dorsal vegetal blastomeres induce neighbouring animal pole blastomeres to become the future mesodermal inducer cells of the organiser. The sperm aster in some way may be the initiator of the first asymmetric movements of material in the egg cytoplasm, before the crescent forms. Gravity can also provide the motive force for the rearrangement of the internal cytoplasm in the fertilized egg, and thus the normal future symmetries can be altered experimentally (Kirschner et al., 1980).

After fertilization, among other morphogenetic and biochemical changes that proceed in the egg, a more lightly pigmented grey crescent forms, because of a shift of cortical pigment towards the animal pole. The grey crescent distinguishes the future upper (dorsal) side of the developing egg, and it ultimately becomes incorporated into the dorsal lip of the blastopore to exert future induction. Goldenberg and Elison (1980) demonstrated regional differences in the degree of cortical granule exocytosis of the *R. pipiens* egg during its activation. They claim that the higher rate of exocytosis (or the ability to react to activating stimuli) of the animal half compared with the vegetal half is a result of differences in Ca^{2+} sensitivity.

The amphibian egg is one of the largest cells to cleave completely. The first cleavage of the zygote is synchronous, but eventually the blastomeres divide at different times independently of one another. The first two meridional cleavages are followed by a horizontal and then a fourth meridional followed by a fifth horizontal cleavage. After the 64-celled stage, a solid morula develops into a spherical blastula containing an extensive cavity the blastocele. Cleavage is uneven, cells of the vegetal half of the blastula dividing more slowly than those of the animal half, probably mainly because of the heavy yolk content of vegetal hemisphere cells. The cytoplasmic constituents of the egg are not displaced to any significant extent during blastulation, and overall they generally retain similar locations in the blastula as in the egg at the beginning of cleavage (see Balinsky, 1970). After fertilization the polysome content of the egg of *Xenopus* increases, but there is little further increase within the cleavage cells until the tailbud stage (NF St.25–30) when organogenesis gets underway and then there is a rapid rise. Most ribosomes are incorporated into polysomes by NF St.42 (Woodland, 1974); indeed it is claimed that the total ribosomal content of the oocyte barely changes until NF St.40 and that there is only a significant increase in the cells after larval hatching (Brown and Littna, 1964). In some support for these results Shih et al. (1978) described the rate of protein synthesis in the oocyte of *R. pipiens* to increase by 70% during its maturation, by another 50% during the 2-celled stage and the rate only doubles, approximately, between the latter stage and the formation of the blastula. Richter et al. (1982) likewise reported a twofold increase in protein synthesis during oocyte maturation of *Xenopus,* primarily because of a recruitment of mRNA rather than to a change in translational efficiency of the ribosomes. Furthermore, as would be expected, there is a considerable degree of change in the type of protein synthesized between cleavage and late blastula stages of *Xenopus,* with a major shift in protein types between gastrula and neurula stages. Actin and tubulin, however, are synthesized *de novo* at all stages of development, though their relative rate of synthesis in pregastrula stages is low (Brock and Reeves, 1978). Bravo and Knowland (1979) have described at least four classes of proteins synthesized differentially in space and time in oocytes, unfertilized eggs, embryos shortly after fertilization, and in later larval stages of *X. laevis.*

Some infrequent tight junctions are first recognized, usually between innermost blastomeres, at all levels of the blastula of *Xenopus* at NF St.7. Deep within the blastula desmosomal-like thickenings also occur between the cells. All the blastomeres have a microfilamentous network located just below the plasma membrane, but extensive bundles of microfilaments and some newly formed microtubules only appear in cells of the blastoporal region at the onset of gastrulation (about NF St.10). Cellular terminal contacts, presumably related to the mechanical and physiological requirements of the cells, by their strong adhesion may assist in the coordination of cellular group movements during gastrulation. Intercellular transport of ions and small molecules may also occur at such "low resistance" junctions (Sanders and Zalik, 1973).

An important feature of a resulting blastula is the streaming movements of intracellular material and shifts of animal and vegetal pole cells that take place, the pregastrulation movements, for even before gastrulation commences some animal pole cells spread over vegetal pole cells. However, notwithstanding these later dynamic cellular activities, a completed blastula consists of gradients of different groups of cells, which at successive circumferential levels differ in their cytoplasmic content, expressed in terms of quantity and distribution (and often individual component size) of pigment, yolk, mitochondria, and doubtless other inorganic and organic constituents, including RNA-containing ribosomal granules, often associated in strands and others containing beta glycogen and RNA (van Gansen, 1967; van Gansen and Schram, 1969). The cellular localization of polysaccharides changes very early during embryogenesis, usually beginning in the animal and dorsal regions of the embryo. A specific area, the grey crescent, has also differentiated in the blastula. Furthermore, various cytoplasmic components show specialized enzyme activities, resulting in varied metabolic functions in different cell groups. Again, in addition to the nucleus the cytoplasm of the oocyte and of the cleaving blastomeres contains various species of RNA and large quantities of DNA, and some synthesis of RNA and DNA proceeds during embryogenesis. Indeed, *de novo* synthesis of DNA begins in the middle blastula stage of *Xenopus laevis* and in the late blastula stage of *Rana temporaria,* which has greater DNA reserves (Vilimkova and Nedvidek, 1962), and RNA polymerase activity already occurs in cells of the blastula of *Xenopus laevis* (Bouloukhère et al., 1980). In fact, dorsal blastomeres begin to synthesize rRNA slightly earlier than ventral ones (Shiokawa and Yamana, 1979). Over two-thirds of the maternal rRNA and poly(A)$^+$RNA (see also Wakahara, 1981) occur in the animal hemisphere cells from the egg through the blastula stage, during cleavage of *Xenopus laevis* eggs, though there is more even distribution of these RNAs between dorsal and ventral halves (Sagata et al., 1981). rRNA synthesis begins during the midblastula stage (Shiokawa et al., 1981a), 4 h earlier (three cell cycles) than previously believed (Misumi et al., 1980). The rate of rRNA synthesis per nucleolated cell is 0.2 pg/h (5.5 × 10^4 molecules/h) at the blastula stage, which rate was constant in later stages. At the blastula stage about 30 molecules of 5S RNA, 10 molecules of capped mRNA, and 900 molecules of 4S RNA were synthesized per molecule of 18S and 28S rRNA, which values were reduced during the gastrula stage. In the neurula, one molecule each of the 5S RNA and capped mRNA, and 10 molecules of the 4S RNA were synthesized per molecule of the 18S and 28S rRNA (Shiokawa et al., 1981b). Gradually, however, the total amount of DNA per cell decreases during embryonic development, as the cytoplasmic reserves of the original zygote are drawn upon. From the tailbud stage to hatching in *Xenopus,* the content of DNA in each individual cell has stabilized and reached that of the normal diploid cell of the adult (Bristow and Deuchar, 1964).

Double-stranded, highly polymerised DNA (Baltus et al., 1968) and high and low molecular weight RNA (Kelley et al., 1971) are bound to yolk platelets, and probably also to mitochondria that contain various enzymes such as cytochrome oxidase, ATPase, acid phosphatase and cathepsins (Weber and Boell, 1955, 1962). In the *Xenopus* oocyte, yolk DNA is linear in its orientation in contrast to that of the mitochondria, which is circular. The yolk platelets here contain ten times the amount found in the mitochondria (Baltus et al., 1968), which hitherto were believed to contain most of the egg DNA (Dawid, 1966).

Thus, ultimately the cells of the different regions of the developed blastula significantly differ in the quality and quantity of a wide range of cytoplasmic constituents that surround their comparable and generally equivalent nuclei. It is these groups of intrinsically different cells of the blastula that are now ready to invaginate, in order to give rise to the next sequential embryological stage, the gastrula.

The Gastrula: Induction and Cellular Differentiation

Programmed embryonic development of amphibian larvae, through gastrulation and neurulation, and so on, ultimately leads to the typical prehatching larva. Since the classic descriptions of amphibian gastrulation by Pasteels (1940) and Holtfreter (1944), the process has been described in numerous textbooks on embryology (see Balinsky, 1970; Deuchar, 1975; Grant, 1978). Indeed, by the use of staining with vital dyes (for example, Nile-blue sulfate, neutral red, and Bismarck brown) soaked in pieces of agar, the future fate of areas of cells from different regions of the blastula surface can be followed and elucidated. In this way "fate maps" of amphibian blastulae have been constructed (Vogt, 1925, 1929; Dalcq, 1938; Pasteels, 1940).

Gastrulation is an orderly movement and rearrangement of groups of cells, which ultimately occupy new topographical relationships with others in the fully formed gastrula. Furthermore, Le Blanc and Brick (1981) reported that during gastrulation the blastula of *R. pipiens* includes superficial and more deeply located populations of cells from different germ layers (i.e., presumptive cells of the head endoderm, notochord, neural ectoderm, and epidermal ectoderm), which even at this early period of development appear to show differences in their spreading and adhesive properties distinct for each region and for different stages of gastrulation. Such differences reflect a precocious and subtle differentiation of early cleavage cells that doubtless is related to early inductive interactions between germ layers, at least by virtue of their spatial rearrangement during gastrulation. Such behavioral differences may also correlate with inductive potentiality. Løvtrup (1965) described the cellular basis of invagination as a transformation of amebocytes, which occupy the dorsal margin of the blastula, into a different type of cell or mechanocyte, whose chemical basis is determined by the animal–vegetal polarity of yolk and cytoplasmic distribution. Polarities in terms of animal–vegetal gradients (originally present in the unfertilized egg (*vide supra*) and a dorsoventral gradient (determined at fertilization by the formation of the grey crescent), have been analyzed by Løvtrup et al. (1978) in relation to the cleavage and gastrulation of the amphibian egg during development. At the base of the blastoporal groove the flask-shaped cells (Ruffini cells) that pull on the embryonic surface as they invaginate are considered to be very early differentiated cells that contain heparin sulfate, possibly a likely candidate as a primary inductor substance in the amphibian embryo. (Løvtrup, (1983). Flickinger (1980) showed that ventral halves of early gastrulae of *R. pipiens* (Shumway stage 10), cultured in a sodium salt of heparin, differentiated dorsal components such as neural, muscle, and pronephric tubule tissues; 60% of the ventral halves of 300 gastrulae responded, whereas < 1% did so when cultured in Niu-Twitty saline solution alone. These results are of interest for Landstrom and Løvtrup (1977) showed that blastoporal flask cells, which contain heparin sulfate, induce mesenchyme and nerve cells in cultured blastular animal hemisphere cells.

Ultimately the dorsal lip of the blastopore (the so-called primary inducer, organizer, or evocator of classical embryologists; see Waddington, 1947), becomes the future chorda mesoderm of specialized cytological and physiological properties. The dorsal lip finally underlies competent ectoderm, the future neural tissue, that is activated to develop into the clearly recognizable neural plate and thence neural tube.

A detailed consideration of causal mechanisms of cellular induction, that is, the response and thence differentiation of target cells after gastrulation and subsequently during embryogenesis (see Saxen and Kohonen, 1969), is beyond the scope of the present work, though the subject is briefly discussed (*vide infra*). However, within the framework of the consideration of amphibian embryogenesis, and hence cellular differentiation, it should be emphasized that during this period cells from different germ layers show differences in affinity, mutual adhesions, and electric discharges between them. Certainly like cells of any particular germ layer differ in their individual behavior from those of other layers; this is revealed when they sort themselves out in vitro (Townes and Holtfreter, 1955; Curtis, 1957). Embryonic induction, first recognized by Spemann (1903) in amphibian embryos in his classic experiments on lens induction and later with the region of the organizer (Spemann, 1921; Spemann and Mangold, 1924), is in all vertebrates a process of great complexity, especially in terms of cellular biochemistry (see, in particular, Yamada, 1958, 1961, 1962; Yamada and Takata, 1961; Nieuwkoop, 1962, 1966; Toivonen and Saxen, 1968; Toivonen et al., 1975; Toivonen et al., 1976; Toivonen and Wartiovaara, 1976; also reviews by Balinsky, 1970; Hamburgh, 1971; Tiedemann, 1976; Løvtrup, 1983). So far, according to the evidence presented, it seems likely that interchange of molecules between inducing and responding cells occurs via surface pores (*vide infra*). Transmissible molecules such as ribonucleoproteins are likely candidates as the main neural inductors in vivo (see Niu, 1958a,b), though other proteins are likely to be concerned. Cellular membrane–membrane contact between inducer and responsive cells is not essential, at least for the initiation of neural induction (Toivonen et al., 1975, 1976; but see Saxen et al., 1976; Toivonen, 1978). Indeed "TA" millipore filters of 0.8 μm average pore size, interposed between competent amphibian ectoderm and inducer tissues, still permit neuralization even over a distance of 20–25 μm (Nyholm et al., 1962). However, intimate membrane contacts between interfaces of cells of the neuroectoderm and chordamesoderm were reported to occur in mid-gastrula stages of *Triturus alpestris* during primary embryonic induction (Grunz and Staubach, 1979).

After neural induction is activated, competent ectoderm shows various ultrastructural changes, which in particular include the development of the RER and alterations in the appearance of mitochondria; in addition the RNA content of the cells increases. Such changes are features of new protein synthesis, which indeed occurs extremely early during embryonic development; intercellular collagen, presumably at least partially synthesized by adjacent cells, is already present in neurulae of *Xenopus* and the rate of synthesis increases 500-fold during development to posthatching stages (Green et al., 1968). Newly formed microfilaments and microtubules are present in neural plate cells of *Hyla* and *Xenopus* (Baker and Schroeder, 1967; Schroeder, 1970; Karfunkel, 1971; Burnside, 1971). Microfilaments (4–6 nm thick) and microtubules (20 nm thick, often in groups of up to 20 in the apical regions of the neural plate cells) influence neurulation either by microfilamentous contractions (Baker and Schroeder, 1967) or by their interdigitation and by increasing overlap. Microtubules may participate in this process, possibly in transport mechanisms (Burnside, 1971). During neurulation of *Hyla regilla, Rana pipiens,* and *Xenopus laevis* the smooth-surfaced facing cells lining the neural groove, which ultimately fuse with their opposite numbers, change in shape to become elongate-polygonal and near their surface intracellular vesicles form; extracellular vesicles are spatially associated with them. Just before cellular fusion, pseudopodia develop and interdigitate with those from opposing cells. Ciliated cells are absent in the fusion area, but are present in other regions of the neurula surface. The glycosaminoglycans content of neural fold cells differs from that of other presumptive epidermal cells, which are cuboidal in shape (Mak, 1978). From a host of examples of cellular differentiation during amphibian embryogenesis, including features of neurulation, it is clear that inducer- or hormonal-stimulated cellular changes also embrace behavioral responses, which are probably a result of and/or certainly influenced by newly activated biosynthetic activity.

Billet and Gould (1971) traced the ultrastructural changes that occur in the epidermal cells of embryos of *Xenopus* during their development and differentiation. The first obvious specific changes, showing real differences from cleavage cells, arise by the end of gastrulation, and significant amounts of RER and polysomes, tonofilaments, and mucus-containing vesicles are recognizable during neurulation (NF St.13–16). Mucus-secreting cells have differentiated after the end of neurulation. During NF St.17–23 (about the end of neurulation to the stage when embryonic elongation commences and the tailbud originates; see Nieuwkoop and Faber, 1956), there is further epidermal cellular differentiation and desmosomes now join adjacent outer cells. Definitive surface-ciliated cells appear from NF St.19, when their ciliary rootlets are clearly distinguishable. Likewise, hatching gland cells of the epidermis of specialized ultrastructure (see section on the skin) are first detected in *Xenopus* at NF St.22 (Yoshizaki, 1973, 1975) and cells of the Anlagen of the cement glands first appear even earlier in *Xenopus* in the early neurula at NF St.15 (Perry and Waddington, 1966).

Probably some protein synthesis that occurs in developing cells during early larval embryogeny is at least in part translated from mRNA originally present in the oocyte. Creation of new mRNA is low in cells during cleavage, but the rate increases after the blastula stage (Tiedemann, 1976). A tadpole synthesizes protein about 50 times more rapidly than the oocyte and 25 times more rapidly than the egg (Woodland, 1974). The amount of rRNA increases rapidly in the neurula cells and 4S tRNA is also synthesized at the late blastula stage to maintain efficient protein synthesis. By the early swimming stage larva probably about 15% of the total cellular RNA is 4S RNA. The 5S RNA is bound to the ribosomes (Brown and Littna, 1966). It is probable that normal nuclear function, in terms of RNA transcription, is not essential for cellular development up to the gastrula stage, though the main limiting factor influencing the rate of protein synthesis to this stage and beyond is doubtless the availability of mRNA. As the cellular polysome content rises (Woodland, 1974), there is a coincident accelerated production of rRNA in the gastrula stage cells (Brown and Littna, 1966), together with the mobilization of maternal oocyte mRNA and thence of other species of newly synthesized mRNA (Tiedemann, 1976). Indeed, mRNA complexity almost doubles from the neurula to early larval stages (Shepherd and Flickinger, 1979). Specific protein synthesis within individual cells is therefore speeded up, resulting in recognizable specialized morphogenesis. For further details on the subject of ribosomes in oogenesis and embryogenesis, the reader is referred to the review by Denis (1974).

RNA synthesis appears to be temperature-dependent. Thus at 21°C (standard temperature), isolated cells of *Xenopus* neurulae labeled in vitro with ^{3}H-uridine actively synthesize 28S and 18S RNA (rRNA) and other RNAs. At 30°C a twofold ($Q_{10} = 2$) increased synthesis of rRNA, 4S, and heterogeneous RNA (broadly distributed around the RNA 40S–18S region) occurs. At 10°C 28S and 18S RNA synthesis is completely inhibited and that of other RNAs reduced by 50%. The absence of 40S rRNA precursor at 10°C suggests that transcription of rRNA is inhibited. All the effects in these cells are reversible within 4 h, when 6-h-treated cells are returned to normal temperatures (Nishio et al., 1978).

Newly synthesized species of mRNA at the level of transcription are essential after gastrulation, for early amphibian embryos, or their tissues, treated with Actinomycin D neither synthesize protein nor do they show any further cellular differentiation. In other words the specialized characters of the marginal zone are newly acquired during cleavage and gastrulation (Nakamura, 1969). Such inhibition by Actinomycin D occurs in clearly recognizable form in the case of early differentiating anuran epidermal hatching gland and ciliary cells (Yoshizaki, 1976), and cement gland cells (Eakin, 1964). Translational products produced by maternal mRNA during early embryogenesis include histones, myosin heavy chain (DeBernardi, 1982), and microtubule proteins; other cell products synthesized on maternal messengers and on the ribosomes during oogenesis are presumably essential for cleavage and for the metabolism of the yolky eggs (Tiedemann, 1976). A useful account of cell differentiation in terms of DNA–RNA activity, among other things, is given by Truman (1974).

During the induction process after gastrulation, responsive tissues become determined, or irrevocably programmed in their structural specificity. As embryogenesis proceeds regional differentiation within individual organ systems begins, initiated by causal mechanisms that are far from being understood but certainly are of great complexity (Toivenen et al., 1976; Tiedemann, 1976). After primary induction mediated by the chorda mesoderm, secondary and tertiary inductive phenomena, and so on, ensue; for example, the formation of the eye lens in relation to the optic cup; the secondary induction by the hind brain to induce the ear vesicle (Model et al., 1981) and tertiary induction by the latter to form the cartilaginous capsule (see Waddington, 1947; Balinsky, 1970). Similarly the pronephric duct induces the mesonephric blastema to differentiate into mesonephric tubules (see Fox, 1963; Gipouloux and Delbos, 1977).

The Mechanism of Induction

The regional nature of the induction process, with its sharply defined demarcated boundaries, led various workers to suggest the existence of neuralizing factor(s) (dorsalizing or activating agents) and mesodermalizing factor(s) (caudalizing or transforming agents) (Nieuwkoop and Grinten, 1961), which by virtue of the balance of concentration gradients between them induce organs at all levels of the embryonic dorsal axis. The terms neuralizing and vegetalizing factors are used by Tiedemann (1976); the vegetalizing factors at least are claimed to penetrate ectoderm cells to cause induction (Tiedemann et al., 1972). The vegetalizing factor (which may also include an inhibiting factor), is prelocalized in the vegetal half of the amphibian embryo. Extracted from chicken embryo trunks, the purified factor is probably a protein of low molecular weight (about 35,000) (Tiedemann et al., 1972). A crude factor with similar inducing properties is present in guinea pig bone marrow (Yamada, 1962) (Table 4). By the use of fluorescently labeled antigenic

Table 4
Method Used to Separate the Active Protein Sample (BMF) from
Guinea Pig Bone Marrow[a]

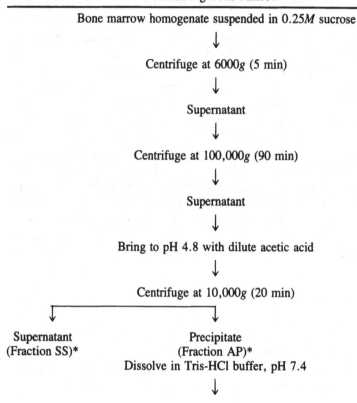

Bone marrow homogenate suspended in 0.25*M* sucrose

↓

Centrifuge at 6000*g* (5 min)

↓

Supernatant

↓

Centrifuge at 100,000*g* (90 min)

↓

Supernatant

↓

Bring to pH 4.8 with dilute acetic acid

↓

Centrifuge at 10,000*g* (20 min)

Supernatant   Precipitate
(Fraction SS)*   (Fraction AP)*
                               Dissolve in Tris-HCl buffer, pH 7.4

↓

(continued)

material from guinea pig bone marrow that passes into the ectoderm of embryonic *Triturus,* conflicting, uncertain, and often confusing evidence nevertheless suggests that during neuralization a variety of inducer molecules, including macromolecules of whole proteins (Vainio et al., 1960; Vainio et al., 1962; Yamada, 1961, 1962), pass from the inducer to influence the differentiation of responsive amphibian ectodermal tissue. How inducing substances released from the archenteron roof operate to initiate cellular transformation of the overlying cells is not known. They may interact initially with the ectodermal cell plasma membrane, or diffuse unchanged further into the responsive cells to associate with carrier molecules, both processes being intermediaries doubtless in a further complex chain of events leading to activities influencing gene expression. Alternatively inducer substances may act directly on the nuclear genes to activate DNA-dependent RNA transcription.

Table 4 (*continued*)

Precipitate with 50% saturation of ammonium sulfate in same buffer
↓
Precipitate
(Fraction ASP)*
↓
Dialyze against the Tris-HCl buffer
↓
Dialyzed sample
↓
Put on a DEAE-cellulose column
↓
Elute sample with increasing concentrations of NACl stepwise or in gradient
↓
Fraction eluted at 0.15*M* NaCl
(Fraction E15)*

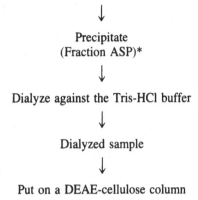

*Asterisks indicate the fractions used after dialysis against the Tris-HCl buffer at pH 7.4. Prospective ectodermal tissue of the early gastrula of *T. pyrrhogaster,* cultured in vitro (Holtfreter solution, pH 7.3, Tris-HCl buffer) including specific active protein sample of guinea pig bone marrow for 3 h, then after transfer to Holtfreter solution forms mesodermal tissues and endodermal tissues, i.e., mainly striated muscle, mesenchyme, and notochord. Occasionally small amounts of neural and mesectoderm tissue differentiate. Guinea pig bone marrow tissue is arranged as a sandwich between two slices of ectoderm (sandwich technique) (method after Yamada, 1962).

The protein-type vegetalizing factor, which determines the endoderm–mesoderm Anlage in amphibian embryos, is probably prelocalized in the vegetal half of the embryo during early cleavage. When prepared in purified form from 11-d-old chicken embryo trunks (smaller quantities of crude extracts of the vegetalizing factor have also been obtained from amphibian cleavage and gastrula stages; Tiedemann, 1976), and tested on gastrula ectoderm of *Triturus alpestris,* implants induced muscle, notochordal, and renal tissues from belly epidermis. A small content of neuralizing factor present induced a lower percentage of small hinder cephalic structures and vesicles (neural induction). A partially purified inhibitor associated with the vegetalizing factor belongs to the acidic glycoprotein fraction. Inhibitor and vegetalizing factor can be partially separated by electrophoresis. Uninhibited vegetalizing factor induces trunk and tail mesodermal tissues, but the inducing capacity is greatly reduced when it is combined with the inhibitor. The chicken vegetalizing factor is also partially inhibited when in combination with chicken DNA from nuclei of 11-d embryo trunks. However, as if the problem was not complicated enough, Asahi et al. (1979) have found that highly purified vegetalizing factor from chicken embryos preferentially induces endoderm in amphibian gastrular ectoderm: in combination with less pure fractions, a high percentage of mesodermal trunks and tails with notochord and somites are induced. These authors speculated that the latter result depended upon the presence of secondary factors (probably protein) acting on the plasma membrane receptors of target cells, which have mesodermal-inducing properties.

Inducers may enter their target cells and conceivably first bind to cytoplasmic membranes and initiate changes in ionic permeability or in the adenyl cyclase–cyclic AMP mechanism. Though various subcellular fractions have a high inducing capacity (see Faulhaber, 1972), the meager evidence available so far supports the view that inducing factors are synthesized by and stored or transported within the ER, and they do not elicit cellular differentiation at the level of translation. During normal induction chromatin of the target cells is probably the significant receptor for newly received inducing substances, which presumably modify gene expression at the level of transcription. The action of the inhibitor in the vegetalizing factor–inhibitor complex is not understood. Perhaps the inhibitor acts as a gene repressor, or it could possibly regulate the system by participating in an inducer–inhibitor complex whose controlled concentration of active inducing factors results in the gradient fields of young embryos (see discussions by Tiedemann et al., 1972; Tiedeman, 1976).

Cytoplasmic penetration of nucleopore filters by inducer tissue, to allow some sort of cellular contact with target tissues, is claimed to be essential for induction to occur. Differences in target cell response achieved by different inducer tissues are considered to be related to their degree of penetrability through the nucleopores (Saxen et al., 1976). Nevertheless, if this is indeed the case, then presumably diffusible inducer substances still pass from inducer to competent target cells to initiate specific cellular differentiation. This assumption is implicit in modern theories on inductive mechanisms during embryogenesis. So far no alternative convincing hypothesis has yet appeared to challenge this view.

Nevertheless, in a recent stimulating review on inductive phenomena, Yamada (1981) has questioned the classical views that induction is purely the result of individual inducing substances acting on competent target tissues. It is known that ectoderm of young amphibian gastrulae can be experimentally channeled into tissues of all three germ layers, that it is omnipotent. Again some heterogeneous inductors (tissues with regional inductive effects, which are related in some manner to chemical factors acting in organizer activity), show mainly significant vegetalizing effects on competent ectoderm. Furthermore, pathways of mesodermal differentiation, including that of the organizer, are elicited in the animal zone of the blastula by interaction with the vegetal zone, and endoderm functions in pregastrula stages to establish patterns of development potency for organizer action. Yamada returns to his earlier ideas of induction and discusses the role of morphogenetic movement—which depends at least in part on an organized system of cell surface filaments and macromolecules, controlled by calcium ions and cyclic nucleotides—in relation to cellular differentiation: the theory of stretch-convergence and transformation of cells during normogenesis. It is heartening to note the optimism still expressed by this eminent embryologist, who pins his faith on modern biochemical technology to deal with these seemingly intractable, albeit fundamental problems.

Whether lectins can act as neuralizing agents with amphibian gastrular ectoderm is problematical, for concanavalin A (Con A) tends to inhibit embryonic morphogenesis (Boucaut et al., 1979). Yamamoto et al. (1981) and Takata et al. (1981), however, suggested that the lectins Con A and *Ulex europeus* glutinin had neural inducing properties and induced changes in the cell surface architecture by binding to the plasma membrane. The concentrations used may, however, have caused some cellular toxicity and cytolysis and thus indirectly led to apparent induction effects. Lectins could well be useful probes to investigate surface glycoconjugates in cell membranes, for they selectively bind to carbohydrate residues at the cell surface to elicit the recognizable phenomena of patching or capping. The experimental results of Duprat et al. (1982), using soybean and garden pea lectins, suggest that in the complex process of neural induction the competent target cells themselves play a specific role, and the structural integrity and conformation of the plasma membrane are either directly or indirectly involved in neurogenesis.

Thus in summary it may be broadly considered that the development of a functionally adapted amphibian larva from the fertilized egg is controlled by basic mechanisms that lead to the segregation and differentiation of a wide variety of organ systems. During embryogenesis the individual intrinsic biochemical properties of different groups of embryonic cells originating basically as a result of egg structure and patterns of cleavage (*vide infra*), and the directive inductive influences that arise, are the main factors evoking the "choice" between the alternative pathways available in organogenesis, so elegantly described by Waddington (1947) in terms of John Piper's pictorial epigenetic landscape. After gastrulation, organogenesis with its specific cellular differentiation is not an independent process, but requires activation by chemical substances derived from other mutually adjacent tissue(s) that are already basically different from responding tissues structurally and functionally. Probably all tissuegenesis and organogenesis during larval development, to at least the level of premetamorphosis (in the sense of Etkin, 1964), result from the chemicophysical reactions of mutually dependent, spatially related tissues, without which the programmed development would not occur.

So-Called "Intrinsic" Properties of Embryonic and Larval Cells

Probably all anuran larval tissues can from an early age respond, in terms of structural and biochemical change, to thyroid hormones, though substantial larval development up to the middle of premetamorphosis at the larval hindlimb bud stage (about NF St.52 of *Xenopus; equivalent to TK St.VI of R. pipiens*), will normally take place in their absence. Furthermore, the circulatory thyroid hormone level is extremely low during premetamorphosis (about NF St.46–55 of *Xenopus* and TK St.I–XI of *R. pipiens*), and it only really starts to rise significantly afterwards during prometamorphosis. The specific effects exerted by thyroid hormones on various individual cell populations, at different periods during the metamorphic cycle, are vaguely described as resulting from their "intrinsic" cellular properties, or their tissue sensitivity, or more broadly and usefully explained as a result of the developmental history of the cells (Tata, 1971) in terms of ontogenetic experience. During anuran embryogenesis and to the beginning of premetamorphosis, organs and tissues, differing in structure, function, and topography are developed. Indeed certain types of skin cells differentiate, function, and thence disappear before the beginning of premetamorphosis. For most of larval life until climax, various component structures of the tail are little different in ultrastructure from comparable ones in the body. Nevertheless, at climax like tissues of the tail and body may suffer quite different fates. As previously described, however, it is clear that the ontogenetic histories of different cell populations, effected in large measure by the patterns of cleavage and early inductive experience, will strongly influence the features of their basic cellular content and architecture. The inductive relationships with other cell groups will result in specific specialization of a variety of cellular structures; doubtless such cells also suffer more subtle influences, expressed for example in terms of regionally specific specialization. The presence of a morphogenetically active "diffusion field" and a "transforming field," conferring regional differences of inducing activity on neural differentiation has previously been postulated (Leussink, 1970). Organs such as the pronephros, liver, and pancreas, different in structure and function, clearly have experienced different ontogenetic patterns of development. They could well be expected to differ in their response to thyroid hormones during metamorphosis, though these facts still do not explain why they respond so differently at climax. Likewise, structurally similar organic structures in a variety of locations in the larva, which ultimately suffer a different fate, are derived from cell populations that were originally in different topographical regions of the blastula. Such different spatial origins presumably would result in differences in mutual relationships with other cell layers and different degrees of response to regionally varied inductive chemical stimuli, perhaps expressed in terms of quality, quantity, or a mixture of both. Thus, although comparable larval tail and body tissues *appear* to be structurally and functionally similar, doubtless they have acquired some specific characters or components (the intrinsic properties), fundamentally different from or not present in their counterparts of other topographical locations, that are at least in part responsible for the contrasting climactic responses to thyroid hormones.

Possible clues to explain some of the phenomena of differential cellular differentiation during larval ontogeny are the facts that (a) embryonic tissues are sensitive to thyroid hormones very early in development and show temporal variability in this response (Knutson and Prahlad, 1971); (b) during larval ontogeny changes occur in the sensitivity of various tissues to thyroid hormones, and there are changes in the capacity and/or affinity of their thyroid hormone binding sites in different kinds of tissues, or within comparable but differently located tissues, during successive stages of ontogeny (see subsequent sections for fuller discussion). Thus T_4-binding in tail cell sap of *R. catesbeiana* only occurs just before tail resorption shortly before climax; in liver cell sap this occurs in much earlier larval stages (Durban and Paik, 1976). Again the number of tail nuclear binding sites for T_3 increases from about 1900 to 3300 during prometamorphosis, as a larger number of the T_3-binding sites are occupied. The enhanced sensitivity of tadpole tail tissue to thyroid hormones near climax may thus result from increased binding capacities for T_3 (Yoshizato and Frieden, 1975). The proportion of T_4-saturable and unsaturable binding sites in the nuclei and cytosol of different tissues of the larval *R. pipiens* likewise is variable (Ergezen and Gorbman, 1977). Therefore, possibly the varied responses to thyroid hormones by different organs and tissues of anuran larvae result, at least in part, from fundamental differences in the affinities and capacities of multiple subcellular thyroid hormone-binding sites (possibly also in their carrier molecules?), which are primarily involved in the mechanism to elicit new cellular biosynthetic activity. However, the fact that there are differences (either in quality or quantity) in hormone binding sites of various cell types still does not explain why, for example, tail tissues such as muscle regress at climax when similar tissues in the body develop and enlarge. The explanation needs to be sought for in the subsequent genetic events that result from the T_3 and T_4 binding at the nuclear receptor sites. The gene activation, initiated by the hormones to result in the transcription of suitable mRNAs, presumably is under some sort of control by factors in the ambient cytoplasm, whose arrangement and distribution were determined to-

gether with those of the receptor sites in the cleavage stages of the blastula and later stages of the gastrula and neurula during regional induction.
It is of interest that mitochondria from different larval tissues of *Xenopus*,
even of the earliest tailbud stages tested, have significant quantitative differences in specific activity of particular enzymes, and thus the "developmental behavior" of such enzymes differs for a given tissue (Weber and
Boell, 1962). Furthermore, the oxidation rates of mitochondria of eggs
and embryos of *Bufo arenarum* change mainly at the time when the larval
opercular folds appear. Perhaps this is bound up with the initiation of different enzyme activities. Coincident with the appearance of thyroidal follicles and colloid, mitochondrial metabolic rate increases gradually to assume that of adults. Thyroid hormones modify mitochondrial activity and
there are different glycolytic pathways via phosphofructase kinase, which
are influenced by different concentrations of ATP and citrate (Legname et
al., 1979). Whether there is any biochemical relationship between, say,
mitochondrial cytochrome oxidase or ATPase activity and hormone binding sites within specific tissues, in terms of, for example, their initiation,
affinities or capacities, and so on, is not known.

Inducers and Hormones

In general inducing substances mainly function during early stages of embryonic and larval development; at least their activity has been investigated mostly during this time. Hormones are active in older larvae and
thereafter. Inducers are manufactured and secreted by inducer cells,
which at least early in embryogeny appear relatively unspecialized in
structure and function. Hormones are the products of specialized and differentiated endocrinal cells. During embryogenesis, inducer substances
are transmitted only short distances to their target cells, measured in
terms of microns between the adjacent cell populations; probably this occurs by diffusion without the aid of carrier molecules. Hormones, in contrast, are conveyed in the blood circulation usually for relatively substantial distances, either dissolved in blood plasma or more usually bound to
some of the plasma proteins, especially in the case of thyroid and steroid
hormones (see Bentley, 1976). The most common carrier for thyroxine is
thyroxine-binding prealbumin (TBPA), but a thyroxine-binding globulin
(TBG) occurs in some mammals including humans (Cohen et al., 1978).
There is no evidence that these plasma protein carrier molecules participate in transferring the hormone into the target cell. Neurotransmitters
such as noradrenaline and acetylcholine may be treated as a special case
and will not be further considered.

Effective regional inductions doubtless occur by virtue of different
gradients of concentration of a number of inducer substances that initiate
varied special cellular differentiation in responsive groups of cells. Thy-

roid hormones of developing amphibian larvae likewise differentially influence target cells as their hormonal circulatory level rises. The cells that react to inducers or hormones appear to alter their response, that is show differences in tissue sensitivity, according to their age or stage of development. Whether inducer substances are effective via threshold concentrations, as in the case of thyroid hormones, is not known. Nor is it clear how inducers function. It is likely, however, that the causal mechanisms of induction are similar to those operating during the hormonal stimulation of target cells; ultimately the inducers either interact with specific surface membrane receptors and adenyl cyclase, with the subsequent release of cyclic AMP (see Sutherland, 1972), and/or interact with individual nuclear or multiple intracellular binding sites to finally influence gene expression. However, cyclic AmP, cyclic GMP, and their mono- and dibutyl derivatives, in various controlled concentrations, do not increase the rate of neural induction, and the neuralizing factor thus may not act by activating membrane-bound AMP or AMP-cyclase (see Tiedemann, 1976). Nevertheless, effective mechanisms within species of similar types of individuals may well vary, for even in groups of chemically related hormonal substances, there is no evidence that their activity bears any specific relationship with their chemistry.

At present little is known about the detailed chemistry of any inducer substance, probably a consequence of the very small quantities available for isolation and purification. It is unlikely that the inducing factors for embryonic amphibian ectoderm, isolated from guinea pig bone marrow (Yamada, 1961, 1962), are identical or indeed even closely similar in chemical composition to the normal amphibian inducers. The crude inducing fractions obtained directly from amphibian embryos (see Tiedemann, 1976) would be more likely to be chemically related to them. A homogeneous factor isolated from 11–13-d-old chick embryos, which in amphibian gastrulae induces ectodermally, endodermally, and mesodermally derived tissues, is proteinaceous, has a molecular weight of 30,000–32,000 daltons, and is inactivated by pepsin and trypsin, but not ribonuclease. Heating at pH 8 and using reagents that influence the stability of S-S linkages also inactivate the factor (Tiedemann et al., 1969). Even less is known about the chemistry of amphibian neuralizing and vegetalizing factors, and highly purified extracts of them are not yet available. The neuralizing factor is not heat-sensitive, nor is it inactivated by disulfide-bond-reducing reagents, but proteolytic enzymes abolish its activity (Tiedemann, 1976).

Extracts of whole *Xenopus* gastrula that include protein, RNA, RN-protein, and deoxy RN-protein showed certain protein and high molecular weight RN-protein fractions to induce hindbrain and spinal cord in amphibian larval ectoderm. Other extracts, for example from microsomes, induced forebrain (Faulhaber, 1972; Faulhaber and Geithe, 1972; see also, Deuchar, 1975). Nevertheless, the contrast between the paucity of our knowledge of the chemistry of embryonic inducing substances with the vast corpus of information accumulated on hormones (see Bentley, 1976), especially those of the thyroid (Pitt-Rivers, 1978), including amphibians (Cohen et al., 1978), is striking. It is not surprising, therefore, that research on hormonal activities involved in the causal mechanisms of the differentiation of cells in various vertebrate embryos, larvae, and fetuses, either in vivo or in vitro, has far outstripped that with inducers in terms of output and or significant information.

The same applies for the juvenile terpenoid hormones and the steroid molting hormone, ecdysone, of insects (Grant, 1978). Until the heroic efforts of those concerned with the processes of vertebrate embryonic induction isolate and determine chemically specific inducing substances in sufficient quantity, this situation is bound to continue.

32. Appraisal of Morphological and Functional Cellular Changes During Metamorphosis

During amphibian larval development and metamorphosis a number of pertinent questions may be posed with regard to the morphological and functional changes that occur. First, do the same clonal populations of cells of different organs and tissues persist throughout larval and adult life? It is known that some larval organs or populations of cells change their functional or biochemical activity. Second, therefore, if clonal persistence occurs during metamorphosis, would the individual cells gradually or indeed suddenly change such activities? Or third, do new clonal populations of cells originate, and do they differ structurally in different organs and tissues to subserve other functional requirements in the adult? Fourth, if these new cells do appear, do they replace the original larval cells that regress or remain in addition to them? Fifth, do entirely new organs develop to substitute for larval ones that then function in the adult? There is clear evidence to answer some of these questions; in other cases the evidence is still equivocal and the problems remain for future solution.

In answer to the first question, it is extremely likely that in some cases at least, organic populations of cells persist, generally unchanged in structure and probably function, throughout ontogeny. Surviving spinal cord neuroblasts, including those of the brachial and lumbar plexuses and their ganglia, once they have fully differentiated cytologically into neurones and assumed a permanent location, integrated synaptic connections with other neurones, and made axonal terminations via the end plates in muscles they could well remain unchanged throughout life; at least from the time the fore- and hindlimb musculature has developed. There would appear to be no reason why there is need for change or replacement, for presumably their functional activities are similar in late metamorphic and postmetamorphic animals. Larval Rohon-Beard and Mauthner cells, however, regress and have disappeared after metamorphosis, having served their larval function, and other neurones assume any of them that could be necessary and required in the adult. New brain centers of neurones do not replace larval ones, but add to the structural complexity of the central nervous system necessary for terrestrial life. Therefore, it is probable that some fully differentiated larval neurones remain throughout adult life, do not divide, and function until some at least show senescence in aging adults. (*see also* Forehand and Farel, 1982a,b).

It is also likely that a single population of larval osteoblasts of the skeleton, including that of the limbs, survives throughout anuran ontoeny, though in contrast to the differentiated neurons of the CNS, some osteoblasts at least may be replaced or continue mitosis.

In answer to the second question, it seems likely that at least some individual clonal populations of cells change their biochemical synthetic activity. Although there are proponents of the view that a new different clone of adult erythrocytes replaces a larval one (see section on blood), other evidence that appears to be more convincing, demonstrates that during metamorphosis individual larval erythrocytes contain both larval and adult hemoglobin. Gradually, within a matter of weeks, the proportion of adult Hb in each cell increases until practically all the postmetamorphic erythrocytes contain adult Hb (Jurd and MacLean, 1970; Benbassat, 1974b). Liver cells may well behave generally in a similar manner with regard to their own specific biosynethsis during metamorphosis (*vide infra*). One may conclude, therefore, that at least in the case of erythrocytes, it would appear that the genes responsible for the production of larval and adult Hb can function simultaneously within the same cell during metamorphosis.

With regard to the third and fourth questions posed, they may be answered categorically in the affirmative, in the case of some larval organs and tissues. For example, in the alimentary canal and pancreas new populations of adult cells, differing from the larval ones in terms of structure and function, originate during the period near or at climax, from nests of undifferentiated cells and they supplant the larval cells that degenerate and disappear (Bonneville and Weinstock, 1970; Beaumont, 1976; Leone et al., 1976; see also section on these organs). Indeed probably the entire larval alimentary canal is wholly remodeled, concomitantly with changes in the mode of feeding practiced by the adult. It is not absolutely clear whether an adult population of liver cells ultimately replaces a larval one, or whether the same population of larval hepatocytes changes its biosynthetic activity (see Cohen, 1970; Frieden and Just, 1970; Cohen et al., 1978; and section on the liver). Certainly larval liver cells change their ultrastructural profile *pari passu* with changes in their biochemical activity (Tata, 1967), and there is little if any evidence of extensive necrosis and thence proliferation of late larval liver cells. Actually hepatocytes mainly hypertrophy during late metamorphosis. Therefore it is more likely that the same population of larval hepatocytes gradually changes its biochemical activity during metamorphosis, and after this period new cells, derived from them by mitosis, continue to function in the same manner in the adults.

In the skin some specific cell populations, either localized closely together as glandular structures or in other cases widely dispersed, persist and function for special purposes only during distinctly limited periods of larval life, some to disappear extremely early in larval development. Other different populations of skin cells originate sequentially in space and time during the metamorphic cycle and persist throughout adult life, probably to be regularly replaced as individual cells are lost, for example, during the sloughing cycle. Replacement, which doubtless occurs in the case of some specialized cells of the epidermis, may well result from individual mitosis or perhaps new ones originate from the epidermal germinal layer. Immigration into the epidermis of nonepidermally derived cells also occurs. Cells of the adult frog epidermal epithelium, though extremely similar in many ways in structure and behavior to those of the larva, especially at climax (Fox, 1977a), do however differ from them in their numbers, morphology, and biochemistry. Nevertheless it seems likely that they are all products of a single cellular population of the basal germinal epithelium. Like the cells of the liver they appear to change their functional activity at climax, accompanied by some modest albeit important structural modifications.

Partial loss and replacement by new muscles and modifications in the structure, topography, and insertion of existing ones occur during metamorphosis. Again, among the endocrine organs the adenohypophysis of the adult frog, for example, has added two new cell types to the existing three of the larva (see Kerr, 1965, 1966).

Examples of the replacement of a larval cell type or of an organ by another different one to subserve generally similar functions in the adult, posed in the fifth question, are well known. Rohon-Beard cells of the nerve cord disappear in the larva, their function superseded by other spinal neurones. The lungs develop gradually during prometamorphosis to replace the external and internal gills. Probably they have equal importance with the skin in respiratory exchange in amphibians enjoying a terrestrial habitat. It is likely that most of the CO_2 diffuses out through the skin and in many forms the lungs are mainly a site of oxygen uptake (Foxon, 1964). The pronephroi have disappeared by the end of climax (Fox, 1970b) and the enlarging and differentiating mesonephroi excrete greater quantities of urea, in relation to the diminishing quantities of ammonia excreted by the pronephroi, as a consequence of the changeover from ammonotelism to ureotelism (Cohen, 1970) during climax (Ashley et al., 1968; see also Frieden and Just, 1970; Dodd and Dodd, 1976). Nevertheless the mesonephroi of the adult aquatic *Xenopus* excrete a high proportion of ammonia (Munro, 1939), though as *Xenopus* ages more urea is excreted, (Deuchar, 1975). Thus the adult mesonephric tubule cells of *Xenopus* modify in function during their lifetime, though doubtless the crucial factor is the function of the liver cells producing enzymes of the urea cycle, and the activity of carbamyl phosphate synthetase.

Molecular Biology of Cellular Differentiation During Metamorphosis

The initiation, maintenance, and completion of metamorphosis of anuran larvae is controlled by thyroid hormones. Thyroid hormones trigger off a predetermined program of cellular morphological and biochemical development; the event are anticipatory and are to prepare an aquatic larva for a future existence as an adult terrestrial animal. Some immature cells like neurones, though they are differentiated cells, acquire adult structure and function, though they probably change very little. Others (or those of the same lineage) that are highly differentiated to function in the larva change their biosynthetic activity for that of the adult, which is probably the case with hepatocytes (Cohen, 1970) and erythrocytes (Jurd and MacLean, 1970; MacLean and Jurd, 1971; see Broyles, 1981). Others are lost and are replaced by new populations of cells (of a different lineage) in the same organ; some cells disappear and are not replaced, having served their larval function.

Embryogenesis and premetamorphosis are important significant phases of anuran ontogeny; the result is a larva at a specifically preconditioned stage of morphological and functional development whose organs, within an integrated regional framework, are individually receptive to respond, each in its own manner, to newly synthesized and gradually increasing concentrations of thyroid hormones released into the circulation. The larval developmental program that thenceforth continues through prometamorphosis and climax is specifically concerned with the major transformation of the aquatic tadpole to a terrestrial froglet. It is known that virtually all types of cells of premetamorphic embryos of *Xenopus* can react, in terms of structural and functional changes, to thyroid hormones early in larval development (Tata, 1970), though larval development is independent of them at least until metamorphosis commences. Probably most larval cells require a thyroid hormonal stimulus during prometamorphosis and climax to elicit the requisite morphogenetic and biochemical changes that occur at this time. The gonads would appear to be the only anuran larval organs that are known to develop independently of the thyroid hormones (Dodd and Dodd, 1976). They develop apparently normally in thyroidectomized larvae, and likewise, in giant larvae of *Xenopus* restricted to NF St.54 by the goitrogen propyl thiouracil for 18 months, their gonads are well-developed; however, the maturation of oocytes in terms of yolk deposition is inhibited (Turner, 1978, and personal communication). Estradiol-17-β induces vitellogenin synthesis by the liver of *Xenopus laevis* larvae of both sexes, but this only begins near the end of prometamorphosis (NF St.57–58), and the response increases during climax. Estrogen receptors (MW 40,000) involved in the synthesis of vitellogenin, have recently been described in adult *Xenopus* liver cell nuclei and cytoplasm. There are 92 ± 18 binding sites in the hepatocyte cytoplasm and 99 ± 19 sites in its nucleus (Westley and Knowland, 1978), far fewer than in other estrogen target cells, such as those of the chick oviduct, or the T_3 and T_4 binding sites in the bullfrog liver and tail cell nuclei (*vide infra*).

Vitellogenin is absorbed by oocytes from the circulation and converted into the yolk proteins lipovitellin and phosvitin (Knowland, 1978; Tata, 1978). During larval development of *Xenopus,* responsiveness to estrogen by the liver for vitellogenin synthesis does not occur in thyrostatic larvae, but this is readily induced by thyroxine treatment. This result therefore suggests that competence by the liver cells for estrogen-induced vitellogenin synthesis is controlled by thyroid hormones. Oocyte maturation is thus a thyroid-controlled activity via the liver (Huber et al., 1979), another new biochemical activity, probably performed by the same larval liver cells, originating during the period of late metamorphosis.

As greater quantities of thyroid hormones are synthesized in the thyroid and thence released into the circulation of the anuran tadpole, the circulatory hormonal level gradually rises during metamorphosis to elicit the orderly sequence of morphogenetic events that occur, terminating at the end of climax. Just (1972) measured the protein-bound iodine (PBI) in plasma samples of larvae of *R. pipiens* during metamorphosis, and found about a 10–15-fold increase in concentration between premetamorphic levels and the highest achieved at climax (30 μg/100 mL plasma), levels substantially lower than those reported by Steinmetz (1954) and Etkin (1964, 1968) from their tadpole immersion experiments. Dodd and Dodd (1976) suggested that the increased synthesis and release of thyroid hormones into the circulation, from about TK St.XV onwards, is restricted to relatively low levels because of the increasing avidity of various tissues for the hormones. Recently, data on plasma binding of thyroid hormones in bullfrog tadpoles have become available (Myauchi et al., 1977; Regard et al., 1978). On the whole their results are in agreement, though Myauchi et al. (1977) emphasize more strongly the more significant binding of T_3 to plasma proteins than that of T_4. Furthermore, essentially they did not detect any significant T_3 and T_4 binding before TK St.XIX–XX, but Regard et al. (1978) described a gradual rise in the plasma level of both hormones through prometamorphic TK ST.XI–XX; at the onset of climax (TK St.XX) both levels sharply increased to peak at the middle of climax (T_3 showed a 15-fold increase and T_4 a 10-fold increase over premetamorphic levels). Both groups of workers reported a reduction of T_3 and T_4 plasma binding levels by up to 80% after climax is completed, with a minimum in young post-climactic froglets and adults. In agreement with these findings the level of PBI also decreases sharply after the completion of climax (Just, 1972). More recently, Suzuki and Suzuki (1981) likewise found the circulating thyroid hormonal level of bullfrog larvae to be very low before TK St.XVIII; thereafter it suddenly increases to reach peak levels at climax. At TK St.XX–XXI the T_4 level was 500 ng/100 mL of plasma and the T_3 level was one-third of this. Afterwards the T_4 level was greatly reduced to be undetectable in a 4-month-old froglet and it was barely 40 ng/100 mL of plasma in the adult. These workers found the plasma PBI to be variable and somewhat lower than that recorded by Just (1972). The sequence of events described by all the groups concerned appears to confirm the previous work by Kaye (1961), that thyroid hormone synthesis is negligible before TK St.XII.

It is of interest that there is an indication of a T_4 5-monodeiodinating system present in the serum of larvae at climax, or soon after, in bullfrog tadpoles, which converts T_4 to T_3. The system is induced by thyroid hormones in premetamorphic tadpoles and is inhibited by propylthiouracil (Galton and Munck, 1981).

The most significant action of the thyroid hormones on amphibian larval cells is the regulation of protein synthesis, that manifests new structure and function. Induced RNA synthesis, presumably mainly at the level of transcription at the nuclear receptor complex, leads ultimately to the origin of various enzymes at the translational level. Thenceforth, together with the synthesis of other new proteins and their assembly into specific cellular organelles, this results in (a) cellular structural differentiation and new biosynthetic activity and (b) new or enhanced catabolic enzyme activity, mainly of acid hydrolases, in structures such as the tail (and also for example, the pronephros, the alimentary tract, neurones of the CNS, and cells of the skin), programmed to regress by the end of climax (Weber, 1969a,b; 1977). Various important cellular changes during metamorphosis involving new protein synthesis and biosynthetic activity have been reported (see Eaton and Frieden, 1969; Tata, 1969, 1971, 1972, 1975; Cohen, 1970; Cohen et al., 1978; Frieden and Just, 1970; Kistler and Weber, 1975; Dodd and Dodd, 1976; Atkinson, 1981; Smith-Gill and Carver, 1981). Probably most anuran embryonic and larval tissues respond to thyroid hormones (Tata, 1970, 1971); specific cells do so in particular ways if thyroxine is applied locally, though the effects may be somewhat widespread if the hormone diffuses away from the site of its application. The type of response by any cell would seem to be determined by its previous developmental history, for similar kinds of tissue in different regions of a larva respond differently to the same hormonal stimulus (*vide supra*).

The initiation and regulation of cellular protein synthesis during anuran metamorphosis, using triiodothyronine (T_3) and thyroxine (T_4), have been extensively studied either in vivo or in vitro, especially with hepatocytes, the tail, and to some extent erythrocytes during their synthesis of larval or adult Hb (see Cohen, 1970; Cohen et al., 1978; Frieden and Just, 1970; Tata, 1971; Benbassat, 1974a,b; Kistler and Weber, 1975; Dodd and Dodd, 1976; Atkinson, 1981). Such work embraces, among other things, the induction of enzymes of the ornithine–urea cycle (Paik and Cohen, 1960; Cohen, 1970; Wixom et al., 1972; Cohen et al., 1978); various catabolic enzymes especially in the tail (Weber, 1967; Frieden and Just, 1970) including collagenase (Smith and Tata, 1976); accumulation of serum albumin (Ledford and Frieden, 1973); activity of DNA-dependent RNA polymerases (Griswold and Cohen, 1972, 1973); changes in the chromatin structure template activity (Kim and Cohen, 1966); stimulus of the synthesis of various species of RNA preceding the appearance of new cell proteins (Finamore and Frieden, 1960; Kim, 1967; Nakagawa et al., 1967; Nakagawa and Cohen, 1967; Eaton and Frieden, 1969; Weber, 1969a; Tata, 1971; Pearson and Paik, 1972; Ryffel and Weber, 1973; Kistler and Weber, 1975); stimulation of DNA synthesis and DNA-polymerase in mitochondria as well as in nuclei after hormonal administration (Campbell et al., 1969; Atkinson et al., 1972) and histone metabolism (Morris and Cole, 1978, 1980). The changeover from larval to adult Hb within the same erythrocytes during climax has been described (Jurd and MacLean, 1970; MacLean and Jurd, 1971). Electronmicroscopy, in particular of hepatocytes, has revealed ultrastructural changes in various organelles, especially of the RER, which indicate the initiation of new protein synthesis, after hormonal administration to anuran larvae (Tata, 1967, 1968a, 1969, 1971; Cohen, 1970; Bennett et al., 1970; Bennett and Glenn, 1970; Spiegel and Spiegel, 1970; Kistler and Weber, 1975; Cohen et al., 1978; Smith-Gill and Carver, 1981).

The mechanism of *de novo* protein synthesis first involves the formation and turnover of various species of nuclear and cytoplasmic RNA. The nuclear synthesis (transcription) and transport into the cytoplasm of various kinds of RNA molecules, of different normally developed embryonic stages of *Xenopus laevis,* was analyzed by Shiokawa et al. (1979). It appears that blastula cell nuclei synthesize snRNA (small nuclear RNAs that migrate on agarose–polyacrylamide gels more slowly than 4S RNA),

4S RNA and 5S RNA, the 4S RNA being transported into the cytoplasm immediately after its formation. The 5S RNA remains in the nucleus and is transported afterwards with 28S rRNA slightly later than 18S rRNA. Nuclear transcription and passage to the cytoplasm, in the blastula cells, of 40S rRNA and poly(A)-RNA was also described. These authors also assumed that the 5S RNA and 28S rRNA comigrate to the cytoplasm and join to form the 60S RNP particle. The examination of the relative amounts and heterogeneity of 10 tRNAs and amino acyl tRNA synthetases, of bullfrog tadpoles and adults, showed that the iso acceptor profiles of seven of the tRNAs (for alanine, arginine, asparagine, leucine, lysine, proline, and threonine) were essentially identical. There were minor differences in the cases of phenyl-alanine and histidine tRNAs, and major quantitative differences for the two species of methionine tRNA (higher levels of tRNA^{met} in tadpoles). Such results may well be expected, for the amino acid compositions of liver protein in tadpoles and adults are generally similar (see Klee et al., 1978).

After the administration of thyroid hormone (T$_4$) to a larva during the early part of metamorphosis, a lag period is followed by the appearance in, for example, liver cells, of new protein (enzymes) of the urea cycle (Cohen, 1970) or an elevation of their biochemical activities (Wixom et al., 1972). During this lag period DNA-dependent mRNA is presumably transcribed in the nucleus, and perhaps cytoplasmic translation of mRNA is sustained by the hormonal stimulus. Actinomycin D, which inhibits RNA production by binding to DNA, inhibits the synthesis of carbamyl phosphate synthetase by hepatocytes, though after the synthesis has commenced, Actinomycin D is ineffective. The probably stable mRNA activity that has occurred is, however, blocked by puromycin at the level of translation (Kim and Cohen, 1968). Nakagawa and Cohen (1967) considered that some of the new RNA may include mRNA synthesized after T$_4$-induced derepression of specific liver cell nuclear DNA templates (Kim, 1967). However, Tata (1971) described only relatively small increases in liver cell nuclear DNA-like mRNA after thyroid hormone administration, though he recognized the fact that this result may well reflect the use of inadequate methods that are not sensitive enough to detect the mRNA among the variety of nuclear species of RNA molecules, or that the mRNA is masked by the far greater increase in the synthesis of rRNA. Indeed in most types of cells the mRNA represents only a small fraction of the cell compared with the rRNA and tRNA. It is of interest that in the essentially nondividing metamorphosing tadpole hepatocytes (Cohen, 1970), there seems to be little DNA synthesis or repair turnover (Pearson and Paik, 1972), which is in line with earlier reports (see Finamore and Frieden, 1960; Paik et al., 1961; Tata, 1965), though Kistler and Weber (1975) reported the DNA concentration per unit of liver weight first to decrease and thence afterwards to increase during late metamorphosis. But preceding the appearance of new cell proteins there is a large increase in the rate of synthesis of ribosomal RNA, which is mainly located in the polysomes of the more complex and lamellated RER. Such enhanced synthesis of rRNA is common to all tissues irrespective of the type of morphogenetic response (Ryffel et al., 1973). The increases in the rate of synthesis of various cell proteins are accompanied by substantial increases in the number of cytoplasmic polysomes and in the amount of associated microsomal membranes in the cells both in vivo and in vitro. Nevertheless, ribosomes do not accumulate appreciably during the initial hormone-induced period (6–10 d), rather they are more often arranged in pockets of high density attached to lamellar membranes, and possibly there is an accelerated turnover of cytoplasmic RNA and ribosomes at this time. Perhaps "new" ribosomes replace "old" ones, though certainly there is a redistribution of ribosomes, recognizable by electronmicroscopy. There is a "heavy rough membrane fraction" where ribosomes are more closely packed on the membranes, and they are more active in protein synthesis in vivo than are the free ribosomes on the "light rough membrane fraction." Possibly "new" ribosomes are active in protein biosynthesis after hormonal stimulation, and it is conceivable that different precoded ribosomes are segregated topographically on the RER and have different synthetic functions (see Tata, 1969, 1971).

Of extreme importance is the question whether mitochondria are significantly involved in protein synthesis during thyroid hormone-induced metamorphosis. Among vertebrates in general it has been suggested that there is activation of energy metabolism in mitochondria after thyroid hormonal stimulation (Sterling, 1977, 1979; see Fig. 59). No changes in the hepatocyte mitochondria of *Xenopus,* either qualitatively or quantitatively, were recognized during T_4-induced metamorphosis by Kistler and Weber (1975). However, mitochondrial swelling and increased uptake and incorporation of amino acids occurred in rat liver hepatocytes after T_4 stimulation, which Buchanan et al. (1970) considered to be interrelated. Recently it has been shown by Nelson et al. (1980) that T_3 increases mitochondrial protein synthesis 2–3-fold in isolated rat hepatocytes, though the treatment appears to have little if any effect on the general synthesis of cytoplasmically translated mitochondrial proteins. However, whether T_3 acts directly on the mitochondria or indirectly by first activating synthesis of cytoplasmic translated proteins is still not clear.

Among amphibians, Cohen (1970) likewise related hepatocyte mitochondrial swelling and their increase in matrix density, after T_4 stimulation of *R. catesbeiana* larvae, as features of protein synthetic activity. In vitro studies have shown that there is an early rise in DNA polymerase activity of mitochondrial and cell sap fractions of hepatocytes after

T_3-induced metamorphosis of *R. catesbeiana*. Incorporation of [^{3}H]-deoxythymidine into DNA of nuclei rose after 5 d of hormone administration. In contrast the specific activity of mitochondrial DNA rose after 1–2 d of similar treatment (Campbell et al., 1969). Likewise there was found to be an increased rate of incorporation of the DNA precursor [^{3}H]-thymidine into DNA of mitochondria, in vivo in *R. catesbeiana* after T_3-induction, which began 1 h after treatment and persisted for 3 d, but thereafter declined to the control level (Atkinson et al., 1972). It is likely therefore that mitochondria could well be included together with other subcellular sites where protein synthesis is stimulated by thyroid hormones.

The process of tissue regression in anuran larvae, exemplified by the behavior of the tail at climax, has been studied in vivo and in vitro (see Shaffer, 1963; Kubler and Frieden, 1964; Weber, 1969a,b; Frieden and Just, 1970; Wang and Frieden, 1973; Fox, 1975, 1977a,b; Atkinson, 1981). The process is accompanied in particular by the increased activity of acid hydrolases. There has been much research on the biochemical aspects of tail resorption of anuran tadpoles, especially of the enzyme activities during climax. Acid hydrolases are the main enzymes concerned (Eeckhout, 1969; Perriard, 1971; Hickey, 1971), probably derived from lysosomes and possibly also from nonlysosomal de novo synthesis by the autolysing cells, and likewise they are present in the heterophagic vacuoles or phagosomes of macrophages that invade and ingest necrotic tissue (Weber, 1977). These enzymes include cathepsins (especially cathepsin D, see Seshimo et al., 1977; Sakai and Horiuchi, 1979), β-glucuronidase (Kubler and Frieden, 1964), β-galactosidase, deoxyribonuclease, ribonuclease, and collagenase (Dodd and Dodd, 1976) and also alanine aminopeptidase (Little et al., 1979).

After thyroid hormone treatment of tadpoles, just before the onset of tail regression, there are bursts of RNA production and thence within 1–2 d protein synthesis ensues. Actinomycin D, which inhibits transcription of RNA, and puromycin or cyclohexamide, which inhibit protein synthesis at the translation level, prevent T_3- or T_4-induction of tail regression. Such inhibition abolishes the hormone-induced activity of the hydrolytic enzymes (Weber, 1965; Eeckhout, 1969; Tata, 1966, 1971). The specific inhibition of T_3-induced tail fin regression of *R. catesbeiana* by pepstatin, a cathepsin D inhibitor, has also been reported (Hiroshi et al., 1977). It is likely, therefore, that tail necrosis (and probably necrosis of other larval tissues also), is the result at least in part of the *de novo* synthesis of lysosomal acid hydrolases, not just to the activation of existing lysosomal enzymes, though Smith and Tata (1976) suggest that the initial T_3-hormone induced changes in the regressing tadpole tail of *Xenopus* might well result from the activation of regulatory ''proteolytic cascades.'' A decreased rate of protein synthesis in the bullfrog tadpole tail muscle tissue is induced early on and sustained by in vivo administration of T_3, and Saleem and Atkinson (1978, 1979) also demonstrated a progressive loss of polyribosomes (and thus less activity overall in their protein synthesis) and a decrease in the proportion of trichloracetic acid-soluble [^{14}C]-leucine, associated with the heavy and medium-sized polyribosomes. They claim that the decreased translational efficiency of these polyribosomes, found to occur in vitro, results from a thermolabile, trypsin-sensitive inhibitory factor(s) found to be present in post-ribosomal supernatant fractions of hormone-treated tadpoles (Saleem and Atkinson, 1980). Administration of T_3 to anuran tadpoles may thus initiate in tail muscle a regulatory effect that operates directly or indirectly at the trans-

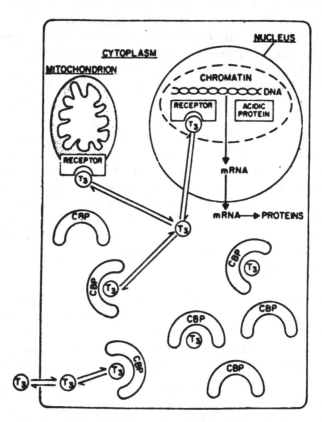

Fig. 59.   Action of thyroid hormones: T_3 (or T_4) within a circle represents unbound hormone that diffuses into the target cell and binds to cytosol binding proteins (CBP). The CBP-T_3 complex is presumed to be in reversible equilibrium with a small quantity of cellular unbound T_3 that can interact at effector binding sites in the nucleus (N) and mitochondria (MIT). There are about 20 million primary sites in the cytosol of rat liver cells, and many CBP molecules thus may not have a bound T_3 molecule.

lational level (apart from any other possible influences acting directly on the nuclei) and influences the rate of protein synthesis. There would appear to be some contradictions with regard to these results and those of previous workers, for hormone-induced tail regression does stimulate the protein (enzyme) synthesis required for cellular autolysis and phagocytosis. Perhaps the reduction in polyribosome numbers is in part the result of the sequential cellular degeneration. Thus various antagonizing activities may occur simultaneously; stimulated protein synthesis, inhibition of protein synthesis, and continuing cellular necrosis. There is some evidence that T_3-stimulation of tail-fin regression in *R. catesbeiana* tadpoles may be mediated via the action of cAMP. Cyclic AMP, like T_3, increases the specific activity of hexosaminidase, a lysosomal enzyme, coincident with tissue reduction. Lithium chloride, which inhibits the activity of adenyl cyclase, also inhibits such changes initiated by T_3 but not when initiated by cAMP (Stuart and Fisher, 1978). However, it is generally believed that thyroid hormones initially do not act at the level of the cell membrane, via the mechanism of adenyl cyclase and cAMP. The results, therefore, suggest that there may be intracellular mechanisms of T_3 action akin to those stimulated by other nonthyroid hormones at the cell surface,

---

Fig. 59 (*cont.*).   Various workers have suggested a variety of mechanisms of action of thyroid hormones on peripheral body tissues (see Frieden, 1981). These include: nuclear transcription, mitochondrial activation, Na-K ATPase (sodium pump) action at the plasma membrane, incorporation into tyrosine pathways, adrenergic receptor sensitivity, membrane action, and a combination of these mechanisms. Mainly from evidence on rat liver cells and clinical results on human subjects, Sterling (1977) proposed a dual action of thyroid hormones (T_3 and T_4) acting on mitochondria and the cell nucleus. The hormones (1) initiate nuclear transcription of mRNA responsible for protein synthesis at the ribosomal level of translation yielding growth, differentiation, and cell maintenance; and (2) activation of mitochondrial energy metabolism, possibly the first observable effect, such as increased oxygen consumption. The mitochondrial membrane protein (MMP), conceived to function as the hormone receptor, is possibly a thermolabile lipoprotein at the inner surface of the mitochondrion. There is no evidence that thyroid hormones together with the CBP are translocated to effector loci in the nucleus and mitochondria. Cytosol binding molecules may hold the hormones within the cell in equilibrium with small quantities of free T_3 and T_4 that bind to the nuclear or mitochondrial receptor sites. Though some workers claim that in amphibians the nucleus is the sole site for reception and action of thyroid hormones, others favor extranuclear binding sites such as mitochondria, as well as nuclei, a view similar to the above hypothesis. (Figure and legend from a review by Sterling, 1977.) A later figure by the same author (Sterling, 1979) incorporated plasma membrane receptors for T_3 at the surface of the target cell.

and/or that alternative chemical pathways mediated by T_3 at various intracellular binding sites are available via cAMP action.

With reference to the various cellular activities of differentiation (and degeneration) influenced by thyroid hormones, Tata (1971, 1972) posed a number of significant questions that still require solution or at least further clarification, including: What is the nature of the replicated RNA and the mechanism of its transfer to and messenger role in the cytoplasm? Again, are new genes transcribed, perhaps by changes in the specificity of the transcriptase enzyme, or is there merely a selective change of the type of RNA transferred to the cytoplasm, which has previously been produced continuously though as yet not selected in early larval stages? Perhaps transcription is influenced by the increased activity of RNase that was found to occur in bullfrog tadpole hepatocytes during metamorphic climax (Tomita et al., 1979). Likewise, do separate classes of cytoplasmic ribosomes synthesize specific proteins, or do the same ribosomes change their biosynthetic activity because of external (extraribosomal) factors influencing translation?

It is at least quite clear that the hormone-stimulated large increase in the rate of synthesis of rRNA (finally to be located in the polysomes of the RER) that precedes protein synthesis (*vide supra*) is a common feature in a variety of different tadpole tissues (Ryffel et al., 1973). A good deal of information, in terms of molecular biology, has now accumulated about the origin, structure, and activity of rRNA cistrons (see Birnstiel et al., 1971). Initially synthesis of new rRNA (as distinct from that already present in ribosomes of the unfertilized egg) begins with the onset of gastrulation when the cell nucleolus first appears (Brown, 1966), and its synthesis gradually increases, though the total content of RNA by the end of embryogenesis is only about double that originally present in the unfertilized egg (Brown and Littna, 1964).

Anucleolate tadpoles of *Xenopus* (minus nucleoli in their cell nuclei) do not synthesize ribosomal RNA, though they can still manufacture mRNA (Brown and Gurdon, 1964). Nucleoli arise at a specific region on the chromosomes, usually distinguishable as a secondary constriction or nucleolar organizer. Mutant *Xenopus* specimens do not possess nucleolar organizers and it is believed that these organizers contain the ribosomal RNA cistrons, which usually have a higher G + C content than the rest of the DNA. In *Xenopus* the nucleolar organizers obey the laws of Mendelian inheritance and are affected by mutation. Likewise in *Bufo* the size of the secondary constrictions generally corresponds to the levels of rDNA, and a mutant axolotl with a small nucleolus in its nuclei was deficient in rDNA.

Birnstiel et al. (1971) provide a 'tentative model' for rDNA of *Xenopus*. The 28S and 18S tandem cistron has a mass of 4.4×10^6 daltons (1 dalton is equivalent to 1/12 of the mass of an atom of carbon-12); 0.6×10^6 daltons code for the nonconserved RNA segment and almost 6×10^6 daltons code for the spacer DNA (which in *Xenopus* is not transcribed), and this segment alternates with the ribosomal transcriptional unit. Generally the basic unit of rDNA consists of about $9–11 \times 10^6$ daltons, which is repeatable many times in the nucleolar organizer DNA. Thus about half of the rDNA cistron codes for precursor rDNA. In *Xenopus* all the ribosomal cistrons of the nucleolar organizer DNA are confined to a ribosomal DNA satellite, which is a discrete separate unit of the genome and constitutes 0.2–0.3% of the genomal DNA. Each nucleolar organizer of a somatic cell nucleus of *Xenopus* weighs about 0.9×10^{-14}g and contains 6×10^9 daltons of rDNA, equivalent to $2.4\times$ the mass of the entire *Escherichia coli* genome.

Ribosomal cistrons, especially of higher organisms, show a high degree of multiplicity. It has been questioned whether such a large number of ribosomal cistrons is constant throughout the life cycle of a differentiated cell, or whether they are synthesized *de novo* and increase in number by amplification from a limited number of "master" genes. Amplification of ribosomal RNA genes has been shown to occur in oocytes of amphibians, fish, and insects by replication of rDNA, and extra copies of the nucleolar organizer region are synthesized selectively in great numbers (see Birnstiel et al., 1971; Collins, 1972). The amplified rDNA of the amphibian oocytes is ultimately contained in the cores of about 1000–1500 nucleoli. Amplified extrachromosomal nucleoli of *Xenopus laevis* oocytes are formed from a protein filamentous complex, a meshwork of 4-nm-thick microfilaments coiled in bundles 30–40 nm in diameter that are densely packed at the nucleolar cortex. They dissolve after treatment with RNase or high salt buffer, and fractions show the presence of a major acidic protein of MW 145,000 (Franke et al., 1981a). Whether amplification occurs in the somatic tissue cells is still not clear. It is claimed to take place in cultured human liver cells (Koch and Cruceanu, 1971) and in the regenerating lens of the adult *Triturus viridescens* (Collins, 1972). However, in liver and tail muscle of *Xenopus* larvae during T_4-induced metamorphosis, when the number of rRNA genes was determined by hybridization, the amount of rDNA remained unchanged and identical with that of adult erythrocyte DNA. In these cases amplification of rDNA templates cannot thus explain the increased synthesis of rRNA. Ryffel et al. (1973) therefore considered that the hormone-induced metamorphic tissue changes may be the result either of a higher rate of transcription, or of the intervention of some kind of post-transcriptional control.

In conclusion it is clear that the model systems of metamorphosis are most complex in anuran amphibians (and insects also), but they provide examples for the analysis of cellular determination and differentiation. The key role played by hormones—especially of the thyroid—as central mechanisms, together with studies on the phenomenon of metamorphic change, possibly provide the most useful means, so far known, for determining the causal mechanisms of molecular biochemical events in cellular differentiation (Ashworth, 1973; Frieden, 1981). Nevertheless, a note of caution is essential, as emphasized by Smith-Gill and Carver (1981). In their extensive review of the biochemistry of amphibian metamorphosis they described only keratin synthesis in the skin and CPS-1 synthesis in the liver to demonstrate quantitative changes in mRNA accumulation in association with metamorphosis; that thyroid hormone action and the level of its control of this accumulation requires further definition. To date, thyroid hormone-induced synthesis of a new class of mRNA has not been demonstrated. We must hope and expect that this will be reported in due course.

Receptor Sites—A Brief Review

Before considering hormone receptor or binding sites in amphibian larval target cells, it would be useful generally to consider, albeit briefly, current knowledge on the subject. A receptor system demands: morphogenetic interactions that demonstrate saturability; that is, a limited or finite number of binding sites, probably of low capacity but high affinity (though the number of sites is claimed to increase in amphibian bullfrog larval tail nuclei during development, Yoshizato and Frieden, 1975); structural and steric specificity, that is, only substances of similar three-dimensional structure can compete for the same binding sites; a tissue specificity related to the biological specificity of the hormone; usually a high affinity related to the concentration of the hormone; and reversibility with kinetics similar to those observed morphologically (see Grant, 1978).

Hormones typically occur in four classes that initially activate their target cells via hormone receptors in at least three ways: by membrane permeability, enzyme activity, or gene expression. Perhaps some hormones utilize more than one mechanism. Catecholamines such as epinephrine and the neural transmitters noradrenaline and actylcholine, act primarily at the level of the cell membrane and affect membrane permeability. Likewise the linear peptide and polypeptide chain hormones, for example, glucagon and the larger polypeptide insulin, and others of the pituitary including growth hormone (somatotropin), FSH (follicle stimulating hormone), vasopressin, prolactin, and thyrotropin all initially interact with receptor molecules at the cell surface (see Cuatrecasas, 1974). Thyroid hormones (like catecholamines, initially derived from the amino acid tyrosine; see Bentley, 1976) and steroid hormones (derived from the fused ring steroid nucleus), enter the target cell directly and interact intracelullarly, possibly at multiple extranuclear receptor protein sites, though ultimately they act at the level of the nucleus to influence gene expression (see Jensen and De Sombre, 1972 on female sex hormones, and Oppenheimer, 1979, and Sterling, 1979, on thyroid hormones). All hormones modulate enzyme activity in a variety of ways, and the cell response to different hormones may include an array of apparently unrelated or pleiotropic changes, possibly because several different sites of action exist within the cell.

In its simplest form a receptor is considered to be a protein with a site for the hormone and an 'executive site', specific for another protein, such as an enzyme or a nucleic acid. The presence of the hormone induces a structural change in the receptor that influences the occupant substance in the 'executive site'; such regulation is expressed in terms of biochemical (enzymic) change or activity. Receptor molecules occur in extremely low concentration in target cells and thus are extremely difficult to investigate, but since Cuatrecasas (1972) described the insulin receptor, reports on other hormone receptors soon followed.

Hormone receptors in target tissues appear to be influenced by the circulating levels of their respective hormones (Raff, 1976). Elevated levels of TSH, which increase the concentration of cAMP, seem to reduce the number (but not the affinity) of TSH binding sites on the plasma membrane of thyroid follicular cells of tapazole-treated rats (Holmes et al., 1980). How TSH regulates the number of sites is not yet understood. Likewise the number (but not the affinity) of binding sites for prolactin increases up to near the end of prometamorphosis and then decreases through climax in the liver, tail, and kidney tissue of the larval bullfrog. The numbers can be regulated by treatment with ovine prolactin and L-T_3, which may be effective by influencing the circulatory endogenous levels of prolactin (Carr et al., 1981).

A typical mechanism of cell surface membrane–hormone receptors is exemplified by the reactive cascade during the intracellular metabolism of glycogen elicited by glucagon and epinephrine (Sutherland, 1972). Interaction of those hormones with their receptors on the *outside* of the cell membrane leads to the membrane-bound enzyme adenyl cyclase producing an amplified amount of cyclic AMP (the nucleotide adenosine-3',5'-monophosphate) from ATP on the *inside* of the membrane. Cyclic AMP has a carrier or 'secondary messenger' function acting within the cell for the hormones. The 'mobile receptor' hypothesis of Cuatrecasas (1974) assumes that the separate individual receptors are not associated with adenyl cyclase in the absence of hormone. Binding of their hormone enables then to bind to cyclase acquired during diffusion through the cell membrane. Different hormonal effects may well depend upon whether the association of the hormone–receptor complex with cyclase stimulates or inhibits the production of cyclic AMP (see Grant, 1978; Blake, 1978). It is likely that the action of the steriod hormones, among the intracellular hormone receptors, has been the most widely investigated and is best understood (Gorski et al., 1969; Jensen and de Sombre, 1972). Probably thyroid hormones function in a similar manner (see Frieden, 1981), though at present it is believed that in contrast to the steroid nuclear receptors, those of the thyroid are not involved in a two-step sequence translocation hormone–receptor complex of the cytosol to the nucleus (see Cohen et al., 1978). Caution with regard to the similarities of steriod and thyroid receptors was advised by Tata (1975), for their intracellular mechanisms may well differ quite substantially. Steroid and thyroid hormones are small molecules that are transported in the blood circulation bound to specific plasma proteins and thence traversing their target cell surface membranes to encounter specific receptors within the cell. Steroid hormonal activity involves a two-step process; they bind to receptor molecules which owing to some conformational change, results in the receptor–hormone complex translocated to the cell nucleus. Here, there is interaction at some specific site in the chromatin, probably the DNA, to elicit DNA-dependent RNA replication via transcriptase enzymes or, among other things, perhaps to influence the selection of specific species of already replicated RNA by changes in permeability at the nuclear–cytoplasmic interface. New protein synthesis is finally initiated at the level of translation in the cytoplasm. The movement of the receptor–hormone complex through the cytoplasm may well influence cytoplasmic functional activity. It is likely, however, that as with the thyroid hormones there are multiple steroid hormone intracellular binding sites, and the mechanism described is almost certainly an oversimplification of the mode of action of these receptor–hormone complexes. (See Alberts et al., 1983).

Thyroid hormones are transported in the blood by prealbumin, a plasma protein of molecular weight 55,000. The protein molecular structure indicates that it contains subunit side chains that define a cylindrical solvent-filled channel, 50 Å long by 8 Å wide, extending through the center of the molecule. Possibly the hormone binds deeply in this channel, which effectively isolates it from the solvent, and the hormone has extensive interactions with the protein side chains extending into the channel (Blake, 1978). In order to equate a rapid rate of binding between a hormone and its receptor site with a rigid 'lock and key' simile, a 'zipper' hypothesis (Burgen et al., 1975) supposed that a single segment of the hormone first binds to a receptor subunit. Then the reciprocal conformational fit of the ligand and the macromolecule rapidly and successively incorporates further segments until the hormone is lined up— zipped up, so to speak—into the receptor.

The factors that control gene transcription in the eukaryotic nucleus are little understood, though molecular systems have been tentatively suggested, for example, by Britten and Davidson (1969) and Georgiev

(1972). The mechanisms regulating gene activity presumably are of greater complexity than are those in the better known prokaryotes, which have been described in some detail for the lac repressor in the operon system of *Escherichia coli* (Jacob and Monod, 1961).

There are about 100 nonhistone acidic proteins (which include the various DNA and RNA polymerases) associated with genomal DNA; they comprise a heterogeneous group of molecular weights ranging from 7000 to 15,000. Most of them are phosphorylated, the amount changing during the cell cycle or during its development. Phosphorylation is also reversible and the conformational changes that occur may influence binding affinities for DNA and/or histone.

The histone proteins (see de Reuck and Knight, 1966) have a molecular weight of 10,000–20,000; usually they are rich in basic amino acids such as arginine and lysine, but they usually lack tryptophan, and only a few molecular species of histones have been identified. In amphibian eggs, for example, the rate of synthesis of histones increases 10-fold in the two-celled stage of *R. pipiens* compared with that of the maturing oocytes, and increases a further three-fold by the blastula stage. The various classes of histones show differing rates of synthesis and they also differ in the time of commencement. Cytoplasmic histone pools are postulated (see Shih et al., 1980). Histones readily combine with the lightly coated eukaryotic DNA, though details of the configurative association are not clearly understood. It is known, however, that histones combine nonspecifically with DNA and appear to regulate its secondary and tertiary structure, possibly the coiling behavior of whole chromosomal sets, and thus DNA activity. Typically, the association of histones with DNA of the chromosomes forms the nucleosomes (or beads) on the chromatin. Probably histones have a nonspecific masking effect on DNA transcription and the nonhistone protein has some masking specificity. It is believed that histones may possibly combine with most of the genomal DNA to cause supercoiling, so that RNA polymerase binding sites are inaccessible and thus only small segments of the DNA are free to be transcribed. Acidic nonhistone protein may protect DNA from histones, permitting RNA polymerase activity to occur at the template (see Watson, 1976; Grant, 1978; Alberts et al., 1983).

Based on X-ray data, Blake and Oatley (1977) showed that the molecular structure of the plasma protein prealbumin dimer has a region in its outer surface that, in structural terms, fulfills the DNA-binding requirements of repressors (Fig. 60). The shape is a cylindrical depression of dimensions sufficient to enclose the double helix DNA molecule to just under half its depth (Fig. 61). The location is between two subunits of protein that give it an exact twofold symmetry. This site is distinct from the hormone-binding site that runs through the center of the molecule, and a pathway of interaction between the two binding sites is conceiva-

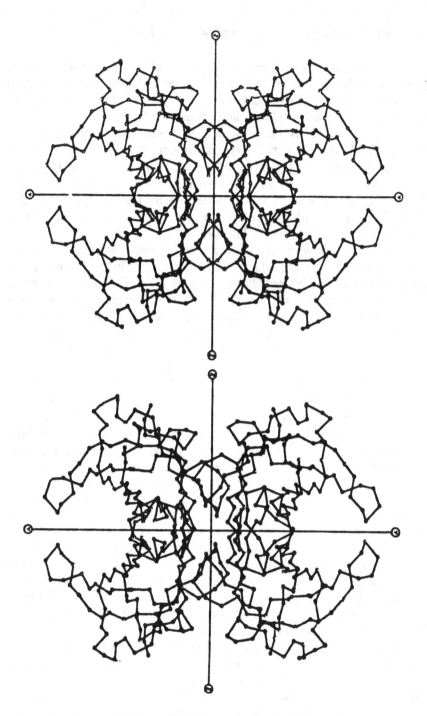

Fig. 60. Stereo drawing from X-ray analysis of the carbon positions of the prealbumin tetramer (MW 55,000) looking down the molecular x axis. A channel runs horizontally through the molecule parallel to the z axis and contains the hormone binding sites. (After Blake and Oatley, 1977.)

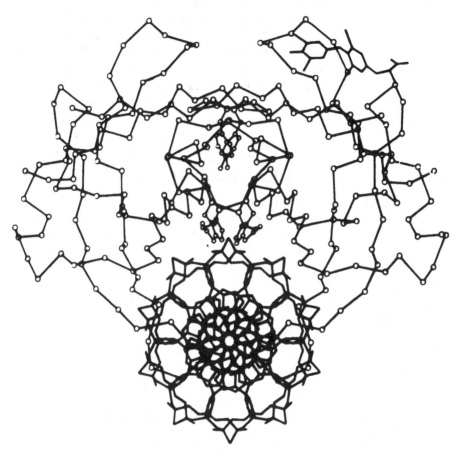

Fig. 61.   The DNA binding sites are considered to be located at the top and bottom of the figure of the tetramer and the axis of the DNA helix lies parallel to the *x* axis.

The T_4-binding plasma protein prealbumin has a DNA site of twofold symmetry. In the groove that lodges the DNA the edges of the nucleotide base pairs are exposed, so as to interact with the protein arms, and thus the nucleotides could permit recognition phenomena of the binding site. The two identical hormone sites in the central channel of the prealbumin molecule are located in such a way that interaction between T_4 and DNA binding sites could allow the hormone to modulate the structure of the DNA site. As yet, however, there is no direct evidence that prealbumin can bind DNA. Nevertheless, the arrangement proposed of the DNA–protein–hormone–binding site interactions suggests a possible biologically significant mechanism, functional within the cell, exemplified by allosteric proteins having a hormone site(s) and the executive sites, containing either another protein, an enzyme or a nucleic acid. (After Blake, 1978.)

ble, enabling the presence of the hormone to modulate the DNA site. Though there is no evidence that prealbumin can bind DNA, such a system could well provide an indication of a system to deal with the problems of DNA–protein–hormone interactions (see Blake, 1978). A model such as this, for a variety of receptors, could well account for the spatial relationships demanded for hormonal modulation of 'executive site' enzymes that in different locations of the cell could then influence either transcription, translation or other cellular biochemical activities.

Receptor (Binding) Sites for Thyroid Hormones in Amphibian Larval Cells

The specificity of different hormones acting on their target cells, indeed the various responses by different individual cell populations to the same hormone, as in the case of amphibian larvae and thyroid hormones, is generally explained by postulating that such differing groups of cells have distinct intrinsic properties. However, more specifically, thyroid hormone-induced cellular responses are considered to act initially at special molecular sites within the cells, where biochemical or some conformational change may occur. In more recent times, especially during the past 10–15 yr, much effort, which has gradually increased, has been expended by workers investigating cellular hormone-binding sites, or receptors, in a variety of classes of vertebrates including amphibian anuran larval cells where, either directly or indirectly, it is assumed that RNA synthesis is ultimately initiated (Kistler et al., 1975a; see also Dratman, 1978). Binding sites are described as being either of high affinity and relatively low capacity (often called specific or saturable) with regard to their favored hormones, or of low affinity and high capacity (nonspecific or virtually nonsaturable) (Tata, 1975). Examination of binding sites by use of $[^{125}I]$-T_3 and $[^{125}I]$-T_4 suggested that in nuclei only of bullfrog tadpole liver cells, high affinity saturable sites for T_3 numbered 12,300 and for T_4 2300 (Kistler et al., 1975a). In tail tissue there are 1500 of these sites for T_3 and 800 for T_4 (Yoshizato et al., 1975a). Probably the nuclear binding sites differ for T_3 and T_4, possibly either in terms of location or molecular structure, or both; they are, however, considered to be part of the nonhistone regions of the nucleus. Saturable binding sites for T_3 and T_4 in liver nuclei of bullfrog larvae were investigated by Galton (1980). She calculated that there were 1100 T_3 saturable sites per liver nucleus, far fewer than the 12,300 sites per liver nucleus described by Kistler et al. (1975a). T_3 binding sites are similar in TK St.X and XV. Stable T_4 and acetic and propionic acid analogs of T_3 competed with $[^{125}I]$-T_3 for sites almost as readily as did stable T_3. The acetic acid analog of T_4, 3,5-diiodothyronine, and (reverse) T_3 were relatively poor competitors. Binding of T_3 to the saturable (but not the nonsaturable) nuclear binding sites was reduced at 4°C.

Oppenheimer et al. (1976) believed that, in mammals at least (i.e., rats), the nuclei alone of liver and kidney cells have high-affinity, limited-capacity receptors for T_3.

Until fairly recently such hormone-binding sites in amphibian cells were also believed to be present only within or at the surface of the nucleus. Indeed Frieden (1981) states that "T_3 and T_4 may be unique in that no significant cytoplasmic receptors or translocation reactions have been found." In contrast to the polypeptide hormones that first interact with receptor molecules on the cell surface, some at least of the small molecular weight hormones (steroid, growth, and thyroid hormones) act directly through intracellular receptors. In mammalian cells they react with a cytosol receptor and the complex is thence transported to the nucleus to initiate or influence in some manner nuclear transcriptional or other biochemical activities (Tata, 1975; Blake, 1978; Alberts et al., 1983, and *vide supra*). The initial sensory estrogen receptor-binding site of a mammalian uterine cell is a receptor protein, analogous to an allosteric feedback control site. Probably a conformational change occurs in this protein, as a result of estrogen binding, that initiates its movement into or onto the nucleus where gene expression is thence regulated, either by direct interaction with DNA involving replication, or perhaps by modifying nuclear membrane control of RNA transfer into the cytoplasm (Gorski et al., 1969).

Easily saturable specific binding sites for $[^{125}I]$-T_3 occur in nuclei of rat liver and kidney cells (see Oppenheimer et al., 1976). Extranuclear cytosol binding sites for T_3 were less easily saturable. Oppenheimer et al. (1972) concluded that perhaps a T_3–cytosol complex interacts with nuclear binding sites to influence gene expression, and that T_4 is only active after its conversion to T_3. Rat liver nuclei were later shown to have about 5000–10,000 specific sites for $[^{125}I]$-T_3. However, apart from the binding of thyroid hormones to the nonhistone nuclear receptor proteins of the nucleus (see Dratman, 1978), binding also appears to occur in the cytosol, plasma membranes, and inner mitochondrial membranes, which have similar high affinity-binding characteristics (see Tata, 1975). Likewise, binding of T_3 (dependent upon the presence of bivalent cations) and T_3 analogs has been found to occur in tail, kidney, and liver cytosol of bullfrog larvae (Yoshizato et al., 1975b; Kistler et al., 1977), and also of T_4 in liver and tail cytosol (cell sap) (Durban and Paik, 1976), whose binding molecule was considered to be protein in nature. Likewise, Ergezen and Gorbman (1977) found unsaturable binding sites for $[^{125}I]$-T_4 in homogenates of liver, tail, and brain tissues of larval *R. pipiens*, and also saturable binding sites in the cytosol, but especially in the nuclear fractions of all three tissues. The brain nuclear and cytosol fractions had the higher proportion of saturable sites than these fractions of other tissues. Saturable binding sites for T_3 were also detected in the nuclei and possibly cytosol, but not in the mitochondria and microsomes, of bullfrog larval liver cells by Toth and Tabachnik (1979). Maximum binding capacity was about 2700 molecules of T_3 per nucleus and the T_3 binding molecules had a molecular weight of between 70,000 and 80,000. The number of binding sites and the dissociation constant for T_3, in liver and in the tail fin cytosol of bullfrog larvae, did not differ from TK Stages I to V and in TK Stage XX. However, significant differences were found between liver and tail fin cytosol in the number of binding sites for T_3 at these stages (Jaffe and Gold, 1977). Nevertheless, so far there is no evidence to support a view that cytosol binding sites for thyroid hormones are involved in the transport of these hormones to the nucleus, and Kistler et al. (1977) suggested that these sites may "moderate" the effect of unbound intracellular hormones.

It is known that the sensitivity of tadpole tail tissue to thyroxine progressively increases during metamorphosis (Derby, 1968). According to Durban and Paik (1976), this change may be related to the increased affinity of T_4 binding in tail cell sap during larval development. T_4-binding protein in tail cell sap was absent in premetamorphic stages of bullfrog larvae, but it arises shortly before climax heralding the changes of tail resorption that ultimately occur; T_4-binding, however, occurs in liver cell sap from early larval stages onwards. The molecular weight of the T_4-binding protein was 60,000 from tail cell sap and 42,000 from that of liver cells. Furthermore, the number of binding sites for $[^{125}I]$-T_3 in bullfrog tail nuclei was reported to increase from about 1900 (at TK St.X) to 3300 (at TK St.XVIII–XIX), and thus a larger fraction of T_3-binding sites are occupied (Yoshizato and Frieden, 1975). There is an increase in nuclear binding capacity, but a concomitant decrease in affinity, and these authors suggest that the increased sensitivity of tadpole tail tissue to thyroid hormones towards climax may result from the increase in nuclear binding capacity for T_3. It should also be noted that T_3 appeared to be 2–5 times more effective in causing tail regression than T_4. Since the initial site of action of thyroid hormones is still not yet clear, and there could well be multiple active subcellular sites (Tata, 1975), perhaps this fact to some extent could reconcile the apparently conflicting results, in terms of recognition of binding sites and their affinities during larval development (*vide supra*), obtained with tail nuclei and T_3 (Yoshizado and Frieden, 1975) and tail cell sap and T_4 (Durban and Paik, 1976). However, more recently Jaffe (1978) found that the binding proteins for T_3 in bullfrog larval liver and tail cytosol had generally similar physicochemical properties (including a molecular weight of about 70,000); probably they are the same protein. Such differences reported in the affinities of liver and tail cytosol to T_3 could thus be caused by a modifying factor influencing their binding to the thyroid hormone.

Evidence from mammalian cells shows that there is an increase in the number of nuclear receptors for T_3 after the exposure of intact cells (GH_1 cells of rat pituitary tumor line) in culture to T_3 (see Samuels and Tsai, 1973). High affinity low capacity binding sites for $[^{125}I]$-T_3 and $[^{125}I]$-T_4 were demonstrated, but only in nuclei, and the maximum binding capacity was identical for both hormones with about 5000 sites per nucleus. Their data suggest that T_3 and T_4 interact with identical nuclear receptors and that conversion of T_4 to T_3 may not be a prerequisite for biological activity. They concluded that receptors increase numerically and the change may be caused by the transfer of unstable cytosol receptors to the nucleus.

Because of the evidence of the involvement of prolactin in the anuran larval metamorphic cycle (see Section I), it is relevant to note that there are receptors binding ovine prolactin in microsomal preparations of tail, gill, and kidney tissues of *R. catesbeiana* larvae. Binding by larval and adult liver preparations and by the larval kidney before TK St.XVII is low or not detectable. However, during normal climax or that induced by thyroid hormones, prolactin binding increases in renal tissue, a feature according to White and Nicoll (1979) that may be related to the hormonal influence on hydromineral homeostasis shifting from the resorbing gills to the kidneys. Saturable binding sites for $[^{125}I]$ ovine prolactin were reported in particulate fractions of liver and tail fin tissue of larval *R. catesbeiana* by Carr and Jaffe (1980). Tail fin tissue showed a 10-fold superior binding capacity over liver tissue, a feature that may be related to the significant prolactin antagonism of T_4-induced tail regression during anuran metamorphosis. Trypsin and phospholipase C enzymes reduce specific binding of prolactin in tail and liver tissues, but neuroaminidase only reduces binding in liver tissue.

When judged by a variety of biological activities, for example, the initial rates of synthesis of nucleic acids, protein, and phospholipids, the time of PO_4 uptake and the induction of acid hydrolases, or in older embryos the characteristic recognizable morphological changes, it was found that embryos of *Xenopus* are unresponsive to exogenous T_3 and T_4 for about 40–60 h after fertilization. After this time competence to respond to these hormones is assumed fairly abruptly, which implies that cells acquire or have sufficiently differentiated specific cell receptors that respond to thyroid hormones to trigger off or initiate in some way the various cellular changes that ultimately occur. Probably all types of cells can respond directly to thyroid hormones (Tata, 1970, 1971). Furthermore, the specific developing cellular biochemistry or morphology probably arises because of a previous ontogenetic history *(vide supra)*, rather than from the result of a diversity of types of thyroid hormone receptor or binding units. After about 60 h (NF St.34–41 of *Xenopus*), cells that become sensitive to thyroid hormones show a binding capacity for radioactive T_3 and T_4 to increase, and simultaneously new cellular differentiation occurs. Beyond NF St.36–40 of *Xenopus* hormone binding is temperature-sensitive, and there is a marked increase in binding at higher temperatures (up to 25°C when compared with that at 5°C) (Tata, 1970). Griswold et al. (1972) concluded that there are two temperature-dependent processes, one transporting T_4 into a liver cell and the second transporting intracellular T_4 into the cell nucleus; the latter process is involved in the low temperature inhibition of T_4-induced metamorphosis. However, Toth and Tabachnick (1980) have shown in *R. catesbeiana* larvae that, though transfer of T_3 from the hepatocyte cytoplasm to its nucleus is slower at 5°C than at 25°C, generally at low and high dosages (200 and 300 pmol/g body wt) there was a high T_3 nuclear uptake at both temperatures. Thus some other mechanism other than a failure of T_3 nuclear binding would seem to be involved in low temperature inhibition of T_3-induced metamorphosis. Temperature sensitivity of hormone-binding sites in cells of amphibian embryos can be reconciled with earlier results on temperature inhibition of T_3- and T_4-induced metamorphic morphological and biochemical changes (Ashley et al., 1968; Fox, 1971), but the mechanism is obviously extremely complex and requires further investigation.

There is a wide range of effects initiated by thyroid hormones in vertebrates (Sterling, 1979; see Fig. 59) including amphibian larvae (Frieden, 1981). It has been suggested by Dratman (1978) that this could possibly be a result of the ability of an organism to produce a diverse group of metabolically relevant iodo compounds, each of which interact with different receptors engaged in specific cellular activities. Such diverse essential iodothyronines may well be amino acid analogs of tyrosine. However, whether such an array of thyroid hormone-type receptors does really exist seems unlikely; perhaps a limited number of them could well show modified responses to a variety of these active iodocompounds.

A Summary: Thyroid Hormones and Larval Metamorphosis

We can now attempt broadly and guardedly to summarize the current state of our knowledge on thyroid hormone action and the resulting cellular differentiation in amphibian development, especially anurans, which undergo dramatic morphological and biochemical changes during their early ontogeny.

Thyroid hormones, especially triiodothyronine, control metamorphosis and without them this process does not occur. Their rate of synthesis and secretion from the thyroid into the circulation and the level maintained there during larval development are determined by a variety of integrated and complex mechanisms. These embrace the activity of other endocrine organs, in particular the pituitary and the hypothalamus, and among other factors are the stage of larval development in relation to hormonal concentration, the type of plasma binding proteins and their affinity for thyroid hormones, temperature, and tissue avidity (often expressed as sensitivity), which probably increases during successive stages of metamorphosis and probably differs between different organs and tissues. The action of prolactin, though not well understood (acting at the peripheral tissue level and possibly depressing the thyroid also), may be a feature of importance and a thyroid hormone antagonist. Competence of cells to respond to thyroid hormones originates early in the development of young larvae of *Xenopus*, about 40–60 h after fertilization. Probably all tissues can respond morphologically and biochemically, but in a manner determined by the previous ontogenetic experience of the individual cells. Thyroid hormones influence cellular changes by first binding to receptor sites of either high affinity and relatively low capacity (specific and saturable), or of low affinity and high capacity (nonspecific or unsaturable), which are located in or at the surface of nuclei and probably also in the cytosol. There are claimed to be about 12,000 specific binding sites for T_3 in bullfrog larval liver cells; probably the number of binding receptor sites is steadily increased to a maximum during larval development. It is also likely that multiple binding sites for thyroid exist, for apart from the nucleus they occur in plasma membranes, microsomes, and inner membranes of mitochondria. Whether the two-step sequence of action of steroid hormones operates in the case of the thyroid hormones, whereby the steroid molecule first binds to a cytosol receptor, which thence by some conformational change is thereby translocated to the nucleus, is not yet confirmed, if it is applicable at all.

Thyroid hormonal stimulation of cells is ultimately expressed by the synthesis and regulation of new cellular proteins that are responsible for changes in structure and biochemical function. The hormones probably act initially (at least at the level of the nucleus), by altering the activity of RNA-polymerase, and by transcriptional changes leading to the replication of a variety of species of RNA, in particular nucleolar ribosomal RNA. Probably the type of cellular differentiation elaborated is basically determined by different species of mRNA coding for specific proteins, but presumably the mechanism requires a large increase in rRNA originating in nucleolar organzers and finally located in the polysomes of the RER, which is seen particularly well in hepatocytes. Perhaps the uptake of amino acids by tRNA is also enhanced by thyroid hormones at the level of translation, which likewise may well be sustained by them. Whether batteries of polyribosomes on the RER are spatially of different nature, are replaced sequentially in space and time, and have different specificities in terms of protein synthesis, is not known. Such important questions are for future solution. Cellular regression can be explained in the same way as cellular differentiation and likewise it operates via new enzyme synthesis, though these are catabolic in nature; probably this process is activated in part by *de novo* synthesis of lysosomal enzymes, but possibly it also involves extralysosomal proteases.

In amphibians (as with other vertebrates) the nature of thyroid (or other) hormone intracellular binding receptors, and their spatial and functional relationship to the hormone and to the nuclear genes (or extranuclear components) are not clearly understood. Thyroid hormones do, however, bind to proteins. These problems are fundamental to our understanding of the nature of hormone-induced cellular differentiation. Receptor sites may well be proteins or protein complexes, which when associated spatially with the hormone are activated to initiate DNA and RNA activity at the requisite cell region, or perhaps regulate other processes such as permeability changes in cell membranes, permitting molecular and ion movements to occur, necessary for specific cellular differentiation.

In principle the concept of hormonal activation of cellular differentiation throughout larval ontogeny may be extended to the action of inducing substances that are present in inducer cells and act on competent responsive target cells during embryogeny. From the earliest successive stages of cleavage of the blastula through gastrulation and neurulation, and so on, different groups of cells, by virtue of their topographic location, are fundamentally different from one another in their cell content. Such intimate spatial relationships of different tissue groupings, which ultimately materialize, mutually influence their developmental fate. Cell contact is not essential for induction to occur, and it is likely that specific substances diffuse from inducer to responsive cells to initiate cellular differentiation. In vertebrates there is little reliable information on the chemistry and mechanism of action of inducer substances on their target cells and current ideas are purely speculative. Presumably they act via membrane or intracellular receptor sites and, like hormones, ultimately stimulate changes in gene expression.

33. Epilog

The preceding account of amphibian metamorphosis began by considering the living adult animals, their larval development, and their phylogenetic origins. Thence gross external development of larvae, their organocellular structure, and the ontogenetic changes that occur in them during metamorphosis were considered and finally we arrived at the subcellular components and their activities, down to the level of molecular structure when organelle self-assembly is initiated, primarily controlled at the level of the nuclear genes; such gene expression is influenced by hormones (and possibly inducer substances particularly during embryogeny), at specific cellular receptor sites in target cells.

According to one's preferences, therefore, big or small is beautiful, for an amphibian, like all other animals, clearly is the sum of its parts, a creature whose functional whole reflects the participation throughout life of all the individual cells. An adult frog, its larva, or the zygote are in reality only different developmental stages of a single individual, basically a genetic 'readout' or blueprint, that has the potential to develop in an adequate ecological niche; epigenesis proceeds in space and time, and is expressed when subject to the pressures of natural selection.

Doubtless, crossopterygians had cells with ultrastructural components such as mitochondria, ribosomes, and hormone receptor sites, like the modern amphibians. It is worth remarking, however, that any real details of most if not all subcellular structures have only been known since the advent of the electron microscope, about 40 years ago, a period of time, when placed in true perspective, that is infinitesimal compared with the several million years since primitive humans appeared, and the 350 million years when the first amphibians trod the earth. Obviously we cannot be complacent and assume that anything like a comprehensive picture of subcellular structure, and especially organelle biosynthetic activity, has yet been achieved, notwithstanding the extensive discoveries during the past few decades. But so much is happening so quickly, as physics and chemistry are unified with cellular biology, that we can face the future with high hopes that a much clearer view of the basic structure and mechanisms of cellular growth, differentiation, and organelle molecular interrelationships will arise, perhaps sooner than once seemed possible.

The succeeding generations of biologists, with the new specialized techniques available, should have a rewarding and exciting future; their predecessors surely will appear so "unworldly" in their earlier understanding of basic biological processes.

References

References

Adler, W. 1901. *Int. Monatschr. Anat. Physiol. Leipzig* **18**, 19–42.

Agarwal, S. K., and I. A. Niazi. 1977. *Proc. Natl. Acad. Sci. India B* **47**, 79–92.

Alberts, B., Bray, D., Lewis, J., Raff, M., Roberts, K., and J. D. Watson. 1983. *Molecular Biology of the Cell.* Garland Publ. Co. New York.

Alcala, A. C. 1962. *Copeia* No. 4, 679–726.

Allen, B. M. 1916. *Q. Rev. Biol.* **4**, 325–372.

Allen, B. M. 1918. *Anat. Rec.* **15**, Suppl. 352–353.

Allen, B. M. 1930. *Univ. Calif. Publ. Zool.* **31**, 53–78.

Allen, E. R. 1978. *Am. Zool.* **18**, 101–111.

Altig, R. 1970. *Herpetologica* **26**, 180–207.

Alvarado, R. H., and A. Moody. 1970 *Am. J. Physiol.* **218**, 1510–1516.

Anderson, E., and J. J. Kollros. 1962. *J. Ultrastruct. Res.* **6**, 35–56.

Anderson, I. G., Briggs, T., Haslewood, G. A. D., Oldham, R. S., Scharen, H., and L. Tokes. 1979. *Biochem. J.* **183**, 507–512.

Anderson, J. D. 1961. *Copeia* No. 4. 371–376.

Anderson, J. D., and P. J. Martino. 1966. *Am. Midl. Nat.* **75**, 257–279.

Anderson, J. D., and R.D. Worthington. 1971. *Herpetologica* **27**, 165–176.

Anderson, J. D., and R. G. Webb. 1978. *J. Herpetol.* **12**, 89–93.

Anderson, P. L. 1943. *Anat. Rec.* **86**, 59–73.

Andres, K. H. 1974. *Symposium on Mechanoreception.* In *Abh. Rhein. Westf. Akad. Wiss.* **53**, 135–152.

Angel, F., and M. Lamotte. 1944. *Ann. Sci. Nat. Zool. Biol. Anim. Ser. II*, **6**, 63–89.

Angelier, E., and M. L. Angelier. 1968. *Ann. Limnol.* **4**, 113–131.

Annandale, N. 1918a. *Rec. Indian Mus.* **15**, 17–23.

Annandale, N. 1918b. *Rec. Indian Mus.* **15**, 59–65.

Annandale, N., and C. R. N. Rao. 1918. *Rec. Indian Mus.* **15**, 25–40.

Anstis, M. 1976. *Trans. R. Soc. S. Aust.* **100**, 192–202.

Archey, G. 1922. *Rec. Canterbury Mus.* **2**, 59–71.

Aron, M. 1925. *C. R. Seances Soc. Biol. Paris* **93**, 678–800.

Aronsson, S. 1976. *Cell. Tissue Res.* **171**, 437–448.

Aronsson, S. 1978. *Gen. Comp. Endocrinol.* **36**, 497–501.

Aronsson, S., and A. Enemar. 1981. *J. Comp. Neurol.* **200**, 315–321.

Asahi, K. I., Born, J., Tiedemann, H., and H. Tiedemann. 1979. *Dev. Biol.* **187**, 231–244.

Ashley, H., and E. Frieden. 1972. *Gen. Comp. Endocrinol.* **8**, 22–31.

Ashley, H., Katti, P., and E. Frieden. 1968. *Dev. Biol.* **17**, 293–307.

Ashton, R. E., and A. L. Braswell. 1979. *Brimleyana* No. 1, 15–22.

Ashworth, J. M. 1973. *Cell Differentiation.* Chapman and Hall. London.

Atkinson, B. G. 1981. In *Metamorphosis: A Problem in Developmental Biology* (eds. L. I. Gilbert and E. Frieden). 2nd. Ed. 397–444. Plenum Press. New York.

Atkinson, B. G., Atkinson, K. H., Just, J. J., and E. Frieden. 1972. *Dev. Biol.* **29,** 162–175.

Atkinson, B. G., and J. J. Just. 1975. *Dev. Biol.* **45,** 151–165.

Atkinson, B. G., and G. H. Little. 1972. *Mech. Age. Dev.* **1,** 299–312.

Baccaglini, P. I., and N. C. Spitzer. 1977. *J. Physiol. (London),* **271,** 93–117.

Bacher, B. E. 1973. *J. Exp. Zool.* **185,** 209–216.

Badenhorst, A. 1978. *Ann. Univ. Stellenbosch Ser.* A2 *(Zoo.),* No. 1, 1–26.

Baffoni, G. M., and G. Catte. 1950. *Rend. Accad. Naz. Lincei.* Ser. 8. **9,** 282–287.

Baffoni, G. M., and G. Catte. 1951. *Riv. Biol.* **43,** 373–397.

Bagnara, J. T. 1964. *Gen. Comp. Endocrinol.* **4,** 299–303.

Bagnara, J. T. 1976. In *Physiology of the Amphibia* (ed. B. Lofts). **3,** 1–52. Academic Press. New York.

Bagnara, J. T., and M. E. Hadley. 1969. *Am. Zool.* **9,** 465–478.

Bagnara, J. T., and M. E. Hadley. 1973. *Chromatophores and Color Change.* Prentice-Hall, Englewood Cliffs, New Jersey.

Bagnara, J. T., Ferris, W., Turner, W. A., and J. D. Taylor. 1978. *Dev. Biol.* **64,** 149–163.

Bagnara, J. T., Matsumoto, J., Ferris, W., Frost, S. K., Turner, W. A., Tchen, T. T., and J. D. Taylor. 1979. *Science* **203,** 410–415.

Baker, P. C., and T. E. Schroeder. 1967. *Dev. Biol.* **15,** 432–450.

Balinsky, B. I. 1970. *An Introduction to Embryology.* W. B. Saunders. Philadelphia.

Baloukhère, M., Thomas, C., Heilporn-Pohl, V., Hanocq, F., and J. Brachet. 1980. *Exp. Cell. Res.* **130,** 291–295.

Baltus, E., Hanocq-Quertier, J., and J. Brachet. 1968. *Proc. Natl. Acad. Sci.* **61,** 419–478.

Barch, S. H. 1953. *Physiol. Zool.* **26,** 223–231.

Barrington, E. J. W. 1946. *Proc. Zool. Soc. London* **116,** 1–21.

Barrio, A. 1976. *Physis. Secc. C. Buenos Aires* **36,** 337–344.

Bartels, W. 1971. *Z. Zellforsch. Mikrosk. Anat.* **116,** 94–118.

Bayard Green, N. 1938. *Copeia* No. 2, 79–82.

Bayard Green, N. 1952. *Diss. Abst.* **17** (1957), Ann Arbor, Michigan.

Bayard Green, N. 1964. *Acc. WV Acad. Sci.* **36,** 34–38.

Beaudoin, A. R. 1954. *Anat. Rec.* **118,** 420.

Beaudoin, A. R. 1956. *Anat. Rec.* **125,** 247–259.

Beaumont, A. 1953a. *C. R. Seances Soc. Biol. Paris* **147,** 56–58.

Beaumont, A. 1953b. *Arch. Anat. Microsc. Morphol. Exp.* **42,** 32–40.

Beaumont, A. 1970. *C. R. Seances Acad. Sci. Paris* **271,** II04–II06.

Beaumont, A. 1977. In *Mècanismes de la Rudimentation des Organes chez les Embryons de Vertébrés* (ed. A. A. Raynaud). Colloq. Int. CNRS No. 266, 113–124.

de Beer, G. R. 1937. *The Development of the Vertebrate Skull.* Oxford University Press.

Belenky, M. A., and V. K. Chetverukhin. 1973. *Dokl. Akad. Nauk. SSSR. Ser. Biol.* **209,** 1232–1234.

Belenky, M. A., Chetverukhin, V. K., and A. L. Polenov. 1973a. *Gen. Comp. Endocrinol.* **21,** 241–249.

Belenky, M. A., Chetverukhin, V. K., and A. L. Polenov. 1973b. *Gen. Comp. Endocrinol.* **21**, 250–261.

Bell, E. T. 1907. *Anat. Anz.* **31**, 283–291.

Benbassat, J. 1970. *Dev. Biol.* **21**, 557–583.

Benbassat, J. 1974a. *J. Cell Sci.* **15**, 347–357.

Benbassat, J. 1974b. *J. Cell Sci.* **16**, 143–156.

Bennett, M. R., and K. Lai. 1981a. *Dev. Biol.* **86**, 212–223.

Bennett, M. R., and K. Lai. 1981b. *Dev. Biol.* **86**, 224–226.

Bennett, T. P., and J. S. Glenn. 1970. *Dev. Biol.* **22**, 535–560.

Bennett, T. P., Glenn, J. S., and H. Sheldon. 1970. *Dev. Biol.* **22**, 232–248.

Benson, D. G. 1972. *Gen. Comp. Endocrinol.* **19**, 129–132.

Bentley, P. J. 1976. *Comparative Vertebrate Endocrinology*. Cambridge University Press.

Bentley, P. J., and G. F. Baldwin. 1980. *Am. J. Physiol.* **239**, R505–508.

di Berardino, M. A., and N. Hoffner. 1970. *Dev. Biol.* **23**, 185–209.

di Berardino, M. A., and T. J. King. 1967. *Dev. Biol.* **15**, 102–128.

Berman, R., Bern, H. A., Nicoll, C. R., and R. C. Strohmann. 1964. *J. Exp. Zool.* **156**, 353–360.

Berns, M. W., and K. S. Narayan. 1970. *J. Morphol.* **132**, 169–180.

Bibb, H. D. 1977. *J. Exp. Zool.* **200**, 265–275.

Bibb, H. D. 1978. *J. Exp. Zool.* **206**, 65–72.

Billet, F. A., and R. P. Gould. 1971. *J. Anat.* **108**, 465–480.

Billings, S. M. 1972. *Z. Anat. Entwicklungsgesch.* **136**, 168–191.

Bird, J. W. C. 1975. In: *Lysosomes in Biology and Pathology* (eds. J. T. Dingle and R. T. Dean). **4**, 75–109. North Holland. Amsterdam.

Birnstiel, M. L., Chipchase, M., and J. Speirs. 1971. In *Progress in Nucleic Acid Research and Molecular Biology* (eds. J. N. Davidson and W. E. Cohn). II, 351–389. Academic Press. New York.

Blackler, A. W. 1958. *J. Embryol. Exp. Morph.* **6**, 491–503.

Blackshaw, S. E., and A. E. Warner. 1976a. *Nature (London)* **262**, 217–218.

Blackshaw, S. E., and A. E. Warner. 1976b. *J. Physiol. (London)* **255**, 209–230.

Blair, A. P. 1964. *Copeia* No. 4, 499.

Blake, C. C. F. 1978. *Endeavour* N. S. **2**, 137–141.

Blake, C. C. F., and S. J. Oatley. 1977. *Nature (London)* **268**, 115–120.

Blatt, L. M., Slickers, K. A., and K. H. Kim. 1969. *Endocrinology* **85**, 1213–1215.

Bles, E. J. 1905. *Trans. R. Soc. Edinburgh* **41**, 789–821.

Blommers-Schlösser, R. M. A. 1975a. *Beaufortia* **24**, No. 309, 7–26.

Blommers-Schlösser, R. M. A. 1975b. *Beaufortia* **23**, No. 296, 15–25.

Blommers-Schlösser, R. M. A. 1979a. *Bijdr. Dierk.* **49**, 261–312.

Blommers-Schlösser, R. M. A. 1979b. *Beaufortia* **29**, No. 352. 1–77.

Blommers-Schlösser, R. M. A. 1982. *Beaufortia* **32**, No. I–II.

Bennett, G. W., Balls, M., Clothier, R. H., Marsden, C. A., Robinson, G., and G. D. Wemyss-Holden. 1981. *Cell Biol. Int. Rpt.* **5**, 151–158.

Boell, E. J., Greenfield, P., and B. Hille. 1963 *Dev. Biol.* **7**, 420–431.

Bollin, E., Carlson, C. A., and K. H. Kim. 1973. *Dev. Biol.* **31**, 185–194.

Boloukhere, M., Thomas, C., Heilporn-Pohl, V., Hanocq, F., and J. Brachet. 1980. *Exp. Cell Res.* **130**, 291–296.

Bonner, J. 1965. *The Molecular Biology of Development*. Oxford University Press.

Bonneville, M. A. 1963. *J. Cell. Biol.* **18**, 579–597.

Bonneville, M. A., and M. Weinstock. 1970. *J. Cell. Biol.* **44**, 151–171.

Bordzilovskaya, N. P., and T. A. Dettlaff. 1979. *Axolotl Newslett.* **7**, 2–22.
Boschwitz, D. 1960a. *Herpetologica* **16**, 91–100.
Boschwitz, D. 1960b. *J. Embryol. Exp. Morph.* **8**, 425–436.
Boschwitz, D. 1961. *Herpetologica* **17**, 192–199.
Boschwitz, D. 1965. *Israel J. Zool.* **14**, 11–23.
Boschwitz, D. 1967. *Israel J. Zool.* **16**, 46–48.
Boschwitz, D. 1969. *Israel J. Zool.* **18**, 277–289.
Boschwitz, D., and H. A. Bern. 1971. *Gen. Comp. Endocrinol.* **17**, 586–588.
Botte, V., and C. Buanano. 1962. *Boll. Zool.* **29**, 471–478.
Boucaut, J. C., Bernard B., Aubery, M., Bourrillon, R., and C. Houillon. 1979. *J. Embryol. Exp. Morph.* **51**, 63–72.
Bovbjerg, A. M. 1963. *J. Morphol.* **113**, 231–243.
Bragg, A. N. 1936. *Copeia* No. 1, 14–20.
Bragg, A. N. 1966. *Wasmann J. Biol.* **24**, 71–73.
Branham, A. E., and J. C. List. 1979. *J. Morphol.* **159**, 311–330.
Bravo, R., and J. Knowland. 1979. *Differentiation* **13**, 101–108.
Bretscher, A. 1949. *Rev. Suisse Zool.* **56**, 34–96.
Briggs, R., and T. J. King. 1952. *Proc. Natl. Acad. Sci.* **38**, 455–463.
Briggs, R., and T. J. King. 1953. *J. Exp. Zool.* **122**, 485–505.
Brink, H. E. 1939. *Proc. Linn. Soc. London* **151**, 120–125.
Bristow, D. A., and E. M. Deuchar. 1964. *Exp. Cell. Res.* **35**, 580–589.
Britten, R. J., and E. H. Davidson. 1969. *Science* **165**, 349–357.
Brock, H. W., and R. Reeves. 1978. *Dev. Biol.* **66**, 128–141.
Brown, A. C., and P. F. Millington. 1968. *Histochemie* **12**, 83–94.
Brown, D., Grosso, A., and R. C. de Sousa. 1981. *J. Cell Sci.* **52**, 197–213.
Brown, D. D. 1966. *Nat. Cancer Inst. Monogr.* No. 23, 293–309.
Brown, D. D., and E. Littna. 1964. *J. Mol. Biol.* **8**, 669–687.
Brown, D. D., and E. Littna. 1966. *J. Mol. Biol.* **20**, 95–112.
Brown, D., D., and J. B. Gurdon. 1964. *Proc. Natl. Acad. Sci.* **51**, 139–146.
Brown, H. A. 1975a. *Comp. Biochem. Physiol.* **50A**, 397–405.
Brown, H. A. 1975b. *Northwest Sci.* **49**, 241–252.
Brown, H. A. 1975c. *Comp. Biochem. Physiol.* **51A**, 863–873.
Brown, H. A. 1976. *Can. J. Zool.* **54**, 552–558.
Brown, M. E. 1946. *Am. J. Anat.* **78**, 79–113.
Brown, P. S., and B. E. Frye. 1969. *Gen. Comp. Endocrinol.* **13**, 126–138.
Broyles, R. H. 1981. In *Metamorphosis: A Problem in Developmental Biology* (eds. L. I. Gilbert and E. Frieden). 2nd. Ed. 461–490.
Broyles, R. H., Johnson, G. M., Maples, P. B., and G. R. Kindell. 1981. *Dev. Biol.* **81**, 299–314.
Bruce, R. C. 1970. *Copeia* No. 4, 776–779.
Bruce, R. C. 1971. *Copeia* No. 2, 234–246.
Bruce, R. C. 1972. *J. Herpetol.* **6**, 43–51.
Bruce, R. C. 1974. *Am. Midl. Nat.* **92**, 173–190.
Bruce, R. C. 1982. *Copeia* No. 1, 117–127.
Brun, R., and H. R. Kobel. 1972. *Rev. Suisse Zool.* **79**, 961–965.
Bruneau, M., and E. Magnin. 1980a. *Can. J. Zool.* **58**, 169–174.
Bruneau, M., and E. Magnin. 1980b. *Can. J. Zool.* **58**, 175–183.
Bruns, R. R., and J. Gross. 1970. *Am. J. Anat.* **128**, 193–224.
Brustis, J. J. 1979. *Arch. Biol.* **90**, 261–272.
Buchanan, J., Primhack, M. P., and D. E. Tapley. 1970. *Endocrinology* **87**, 993–999.
Budtz, P. E. 1979. *J. Zool.* **189**, 57–92.

Burgen, A. S. V., Feeney, J., and G. C. K. Roberts. 1975. *Nature (London)* **253**, 753–755.

Burnside, B. 1971. *Dev. Biol.* **26**, 416–441.

Busak, S. D., and G. R. Zug. 1976. *Herpetologica* **32**, 130–137.

Busolati, G., and A. G. E. Pearse. 1967. *J. Endocrinol.* **37**, 205–209.

Cambar, R. 1947. *C. R. Seances Soc. Biol. Paris* **141**, 752–754.

Cambar, R., and J. D. Gipouloux. 1956. *Bull. Biol. Fr. Belg.* **90**, 198–217.

Cambar, R., and B. Marrot. 1954. *Bull. Biol. Fr. Belg.* **88**, 168–177.

Cambar, R., and S. Martin. 1959. *Actes Soc. Linn. Bordeaux* **98**, 1–20.

Campbell, A. M., Corrance, M. H., Davidson, J. N., and H. M. Kerr. 1969. *Proc. R. Soc. Edinburgh* Sect. A. **70**, 295–310.

Campantico, E., Giunta, C., Guardabassi, A., and M. Vietti. 1970. *Gen. Comp. Endocrinol.* **18**, 396–399.

Caramaschi, U. 1979. *Rev. Bras. Biol.* **39**, 169–171.

Caramaschi, U., Jim, J., and C. M. de Carvalho. 1980. *Rev. Bras. Biol.* **40**, 405–408.

Carpenter, K. L., and J. B. Turpen. 1979. *Differentiation* **14**, 167–174.

Carr, F. E., and R. C. Jaffe. 1980. *Mol. Cell. Endocrinol.* **17**, 145–155.

Carr, F. E., Jacobs, P. J., and R. C. Jaffe. 1981. *Mol. Cell. Endocrinol.* **23**, 65–76.

Carroll, E. J., and J. L. Hedrick. 1974. *Dev. Biol.* **38**, 1–13.

Carver, V. H., and E. Frieden. 1977. *Gen. Comp. Endocrinol.* **31**, 202–207.

Cei, J. M. 1968. *Herpetologica* **24**, 141–146.

Chacko, T. 1965a. *Acta Zool. (Stockholm)* **46**, 83–108.

Chacko, T. 1965b. *Acta Zool. (Stockholm)* **46**, 311–328.

Champy, C. 1922. *Arch. Morph. Gen. Exp.* **4**, 1–58.

Channing, A. 1972. *Ann. Natal Mus.* **21**, 509–511.

Channing, A. 1973. *Zool. Afr.* **8**, 153–156.

Channing, A. 1976a. *Arnoldia Misc. Publi. (Rhodesia)* **8**, 1–4.

Channing, A. 1976b. *Zool. Afr.* **11**, 299–312.

Channing, A. 1978. *Herpetologica* **34**, 394–397.

Channing, A., and D. E. van Dyke. 1976. *A Guide to the Frogs of South West Africa*. Univ. Durban. Westville Press.

Chapman, D. 1975. In *Cell Membranes: Biochemistry, Cell Physiology and Pathology* (eds. G. Weissmann and R. Claiborne). 13–22. H. P. Publ. Co. New York.

Chapman, D., and D. F. H. Wallach. 1968, 1973, 1976. *Biological Membranes*. Vols. 1–3, Academic Press. London.

Chapman, G. B., and A. B. Dawson. 1961. *J. Biophys. Biochem. Cytol.* **10**, 425–435.

Cherry, R. J. 1976. In *Biological Membranes* (eds. D. Chapman and D. F. H. Wallach). Vol. 3, 47–102. Academic Press. London.

Choi, J. K. 1963. *J. Cell. Biol.* **16**, 53–72.

Chopra, D. P., and J. D. Simnett. 1969a. *J. Embryol. Exp. Morph.* **21**, 539–548.

Chopra, D. P., and J. D. Simnett. 1969b. *Exp. Cell. Res.* **58**, 319–322.

Chopra, D. P., and J. D. Simnett. 1970. *J. Embryol. Exp. Morph.* **24**, 525–533.

Chopra, D. P., and J. D. Simnett. 1971. *J. Embryol. Exp. Morph.* **25**, 321–329.

Chou, H. T. I., and J. J. Kollros. 1974. *Gen. Comp. Endocrinol.* **22**, 255–260.

Christensen, A. K. 1964. *Am. J. Anat.* **115**, 257–278.

Chu-Wang, I. W., Oppenheim, R. W., and P. B. Farel. 1981. *Brain Res.* **213**, 307–318.

Claes, H. 1964. *Acta Anat.* **59**, 229–281.

Claes, H. 1965. *Acta Anat.* **62**, 104–156.

Clergue-Gazeau, M., and R. Thorn. 1976. *Ann. Spéléol.* **36,** 169–174.

Cleveland, D., and M. Kirschner. 1982. In *Cold Spring Harbor Symposium* **46,** 171–183.

Clothier, R. H., Balls, M., Water, A. D., Marsden, C. A., and G. W. Bennett. 1983. In *Thyrotropin-Releasing Hormones* (eds. E. C. Griffiths and G. W. Bennett). 203–215. Raven Press, New York.

Cochran, D. M. 1972. *Living Amphibians of the World.* Hanover House, New York (Hamish Hamilton, London).

Cochran, S. L., Hackett, J. H., and D. L. Brown. 1980. *Neuroscience* **5,** 1629–1646.

Coghill, G. E. 1914. *J. Comp. Neurol.* **24,** 161–234.

Cohen, P. P. 1970. *Science* **168,** 533–543.

Cohen, P. P., Brucker, R. F., and S. M. Morris. 1978. In *Hormonal Proteins and Peptides* (ed. C. H. Li). Vol. 6, 273–381. Academic Press. New York.

Cohen, W. D. 1982. *Protoplasma* II3, 23–32.

Coleman, R. 1970. In *Calcitonin: Proc. 2'nd. Int. Symp.* (ed. S. Taylor). 348–358. W. Heineman. London.

Coleman, R. 1975. *Cell. Tissue Res.* **164,** 215–232.

Coleman, R., Evennett, P. J., and J. M. Dodd. 1967. *Histochemie* **10,** 33–43.

Coleman, R., Evennett, P. J., and J. M. Dodd. 1968. *Gen. Comp. Endocrinol.* **10,** 34–46.

Coleman, R., and A. D. Phillips. 1974. *Cell. Tissue Res.* **148,** 69–82.

Collins, J. M. 1972. *Biochemistry* **11,** 1259–1264.

Collins, S. 1961. MS. Thesis, Univ. Arizona. Quoted in Bagnara (1976).

Colombo, R. 1982. *Experientia* **38,** 408–409.

del Conte, E., and J. L. Sirlin. 1952. *Anat. Rec.* **112,** 125–135.

Cooper, E. L. 1967. *J. Morphol.* **122,** 381–397.

Cooper, E. L. 1976. In *Physiology of the Amphibia* (ed. B. Lofts). Vol. 3, 163–272. Academic Press. New York.

Cooper, E. L., Brown, B. A., and B. S. Baculi. 1971. In *Morphological and Functional Aspects of Immunity* (eds. K. L. Lindahl-Kiessling, G. Alm and M. Hanna). 1–10. Plenum Press. New York.

Copp, D. H. 1969. *J. Endocrinol.* **43,** 137–161.

Cortelyou, J. R. 1967. *Gen. Comp. Endocrinol.* **9,** 234–240.

Crim, J. W. 1975. *J. Exp. Zool.* **192,** 355–362.

da Cruz, C. A. G., and O. L. Peixoto. 1978. *Rev. Bras. Biol.* **38,** 297–299.

Cuatrecasas, P. 1972. *Proc. Natl. Acad. Sci.* **69,** 1277–1281.

Cuatrecasas, P. 1974. *Annu. Rev. Biochem.* **43,** 169–214.

Curtis, A. S. G. 1957. *Proc. R. Phys. Soc. Edinburgh Sect. B.* **26,** 25–32.

Czolowska, R. 1969. *J. Embryol. Exp. Morph.* **22,** 229–251.

Dabagyan, N. V., and L. A. Sleptsova. 1975. *Herbaceous (Grass) Frogs, Rana temporaria. Objects of Biological Development.* M. Nauka. pp. 442–462. (In Russian.)

Dacke, C. G. 1979. *Calcium Regulation in Sub-Mammalian Vertebrates.* Academic Press. London.

Dalcq, A. M. 1938. *Form and Causality in Early Development.* Cambridge University Press.

Danielli, J. F. 1975. In *Cell Membranes: Biochemistry, Cell Biology and Pathology* (eds. G. Weissmann and R. Claiborne). 1–11. H. P. Publ. Co. New York.

Davies, S. N., Kitson, D. L., and A. Roberts. 1982. *J. Embryol. Exp. Morph.* **70**, 215–224.

Davis, B. P., Jeffrey, J. J., Eisen, A. Z., and A. Derby. 1975. *Dev. Biol.* **44**, 217–222.

David, I. B. 1966. *Proc. Natl. Acad. Sci.* **56**, 269–276.

Davson, H., and J. F. Danielli. 1943. *The Permeability of Natural Cell Membranes.* Cambridge University Press.

Dawid, T. B. 1966. *Proc. Natl. Acad. Sci.* **56**, 269–276.

De Bernardi, F. 1982. *Exp. Cell. Biol.* **50**, 281–290.

Decker, R. S. 1974a. *Dev. Biol.* **41**, 146–161.

Decker, R. S. 1974b. *J. Cell. Biol.* **61**, 599–612.

Decker, R. S. 1976. *Dev. Biol.* **49**, 101–118.

Decker, R. S. 1978. *J. Comp. Neurol.* **180**, 635–660.

Decroly, M., Goldfinger, M., and N. Six-Tondeur. 1979. *Biochim. Biophys. Acta* **587**, 567–578.

Dempster, W. T. 1933. *J. Exp. Zool.* **64**, 495–511.

Denis, H. 1974. In *Chemical Zoology* (eds. M. Florkin and B. T. Scheer). **9**, 3–22. Academic Press. New York.

Dent, J. N. 1942. *J. Morphol.* **71**, 577–601.

Dent, J. N. 1956. *Copeia* No. 4, 207–210.

Dent, J. N. 1968. In *Metamorphosis: A Problem in Developmental Biology* (eds. W. Etkin and L. I. Gilbert). 271–311. Appleton-Century-Crofts. New York.

Derby, A. 1968. *J. Exp. Zool.* **168**, 147–156.

Derby, A., Jeffrey, J. J., and A. Z. Eisen. 1979. *J. Exp. Zool.* **207**, 391–398.

Deuchar, E. M. 1966. *Biochemical Aspects of Amphibian Development.* Methuen. London.

Deuchar, E. M. 1975. *Xenopus: The South African Clawed Frog.* J. Wiley. England.

Dhanarajan, Z. C., and B. G. Atkinson. 1981. *Dev. Biol.* **82**, 317–328.

Dijkgraaf, S. 1963. *Biol. Rev.* **38**, 51–105.

Dmytrenko, G. M., and G. S. Kirby, 1981. *J. Exp. Zool.* **215**, 179–182.

Dodd, J. M. 1950. *Nature (London)* **165**, 283.

Dodd, M. H. I., and J. M. Dodd. 1976. In *Physiology of the Amphibia* (ed. B. Lofts). **3**, 467–599. Academic Press. London.

Doerr-Schott, J. 1968. *Z. Zellforsch. Mikrosk. Anat.* **90**, 616–645.

Dorn, A. R., and R. H. Broyles. 1982. *Proc. Natl. Acad. Sci.* **79**, 5592–5596.

Douarin, N. 1966. *Ann. Biol.* **5**, 105–171.

Douglas, W. H. J., Ripley, R. C., and R. A. Ellis. 1970. *J. Cell. Biol.* **44**, 211–215.

Dournon, C., and P. Chibon. 1974. *Arch. Entwicklungsmech. Org.* **175**, 27–47.

Dowling, J. T., and D. Razevska. 1966. *Gen. Comp. Endocrinol.* **6**, 162–169.

Dratman, M. B. 1978. In *Hormonal Proteins and Peptides* (ed. C. H. Li). **6**, 205–271. Academic Press. New York.

Driesch, H. 1891. *Z. Wiss. Zool.* **53**, 160–182.

Ducibella, T. 1974a. *Dev. Biol.* **38**, 175–186.

Ducibella, T. 1974b. *Dev. Biol.* **38**, 187–194.

Duellman, W. E. 1969. *Herpetologica* **25**, 241–247.

Duellman, W. E., and J. D. Lynch. 1969. *Herpetologica* **25**, 231–240.

Duellman, W. E., and S. C. Maness. 1980. *J. Herpetol.* **14**, 213–222.

Dunlap, D. G. 1966. *J. Morphol.* **119**, 241–258.

Dunn, E. R. 1915. *Copeia* No. 21, 28–30.

Duprat, A. M., Gualandris, L., and P. Rouge. 1982. *J. Embryol. Exp. Morph.* **70**, 171–187.

Durand, J. P. 1971. Thesis: Faculté des Sciences, Université Paul Sabatier de Toulouse. France.

Durban, E., and W. K. Paik. 1976. *Biochim. Biophys. Acta* **437**, 175–189.

Dushane, G. P. 1938. *J. Exp. Zool.* **78**, 485–503.

Eakin, R. M. 1947. *Univ. Calif. Publ. Zool.* **51**, 245–287.

Eakin, R. M. 1963. *Dev. Biol.* **7**, 169–179.

Eakin, R. M. 1964. *Z. Zellforsch. Mikrosk. Anat.* **63**, 81–96.

Eakin, R. M., and F. E. Bush. 1957. *Anat. Rec.* **129**, 279–295.

Eaton, J. E., and E. Frieden. 1969. *Gen. Comp. Endocrinol.* Suppl. 2, 398–407.

Eaton, T. H., and R. M. Imagawa. 1948. *Copeia* No. 4, 263–266.

Eberth, C. J. 1866. *Arch. Mikrosk. Anat. Entwicklungsmech.* **2**, 490–506.

Edds, M. V., and P. R. Sweeney. 1961. In *Synthesis of Molecular and Cellular Structure* (ed. D. Rudnick). III–138. Ronald Press.

Eddy, L., and H. Lipner. 1975. *Gen. Comp. Endocrinol.* **25**, 462–465.

Eeckhout, Y. 1969. *Mém. Acad. R. Med. Belg.* **38**, 1–113.

Eichelberg, H., and H. Schneider. 1973. *Z. Zellforsch. Mikrosk. Anat.* **141**, 223–233.

Eichler, V. B., and R. A. Porter. 1981. *J. Comp. Neurol.* **203**, 121–130.

Elias, H., and J. C. Sherrick. 1969. *Morphology of the Liver*. pp. 192–195. Academic Press. New York.

Emmett, A. D., and F. P. Allen. 1919. *J. Biol. Chem.* **38**, 325–344.

Enemar, A. 1978. *Gen. Comp. Endocrinol.* **34**, 211–218.

Enemar, A., Essvik, B., and R. Klang. 1968. *Gen. Comp. Endocrinol.* **11**, 328–331.

Epperlein, H. H., and M. Junginger. 1982. *Amphibia-Reptilia* **2**, 295–308.

Epple, A. 1966. *Gen. Comp. Endocrinol.* **7**, 207–214.

Ergezen, S., and A. Gorbman. 1977. *Gen. Comp. Endocrinol.* **31**, 492–494.

Etkin, W. 1963. *Science* **139**, 810–814.

Etkin, W. 1964. In *Physiology of the Amphibia* (ed. J. A. Moore). **1**, 427–468. Academic Press. New York.

Etkin, W. 1965. *J. Morphol.* **116**, 371–378.

Etkin, W. 1968. In *Metamorphosis: A Problem in Developmental Biology* (eds. W. Etkin and L. I. Gilbert). 313–348. Appleton-Century-Crofts. New York.

Etkin, W. 1970. In *Hormones and the Environment* (eds. G. K. Benson and J. G. Phillips). *Mem. Soc. Endocrinol.* No. 18, 137–155. Cambridge University Press.

Etkin, W., and A. G. Gona. 1967. *J. Exp. Zool.* **165**, 249–258.

Etkin, W., and A. G. Gona. 1974. In *Handbook of Physiology-Endocrinology*. **3**, 5–20.

Etkin, W., and R. Lehrer. 1960. *Endocrinology* **67**, 457–466.

Ewer, D. W. 1959. *Nature (London)* **183**, 271.

Exbrayat, J. M., Delsol, M., and J. Flatin. 1981. *C. R. Seances Acad. Sci. Paris* **292**, 417–420

Eycleshymer, A. C., and J. M. Wilson. 1910. In *Keibel's Normentafeln zur Entwicklungsgeschichte der Wirbeltiere*. **11**, 1–50. Gustav Fischer. Jena.

Fährmann, W. 1971. *Z. Mikrosk. Anat. Forsch.* **84**, 1–25.

Fales, D. E. 1935. *J. Exp. Zool.* **72**, 147–174.

Farquhar, M. G., and G. E. Palade. 1965. *J. Cell. Biol.* **26**, 263–291.

Farrar, E., and B. E. Frye. 1979. *Gen. Comp. Endocrinol.* **39**, 372–380.

Faulhaber, I. 1972. Arch. *Entwicklungsmech. Org.* **171,** 87–108.

Faulhaber, I., and H. P. Geithe. 1972. *Rev. Suisse Zool.* **70** (suppl.), 103–117.

Ferber, E. 1973. In *Biological Membranes* (ed. D. Chapman and D. F. H. Wallach). Vol. **2,** 221–252. Academic Press. London.

Ferigo, E., Cardellini, P., Rodino, E., Sala, M., and B. Salvato. 1977. *Acta Embryol. Exp.* **2,** 137–154.

Ferrier, V. 1974. *Ann. Embryol. Morph.* **7,** 407–416.

Field, H. H. 1891. *Bull. Mus. Comp. Zool. Harv. Univ.* **21,** 201–340.

Finamore, F. J., and E. Frieden. 1960. *J. Biol. Chem.* **235,** 1751–1755.

Fitzgerald, K. T., Guillette, L. J., and D. Duvall. 1979. *J. Herpetol.* **13,** 457–460.

Flavin, M. 1973. Thèse Doctorate de Specialité. Université Paul Sabatier, Toulouse, 1532. (see Flavin, Duprat and Rosa, 1979.)

Flavin, M., Duprat, A. M., and J. Rosa. 1979. *Cell. Diff.* **8,** 405–410.

Flavin, M., Duprat, A. M., and J. Rosa. 1982. *Cell. Differ.* **11,** 27–33.

Flavin, M., Blouquit, Y., Duprat, A. M., and J. Rosa. 1978. *Comp. Biochem. Physiol.* **61B,** 539–544.

Flickinger, R. A. 1964. *Gen. Comp. Endocrinol.* **4,** 285–289.

Flickinger, R. A. 1980. *Arch. Dev. Biol.* **188,** 9–11.

Forbes, M. S., Zaccaria, R. A., and J. N. Dent. 1973. *Am. J. Anat.* **138,** 37–72.

Forehand, C. J., and P. B. Farel. 1982a. *J. Comp. Neurol.* **209,** 386–394.

Forehand, C. J., and P. B. Farel. 1982b. *J. Comp. Neurol.* **209,** 395–408.

Forge, P., and R. Barbault. 1977. *Terre Vie* **31,** 117–125.

Forman, L. J., and J. J. Just. 1976. *Dev. Biol.* **50,** 537–540.

Forman, L. J., and J. J. Just. 1981. *Gen. Comp. Endocrinol.* **44,** 1–12.

Formas, J. R. 1976. *J. Herpetol.* **10,** 221–225.

Formas, J. R., and E. Pugin. 1978. *Herpetologica* **34,** 335–338.

Fortune, J. E., and A. W. Blackler. 1976. *J. Embryol. Exp. Morph.* **36,** 453–468.

Fox, H. 1956. *J. Embryol. Exp. Morph.* **4,** 139–152.

Fox, H. 1957. *J. Embryol. Exp. Morph.* **5,** 274–282.

Fox, H. 1960. *J. Embryol. Exp. Morph.* **8,** 495–504.

Fox, H. 1962a. *J. Embryol. Exp. Morph.* **10,** 103–114.

Fox, H. 1962b. *J. Embryol. Exp. Morph.* **10,** 224–230.

Fox, H. 1963. *Q. Rev. Biol.* **38,** 1–25.

Fox, H. 1966. *J. Embryol. Exp. Morph.* **16,** 487–496.

Fox, H. 1970a. *Arch. Biol.* **81,** 1–20.

Fox, H. 1970b. *J. Embryol. Exp. Morph.* **24,** 139–157.

Fox, H. 1971. *Exp. Gerontol.* **6,** 173–177.

Fox, H. 1972a. *Arch. Biol.* **83,** 373–394.

Fox, H. 1972b. *Arch. Biol.* **83,** 395–405.

Fox, H. 1972c. *Arch. Biol.* **83,** 407–417.

Fox, H. 1973a. *Z. Zellforsch. Mikrosk. Anat.* **138,** 371–386.

Fox, H. 1973b. *J. Embryol. Exp. Morph.* **30,** 377–396.

Fox, H. 1974. *J. Zool.* **174,** 217–235.

Fox, H. 1975. *J. Embryol. Exp. Morph.* **34,** 191–207.

Fox, H. 1976. *J. Microsc. (Paris)* **26,** 43–46.

Fox, H. 1977a. In *Comparative Anatomy of the Skin* (ed. R. I. C. Spearman). *Symp. Zool. Soc. Lond.* No. 39. 269–289. Academic Press. London.

Fox, H. 1977b. In *Mécanismes de la Rudimentation des Organes chez les Embryons de Vertébrés* (ed. A. A. Raynaud). Colloq. Int. CNRS No. 266. 93–112.

Fox, H. 1977c. In *Biology of the Reptilia* (eds. C. Gans and T. S. Parsons). 1–155. Academic Press. London.

Fox, H. 1978. In *XIX Morphological Congress Symposium*. (ed. E. Klika). Charles University, 559–567. Prague. Sept. 6–9, 1976.

Fox, H. 1981. In *Metamorphosis: A Problem in Developmental Biology* (eds. L. I. Gilbert and E. Frieden). 2nd. Ed. 327–362. Plenum Press. New York.

Fox, H. 1983. *J. Zool.* **199**. 223–248.

Fox, H., Bailey, E., and R. Mahoney. 1972. *J. Morphol.* **138**, 387–406.

Fox, H., and L. Hamilton. 1971. *J. Embryol. Exp. Morph.* **26**, 81–98.

Fox, H., Mahoney, R., and E. Bailey. 1970. *Arch. Biol.* **81**, 21–50.

Fox, H., and J. M. Moulton. 1968. *Arch. Anat. Microsc. Morphol. Exp.* **57**, 107–119.

Fox, H., and S. C. Turner. 1967. *Arch. Biol.* **78**, 61–90.

Fox, H., and M. Whitear. 1978. *Biol. Cellulaire.* **32**, 223–232.

Fox, H., Lane, E. B., and M. Whitear. 1980. In *The Skin of Vertebrates (eds. R. I. C. Spearman and P. A. Riley). Linn. Soc. Symp. Ser.* No. 9. 271–281. Academic Press. London.

Foxon, G. E. H. 1964. In *Physiology of the Amphibia* (ed. J. A. Moore). **1**, 151–209. Academic Press. New York, London.

Francois-Krassowska, A. 1978. *Acta Biol. Cracov. Ser. Zool.* **2**, 1–44.

Frangioni, G., and G. Borgioli. 1978. *Monit. Zool. Ital.* **12**, 181–198.

Franke, W. W., Kleinschmidt, J. A., Spring, H., Krohne, G., Grund, C., Trendelenburg, M. F., Stoehr, M., and U. Scheer. 1981a. *J. Cell. Biol.* **90**, 289–299.

Franke, W. W., Scheer, U., Krohne, G., and E. Dieter-Jarasch. 1981b. *J. Cell. Biol.* **91**, 39–50.

Franke, W. W., Schmid, E., Schiller, D. L., Winter, S., Jarasch, E. D., Moll, R., Denk, H., Jackson, B. W., and K. Illmensee. 1982. In *Cold Spring Harbor Symposium,* **46**, 431–453.

Franzini-Armstrong, C., Landmesser, L., and G. Pilar. 1975. *J. Cell. Biol.* **64**, 493–497.

Frieden, E. 1981. In *Metamorphosis: A Problem in Developmental Biology* (eds. L. I. Gilbert and E. Frieden). 2nd. Ed. 545–563. Plenum Press. New York.

Frieden, E., and J. J. Just. 1970. In *Biochemical Actions of Hormones* (ed. G. Litwak). **1**, 1–52. Academic Press. New York.

Fritzsch, B., and U. Wahnschaffe. 1983. *Cell. Tissue Res.* **229**, 483–503.

Frohlich, J., Aurin, H., and P. Kemnitz. 1977. *Verh. Anat. Ges.* **71**, 1171–1175.

Fry, A. E., Leius, V. K., Bacher, B. E., and J. C. Kaltenbach. 1973. *Gen. Comp. Endocrinol.* **21**, 16–29.

Frye, B. E. 1964. *J. Exp. Zool.* **155**, 215–224.

Gage, S. H. 1891. *Am. Nat.* **25**, 1084–1110.

Gahan, P. B. 1967. In *Int. Rev. Cytol.* (eds. G. H. Bourne and J. F. Danielli). **21**, 1–63.

Gallien, L., and O. Bidaud. 1959. *Bull. Soc. Zool. Fr.* **84**, 22–32.

Gallien, L., and M. Durocher. 1957. *Bull. Biol. Fr. Belge.* **91**, 97–114.

Gallien, L., and C. H. Houillon. 1951. *Bull. Biol. Fr. Belge.* **85**, 373–375.

Galton, V. A. 1980. *Endocrinology* **106**, 859–866.

Galton, V. A., and K. Munck. 1981. *Endocrinology* **109**, 1127–1131.

van Gansen, P. 1967. *Exp. Cell. Res.* **47**, 157–166.

van Gansen, P., and A. Schram. 1969. *J. Embryol. Exp. Morph.* **22**, 69–78.

Garton, J. S., and R. A. Brandon. 1975. *Herpetologica* **31**, 150–161.

Gartz, R. 1970. *Cytobiol. Z. Exp. Zellforsch.* **2**, 220–234.

Gasche, P. 1944. *Helv. Physiol. Pharmacol. Acta* **2**, 607–626.
Gasser, F. 1964. *Bull. Soc. Zool. Fr.* **89**, 423–428.
Gaudon, A. J. 1965. *Herpetologica* **21**, 117–130.
Geigy, R. 1941. *Rev. Suisse Zool.* **48**, 483–494.
George, I. D. 1940. PhD Thesis. University of Michigan, Ann Arbor, Michigan.
Georgiev, G. P. 1972. In *Current Topics in Developmental Biology (eds. A. A. Moscona and A. Monroy).* **7**, 1–60. Academic Press. New York.
Gibley, C. W., and J. P. Chang. 1966. *Am. Zool.* **6**, 610.
Gillois, M., and A. Beaumont. 1964. *C. R. Seances Soc. Biol. Paris* **158**, 8–10.
Gipouloux, J. D., and M. Delbos. 1977. *J. Embryol. Exp. Morph.* **41**, 259–268.
Gipouloux, J. D., and J. Hakim. 1976. *Bull. Biol. Fr. Belge.* **110**, 283–298.
Giunta, C., Campantico, E., Vietti, M., and A. Guastalla. 1972. *Gen. Comp. Endocrinol.* **18**, 568–571.
Gläesner, L. 1925. In *Kiebel's Normentafeln zur Entwicklungsgeschichte der Wirbeltiere.* **14**, 1–49. Gustav Fischer. Jena.
Glucksohn, S. 1931. *Arch. Entwicklungemech. Org.* **125**. 341–405.
Goin, C. J. 1947. *Nat. Hist. Misc. Chicago Acad. Sci.* No. 10, 1–4.
Goldenberg, M., and R. P. Elison. 1980. *Dev. Growth. Differ.* **22**, 345–356.
Goldin, G., and B. Fabian. 1971. *Acta Embryol. Morphol. Exp.* **1**, 31–39.
Goldin, G., and B. Fabian. 1972. *S. Afr. Med. J.* **46**, 778.
Gomperts, B. D. 1977. *The Plasma Membrane: Models for Structure and Function.* Academic Press. London.
Gona, A. G. 1967. *Endocrinology* **81**, 748–754.
Gona, A. G. 1969. *Z. Zellforsch. Mikrosk. Anat.* **95**, 483–494.
Gona, A. G. 1972. *J. Comp. Neurol.* **146**, 133–142.
Gona, A. G., and O. D. Gona. 1977. *Exp. Neurol.* **57**, 581–587.
Goniakowska-Witalinska, L. 1982. *Anat. Embryol.* **164**, 113–137.
Goodrich, E. S. 1930. *Studies on the Structure and Development of Vertebrates.* Macmillan. London.
Goos, H. J. T. 1969. *Z. Zellforsch. Mikrosk. Anat.* **97**, 449–458.
Goos, H. J. T. 1978. *Am. Zool.* **18**, 401–410.
de Graf, A. R. 1957. *J. Exp. Biol.* **34**, 173–176.
Gorski, J., Shyamala, G., and D. Toft. 1969. In *Current Topics in Developmental Biology* (eds. A. A. Moscona and A. Monroy). **4**, 149–166.
Gosner, K. L. 1960. *Herpetologica* **16**, 183–190.
Gosner, K. L., and I. H. Black. 1954. *Copeia* No. 4, 251–255.
Gosner, K. L., and D. A. Rossman. 1960. *Herpetologica* **16**, 225–232.
Gottschaldt, K. M., and C. Vahle-Hinz. 1981. *Science* **214**, 183–186.
del Grande, P., and V. Franceschini. 1982. *Anat. Embriol.* **87**, 53–70.
Grant, M. P. 1930a. *Anat. Rec.* **45**, 1–25.
Grant, M. P. 1930b. *Anat. Rec.* **47**, 330.
Grant, P. 1978. *Biology of Developing Systems.* Holt, Reinhart and Winston. New York.
Green, H., Goldberg, G. B., Schwartz, M., and D. D. Brown. 1968. *Dev. Biol.* **18**, 391–400.
Greven, H. 1980. *Z. Mikrosk.-Anat. Forsch.* **94**, 196–208.
Griffiths, I. 1961. *Proc. Zool. Soc. London* **137**, 249–283.
Griswold, M. D., and P. P. Cohen. 1972. *J. Biol. Chem.* **247**, 353–359.
Griswold, M. D., and P. P. Cohen. 1973. *J. Biol. Chem.* **248**, 5854–5860.
Griswold, M. D., Fischer, M. S., and P. P. Cohen. 1972. *Proc. Natl. Acad. Sci.* **69**, 1468–1489.

Gross, J. 1966. *J. Invest. Dermatol.* **47**, 274–277.

Grunz, H., and J. Staubach. 1979. *Differentiation* **14**, 59–66.

Guardabassi, A. 1959. *Arch. Ital. Anat. Embriol.* **64**, 105–127.

Guastalla, A., and E. Campantico. 1979. *Monit. Zool. Ital.* **13**, 11–23.

Gudernatsch, J. R. 1912. *Arch. Entwicklungsmech. Org.* **35**, 457–483.

Günther, R. 1978. *Mitt. Zool. Mus. Berlin* **54**, 161–179.

Gurdon, J. B. 1962. *Dev. Biol.* **4**, 256–273.

Gurdon, J. B. 1968. *J. Embryol. Exp. Morph.* **20**, 401–411.

Gurdon, J. B., and R. A. Laskey. 1970. *J. Embryol. Exp. Morph.* **24**, 227–248.

Gurdon, J. B., Lane, C. D., Woodland, H. R., and G. Marbaix. 1971. *Nature (London)* **233**, 177–182.

Gurdon, J. B. and V. Uelinger. 1966. *Nature (London)* **210**, 1240–1241.

Hadley, M. E., and J. M. Goldman. 1972. In *Pigmentation: Its Genesis and Biological Control* (ed. V. Riley). 225–245. Appleton-Century-Crofts. New York.

Hamada, K., Sakai, Y., Tsushema, K., and R. Shukuya. 1966. *J. Biochem. (Tokyo)* **60**, 37–41.

Hamada, K., and R. Shukuya, 1966. *J. Biochem. (Tokyo)* **59**, 397–403.

Hamburger, V. 1950. *A Manual of Experimental Embryology.* University of Chicago Press.

Hamburgh, M. 1971. *Theories of Differentiation.* Edward Arnold. London.

Hamilton, L. 1969. *J. Embryol. Exp. Morph.* **22**, 253–264.

Hammerman, D. L. 1969. *Acta. Zool.* **50**, 11–33.

Hanaoka, Y. 1966. *J. Fac. Sci. Hokkaido Univ.* **16**, 106–112.

Hanaoka, Y., Koya, S. M., Kondo, Y., Kobayashi, Y., and K. Yamamoto. 1973. *Gen. Comp. Endocrinol.* **21**, 410–423.

Hanke, W. 1974. In *Chemical Zoology* (eds. M. Florkin and B. T. Scheer). **9**, 123–159. Academic Press. New York.

Hara, K., and E. C. Boterenbrood. 1977. *Arch. Dev. Biol.* **181**, 89–93.

Harrison, G. R. 1925. *J. Exp Zool.* **41**, 349–427.

Hay, E. D. 1961a. *Anat. Rec.* **139**, 236.

Hay, E. D. 1961b. *J. Biophys. Biochem. Cytol.* **10**, 457–463.

Hay, E. D. 1963. *Z. Zellforsch. Mikrosk. Anat.* **59**, 6–34.

Hazard, E. S., and V. H. Hutchinson. 1978. *J. Exp. Zool.* **206**, 109–118.

Heady, J. E., and J. J. Kollros. 1964. *Gen. Comp. Endocrinol.* **4**, 124–131.

Hedeen, S. E. 1971. *Herpetologica* **27**, 160–165.

Hedeen, S. E. 1975. *Ohio J. Sci.* **75**, 182–183.

Hemme, L. 1972. *Z. Zellforsch. Mikrosk. Anat.* **125**, 353–377.

Henle, K. 1981. *Amphibia-Reptilia* **2**, 123–132.

Henry, M., and J. Charlemagne. 1981. *Dev. Comp. Immunol.* **5**, 449–460.

Herner, A. E., and E. Frieden. 1960. *J. Biol.* **235**, 2845–2851.

Herner, A. E., and E. Frieden. 1961. *Arch. Biochem. Biphys.* **95**, 25–35.

Herreid, C. F., and S. Kinney. 1967. *Ecology* **48**, 579–590.

Hetherington, T. E., and M. H. Wake. 1970. *Zoomorphologie* **93**, 209–225.

Heyer, R. W. 1979. *Smithson. Contrib. Zool.* No. 301, 1–43.

Hickey, E. D. 1971. *Arch. Entwicklungsmech. Org.* **166**, 303–330.

Highton, R. 1956. *Copeia* No. 2, 75–93.

Hillier, A. P. 1970. *J. Physiol. (London)* **211**, 585–597.

Hillis, D. M. 1982. *Copeia* **No. I**, 168–174.

Hing, L. K. 1959. *Treubia* **25**, 89–111.

Hiroshi, S., Ryuzaki, M., and K. Yoshizato. 1977. *Dev. Biol.* **59**, 96–100.

Hoffman, C. W., and J. N. Dent. 1977. *Gen. Comp. Endocrinol.* **32,** 512–521.
Hollyfield, J. G. 1966. *Dev. Biol.* **14,** 461–480.
Hollyfield, J. G. 1967. *J. Morphol.* **119,** 1–9.
Holmes, R. L., and J. N. Ball. 1974. *The Pituitary Gland: A Comparative Account.* Cambridge University Press. London.
Holmes, S. D., Gitlin, J., Titus, G., and J. R. Field. 1980. *Endocrinology* **106,** 1892–1899.
Holmgren, N. 1933. *Acta Zool. (Stockholm)* **14,** 185–295.
Holtfreter, J. 1944. *J. Exp. Zool.* **95,** 171–212.
Holtzer, H., Bennett, G. S., Tapscott, S. J., Croop, J. M., and Y. Toyama. 1982. In *Cold Spring Harbor Symposium* **46,** 317–329.
Hora, S. L. 1933. *Trans. R. Soc. Edinburgh* **57,** 469–472.
Horstadius, S. 1950. *The Neural Crest.* Oxford University Press.
Horstadius, S., and A. Wolsky. 1936. *Arch. Entwicklungsmech. Org.* **135,** 69–113.
Horton, J. D. 1971. *Am. Zool.* **11,** 219–228.
Hourdry, J. 1971a. *J. Microsc. (Paris)* **10,** 41–58.
Hourdry, J. 1971b. *Histochemie* **26,** 126–141.
Hourdry, J. 1971c. *Histochemie* **26,** 142–159.
Hourdry, J. 1972. *J. Ultrastruct. Res.* **39,** 327–344.
Hourdry, J. 1973. *J. Microsc. (Paris)* **18,** 45–54.
Hourdry, J. 1974. *J. Microsc. (Paris)* **20,** 165–182.
Hourdry, J. 1977. In *Mécanismes de la Rudimentation des Organes chez les Embryons de Vertébrés* (ed. A. A. Raynaud). Colloq. Int. CNRS No. 266. 125–136.
Hourdry, J., and M. Dauca. 1977. *Int. Rev. Cytol. Suppl.* **5,** 337–385.
Housley, M. D., and K. K. Stanley. 1982. *Dynamics of Biological Membranes: Influence on Synthesis, Structure and Function.* Wiley. London.
Hsu, T. W., and T. A. Lyerla. 1977. *J. Exp. Zool.* **199,** 25–32.
Huber, S., Ryffel, G. U., and R. Weber. 1979. *Nature (London)* **278,** 65–67.
Hughes, A. F. W. 1959. *J. Embryol. Exp. Morph.* **7,** 22–38.
Hughes, A. F. W. 1961. *J. Embryol. Exp. Morph.* **9,** 269–284.
Hughes, A. F. W. 1966. *J. Embryol. Exp. Morph.* **16,** 401–430.
Hughes, 1968. *Aspects of Neural Ontogeny.* Logos Press. London.
Hulse, A. C. 1979. *J. Herpetol.* **13,** 153–156.
Hunt, R. K., and M. Jacobson. 1971. *Dev. Biol.* **26,** 100–124.
Hurley, M. P. 1958. *Growth* **22,** 125–166.
Huxley, J. S., and G. R. de Beer. 1934. *The Elements of Experimental Embryology.* Cambridge University Press. London.

Ichikawa, A., and M. Ichikawa. 1969. *Acta Anat. (Japon)* **44,** 80.
Ide, H. 1973. *Gen. Comp. Endocrinol.* **21,** 390–397.
Ide, H. 1982. *Gen. Comp. Endocrinol.* **47,** 340–345.
Ilic, V., and D. Brown. 1980. *Anat. Rec.* **196,** 153–161.
Inger, R. F. 1954. *Fieldiana Zool.* **33,** 183–531.
Inger, R. F. 1956. *Fieldiana Zool.* **34,** 389–424.
Ireland, P. H. 1974. *Herpetologica* **30,** 338–343.
Ireland, P. H. 1976. *Herpetologica* **32,** 233–238.
Isenberg, G., Leonard, K., and B. M. Jockusch. 1982. *J. Mol. Biol.* **158,** 231–249.
Ito, R. 1980. *Arch. Histol. (Japon)* **43,** 231–240.
Iwama, H. 1968. *Normal Table of Megalobatrachus japonicus.* University of Nagoya. 1–28. Tasaki Publ. Company, Nagoya.
Iwasawa, H., and N. Kawasaki. 1979. *Jap. J. Herpetol.* **8,** 22–35.

Iwasawa, H., and Y. Kera. 1980. *Jap. J. Herpetol.* **8,** 73–89.

Iwasawa, H., and Y. Morita. 1980. *Zool. Mag. (Tokyo)* **89,** 65–75.

Iwasawa, H., Maruyama, T., and S. Tanaka. 1978. *Sci. Rep. Niigata Univ.* Ser.D. No.15, 1–8.

Izecksohn, E., da Cruz, C. A. G., and O. L. Peixoto. 1979. *Rev. Bras. Biol.* **39,** 233–236.

Jackson, I. M. D. 1978. *Am. Zool.* **18,** 385–399.

Jackson, I. M. D., and S. Reichlin. 1977. *Science* **198,** 414–415.

Jackson, I. M. D., and S. Reichlin. 1979. *Endocrinology* **104,** 1814–1821.

Jacob, F., and J. Monod. 1961. *Cold Spring Harbor Symposium* **26,** 193–211.

Jacoby, J., and C. B. Kimmel. 1982. *J. Comp. Neurol.* **204,** 364–376.

Jae Chung Hah. 1974. *Korean J. Zool.* **17,** 177–184.

Jae Chung Hah. 1975. *J. Korean Univ.* **20,** 31–36.

Jaffe, R. C. 1978. *Mol. Cell. Endocrinol.* **11,** 205–211.

Jaffe, R. C., and I. I. Geschwind. 1974a. *Gen. Comp. Endocrinol.* **22,** 289–295.

Jaffe, R. C., and I. I. Geschwind. 1974b. *Proc. Soc. Exp. Biol. Med.* **146,** 961–966.

Jaffe, R. C., and M. C. Gold. 1977. *Mol. Cell. Endocrinol.* **8,** 1–13.

Jaffee, O. C. 1954. *J. Morphol.* **95,** 109–123.

Jaffee, O. C. 1963. *Anat. Rec.* **145,** 179–182.

Janes, R. G. 1934. *J. Exp. Zool.* **67,** 73–91.

Janes, R. G. 1937. *J. Morphol.* **61,** 581–611.

Jarvik, E. 1960. *Théories de l'Évolution des Vértébres. Reconsiderees à la Lumiere des Recentes Découvertes sur les Vértébres Inférieurs.* Masson. Paris.

Jarvik, E. 1981. *Syst. Zool.* **30,** 378–384.

Jayatilaka, A. D. P. 1978. *J. Anat.* **125,** 579–591.

Jensen, E. V., and E. R. de Sombre. 1972. *Annu. Rev. Biochem.* **41,** 203–230.

de Jongh, H. T. 1968. *Neth. J. Zool.* **18,** 1–103.

Jordan, E. C. 1893. *J. Morphol.* **8,** 269–366.

Jordan, H. E. 1938. In *Handbook of Hematology* (ed. H. Downey). **2,** Sect. XII. 704–862. Harper (Hoeber). New York.

Jordan, H. E., and C. C. Speidel. 1923. *Anat. Rec.* **25,** 137.

Jordan, H. E., and C. C. Speidel. 1929. *Proc. Soc. Exp. Biol. Med.* **27,** 67–68.

Jorquera, B., and L. Izquierdo. 1964. *Biologica (Santiago)* **36,** 43–53.

Jorquera, B., and E. Pugin. 1975. *Mus. Nac. Hist. Nat. (Santiago) Publ. Occas.* **20,** 1–18.

Jorquera, B., Pugin, E., and O. Y. Goicoechea. 1972. *Arch. Med. Vet.* **4,** 5–19.

Jorquera, B., Pugin, E., and O. Y. Goicoechea. 1974. *Biol. Soc. Biol. Concepcion* **48,** 127–146.

Jurand, A. 1962. *J. Embryol. Exp. Morph.* **10,** 602–621.

Jurand, A. 1974. *J. Embryol. Exp. Morph.* **32,** 1–33.

Jurd, R. D., and N. Maclean. 1970. *J. Embryol. Exp. Morph.* **23,** 299–309.

Jurgens, J. D. 1979. *Madoqua* **11,** 185–208.

Just, J. J., 1972. *Physiol. Zool.* **45,** 143–152.

Just, J. J., and B. G. Atkinson. 1972. *J. Exp. Zool.* **182,** 271–280.

Just, J. J., Gartz, R. N., and E. C. Crawford. 1973. *Respir. Physiol.* **17,** 276–282.

Just, J. J., Kraus-Just, J., and D. A. Check. 1981. In *Metamorphosis: A Problem in Developmental Biology* (eds. L. I. Gilbert and E. Frieden). 2nd Ed. 265–326. Plenum Press. New York.

Just, J. J., Schwager, J., and R. Weber. 1977. *Arch. Devel. Biol.* **183,** 307–323.

Just, J. J., Schwager, J., Weber, R., Fey, H., and H. Pfister. 1980. *Arch. Devel. Biol.* **188,** 75–80.
Juszczyk, W., and M. Zakszewski. 1981. *Acta Biol. Cracov Ser. Zool.* **23,** 127–136.

Kaltenbach, J. C. 1953a. *J. Exp. Zool.* **122,** 21–39.
Kaltenbach, J. C. 1953b. *J. Exp. Zool.* **122,** 449–467.
Kaltenbach, J. C. 1968. In: *Metamorphosis: A Problem in Developmental Biology* (eds. W. Etkin and L. I. Gilbert). 399–441. Appleton-Century-Crofts. New York.
Kaltenbach, J. C. 1971. In *Hormones in Development* (eds. M. Hamburgh ang E. J. W. Barrington). 281–297. Appleton-Century-Crofts. New York.
Kaltenbach, J. C., and K. L. Lee. 1977 *Am. Zool.* **17,** 454.
Kaltenbach, J. C., Lipson, M. L., and C. H. K. Wang. 1977. *J. Exp. Zool.* **202,** 103–120.
Kaltenbach, J. C., Wang, C. H. K., and M. L. Lipson. 1981. *J. Exp. Zool.* **216,** 247–260.
Karfunkel, P. 1971. *Dev. Biol.* **25,** 30–56.
Katagiri, C. 1975. *J. Exp. Zool.* **193,** 109–118.
Kaung, H. C. 1975. *Anat. Rec.* **182,** 401–414.
Kaung, H. C., and J. J. Kollros. 1977. *Anat. Rec.* **188,** 361–370.
Kawada, J., Taylor, R. E., and S. B. Barker. 1969. *Comp. Biochem. Physiol.* **30,** 965–975.
Kawada, J., Taylor, R. E., and S. B. Barker. 1972. *Endocrinol. Japon* **19,** 53–57.
Kaye, N. W. 1961. *Gen. Comp. Endocrinol.* **1,** 1–19.
Kaywin, L. 1936. *Anat. Rec.* **64,** 413–441.
Kelley, R. O., Nakai, G. S., and M. E. Guganig. 1971. *J. Embryol. Exp. Morph.* **26,** 181–193.
Kelly, D. E. 1966a. *Anat. Rec.* **154,** 685–700.
Kelly, D. E. 1966b. *J. Cell. Biol.* **28,** 51–72.
Kemp, N. E. 1963. *Dev. Biol.* **7,** 244–254.
Kemp, N. E., and J. A. Hoyt. 1969. *Dev. Biol.* **20,** 387–410.
Kenny, J. S. 1968. *Caribb. J. Sci.* **8,** 35–45.
Kenny, J. S. 1969. *J. Zool.* **157,** 225–246.
Kerr, T. 1965. *Gen. Comp. Endocrinol.* **5,** 232–240.
Kerr, T. 1966. *Gen. Comp. Endocrinol.* **6,** 303–311.
Kerr, F. R., Harmon, B., and J. Searle. 1974. *J. Cell. Sci.* **14,** 571–585.
Kessel, R. G., and L. S. Ganion. 1980. *J. Submicrosc. Cytol.* **12,** 647–654.
Kezer, J. 1952. *Copeia* No. 4, 234–237.
Khan, M. S. 1965. *Biologia (Lahore)* **11,** 1–39.
Khan, M. S. 1968. *Pak. J. Sci. Res.* **20,** 93–106.
Kielbówna, L. 1966. *Zool. Pol.* **11,** 247–255.
Kielbówna, L. 1975. *Cell. Tissue Res.* **159,** 277–286.
Kielbówna, L. 1981. *J. Embryol. Exp. Morph.* **64,** 295–304.
Kielbówna, L., and B. Koscielski. 1979. *Arch. Devel. Biol.* **185,** 295–303.
Kikuyama, S., Miyakawa, M., and Y. Arai. 1979. *Cell. Tissue Res.* **198,** 27–33.
Kikuyama, S., Yamamoto, K., and M. Mayumi. 1980a. *Gen. Comp. Endocrinol.* **41,** 212–216.
Kikuyama, S., Yamamoto, K., and T. Seki. 1980b. *Gunma Symp. Endocrinology* **17,** 3–13.

Kilarski, W., and M. Kozlowska. 1981. *Z. Mikrosk.-Anat. Forsch. (Leipzig)* **95,** 963–978.

Kim, K. H. 1967. *Fed. Proc. Fed. Am. Soc. Exp. Biol.* **26,** 603.

Kim, K. H., and P. P. Cohen. 1966. *Proc. Natl. Acad. Sci.* **55,** 1251–1255.

Kim, K. H., and P. P. Cohen. 1968. *Biochim. Biophys. Acta* **166,** 574–577.

Kim, K. H., and K. A. Slickers. 1971. In *Hormones in Development* (eds. M. Hamburgh and E. J. K. Barrington). 321–334. Appleton-Century-Crofts. New York.

King, J. A., and R. P. Millar. 1981. *Gen. Comp. Endocrinol.* **44,** 20–27.

King, T. J., and R. Briggs. 1954. *J. Embryol. Exp. Morph.* **2,** 73–80.

Kirschner, M., Gerhart, J. C., Hara, K., and G. A. Ubbels. 1980. In *The Cell Surface: mediator of developmental processes* (eds. S. Subtelny and N. K. Wessells). 187–215. Academic Press. New York.

Kistler, A., and R. Weber. 1975. *Mol. Cell. Endocrinol.* **2,** 261–288.

Kistler, A., Yoshizato, K., and E. Frieden. 1975a. *Endocrinology* **97,** 1036–1042.

Kistler, A., Yoshizato, K., and E. Frieden. 1975b. *Dev. Biol.* **46,** 151–159.

Kistler, A., Yoshizato, K., and E. Frieden. 1977. *Endocrinology* **100,** 134–137.

Klee, J. H., Dipietro, D., Fournier, M. J., and M. S. Fischer. 1978. *J. Biol. Chem.* **253,** 8074–8080.

Knight, F. C. E. 1938. *Arch. Entwicklungsmech. Org.* **137,** 461–473.

Knowland, J. 1978. *Differentiation* **12,** 47–51.

Knutson, T. L., and K. V. Prahlad. 1971. *J. Exp. Zool.* **178,** 45–57.

Koch, J., and A. Cruceanu. 1971. *Physiol. Chem.* **352,** 137–142.

Kollros, J. J. 1940. *J. Exp. Zool.* **85,** 33–52.

Kollros, J. J. 1957. *Proc. Soc. Exp. Biol. Med.* **95,** 138–141.

Kollros, J. J. 1961. *Am. Zool.* **1,** 107–114.

Kollros, J. J. 1968. In *Ciba Foundation Symposium. Growth of the Nervous System* (eds. G. E. W. Wolstenholme and M. O'Connor). 179–192.

Kollros, J. J. 1972. In *7th. Mid-West Conference on Endocrinology and Metabolism.* 83–96.

Kollros, J. J. 1981. In *Metamorphosis: A Problem in Developmental Biology* (eds. L. I. Gilbert and E. Frieden). 2nd. Ed. 445–459. Plenum Press. New York.

Kollros, J. J., and J. C. Kaltenbach. 1952. *Physiol. Zool.* **25,** 163–170.

Kollros, J. J., and V. M. McMurray. 1955. *J. Comp. Neurol.* **102,** 47–63.

Kollros, J. J., and V. M. McMurray. 1956. *J. Exp. Zool.* **131,** 1–26.

Kollros, J. J., and V. Pepernik. 1952. *Anat. Rec.* **113,** 527.

Kollros. J. J., and J. Race. 1960. *Anat. Rec.* **136,** 224.

Kopec, J. 1970. *Zool. Pol.* **20,** 199–216.

Kopsch, B. 1952. *Die Entwicklung des braunen Grasfrosches, Rana fusca Roessel.* Thieme. Stuttgart.

Kordylewski, L. 1978. *J. Embryol. Exp. Morph.* **45,** 215–228.

Korky, J., and R. G. Webb. 1973. *J. Herpetol.* **71,** 47–49.

Koskela, P. 1971. *Ann. Zool. Fenn.* **10,** 414–418.

Kramer, B. 1972. *Dev. Biol.* **29,** 220–226.

Kramer, B. 1978. *S. Afr. J. Sci.* **74,** 462–465.

Kramer, B. 1980. *J. Anat.* **130,** 809–820.

Kubler, H., and E. Frieden. 1964. *Biochim. Biphys. Acta* **93,** 635–643.

Kudô, T. 1938. In *Keibel's Normentafeln zur Entwicklungsgeschichte der Wirbeltiere.* **16,** 1–98. Gustav Fischer. Jena.

Kujat, R. 1981. *Fol. Biol. (Kraków)* **29,** 275–278.

Kullberg, R. W., Lentz, T. L., and M. W. Chen. 1977. *Dev. Biol.* **60,** 101–129.

Kunzenbacher, I., Bereiter Hahn, J., Osborn, M., and K. Weber. 1982. *Cell. Tissue Res.* **222,** 445–457.
Kusa, M., Matsuura, K., and S. Yahagi. 1976. *Zool. Mag. (Tokyo)* **85,** 70–72.

Lamb, A. H. 1974. *Brain Res.* **67,** 527–530.
Lambertini, G. 1928. *Ric. Morfol.* **9,** 71–88.
Lamborghini, J. E. 1980. *J. Comp. Neurol.* **189,** 323–334.
Lamborghini, J. E., Ravenaugh, M., and N. C. Spitzer. 1979. *J. Comp. Neurol.* **183,** 741–752.
Lamotte, M., and F. Xavier. 1972a. *Ann. Embryol. Morph.* **5,** 315–340.
Lamotte, M., and F. Xavier. 1972b. *Bull. Soc. Fr. Belge.* **97,** 413–428.
Lamotte, M., and M. Zuber-Vogeli. 1954. *Bull. Inst. Franc. Afr. Noire* **16A,** 940–953.
Lamotte, M., and J. Lesure. 1977. *Terre Vie* No. 2, 225–312.
Landstrom, U. 1977. *J. Embryol. Exp. Morph.* **41,** 23–32.
Landstrom, U., and S. Løvtrup. 1977. *Acta Embryol. Exp.* **2,** 171–178.
Lane, E. B., and M. Whitear. 1980. *Can. J. Zool.* **58,** 450–455.
Lanot, R. 1962. *Bull. Biol. Fr. Belge.* **96,** 703–721.
Lapiere, C. M., and J. Gross. 1963. In *Mechanisms of Hard Tissue Destruction* (ed. R. F. Sognnaes). 663–694. Am. Assn. Adv. Sci. Washington.
Larsen, J. H. 1968. *J. Ultrastruct. Res.* **24,** 190–209.
Larsen, L. O. 1976. In *Physiology of the Amphibia* (ed. B. Lofts). **3,** 53–100. Academic Press. London.
Lasek, R. J., and S. T. Brady. 1982. In *Cold Spring Harbor Symposium,* **46,** 113–124.
Lavker, R. M. 1974. *J. Morphol.* **142,** 365–377.
Lazarides, E. 1980. *Nature (London)* **283,** 249–256.
Lazarides, E., Granger, B. L., Gard, D. L., O'Connor, C. M., Breckler, J., Price, M., and S. I. Danto. 1982. In *Cold Spring Harbor Symposium* **46,** 351–378.
Leavitt, N. 1948. In *Experimental Embryology* (R. Rugh). Burgess. Minneapolis.
Leblanc, J., and I. Brick. 1981. *J. Embryol. Exp. Morph.* **61,** 145–163.
Ledford, B. E., and E. Frieden. 1973. *Dev. Biol.* **30,** 187–197.
Lee, A. K., and E. H. Mercer. 1967. *Science* **159,** 87–88.
Leeson, T. S. 1977. *Am. J. Anat.* **150,** 185–192.
Leeson, C. R., and L. T. Threadgold. 1960. *Acta Anat.* **43,** 298–302.
Leeson, C. R., and L. T. Threadgold. 1961. *Acta Anat.* **44,** 159–173.
Legname, H. S., Toledo, C. L. V., and A. H. Legname. 1979. *Acta Embryol. Exp.* No. (1). 17–27.
Lehmann, R. 1967. *Naturwissenschaften* **54,** 25–26.
Lemanski, L. F., and R. Aldoroty. 1977. *J. Morphol.* **153,** 419–426.
Leone, F., Lambert-Gardini, S., Sartori, C., and S. Scapin. 1976. *J. Embryol. Exp. Morph.* **36,** 711–724.
Lescure, J. 1976. *Bull. Soc. Fr. Belge.* **101,** 299–306.
Leussink, J. A. 1970. *Neth. J. Zool.* **20,** 1–79.
Lewinson, D., Rosenberg, M., and M. R. Warburg, 1982. *Biol. Cell.* **46,** 75–84.
Lewis, S., and C. Straznicky. 1979. *J. Comp. Neurol.* **183,** 633–646.
Licht, L. E. 1975. *Can. J. Zool.* **53,** 1254–1257.
Liem, R. K. H., Keith, C. H., Leterrier, J. F., Trenkner, E., and M. L. Shelanski. 1982. In *Cold Spring Harbor Symposium* **46,** 341–350.
Limbaugh, B. A., and E. P. Volpe. 1957. *Am. Mus. Novit.* No. 1842, 1–32.
Lind, S. E., Yin, H. L., and T. P. Stossel. 1982. *J. Clin. Invest.* **69,** 1384–1387.

Ling, B. Y. F., and T. A. Lyerla. 1976. *J. Exp. Zool.* **195,** 191–197.

Lipson, M. J., Cerskus, R., and J. R. Silbert. 1971. *Dev. Biol.* **25,** 198–208.

Lipson, M. S., and J. C. Kaltenbach. 1965. *Am. Zool.* **5,** 212.

Little, G. H., Garner, C. W., and J. W. Pelley. 1979. *Comp. Biochem. Physiol.* **62,** 163–165.

Liu, C. C. 1950. *Fieldiana Zool. Mem.* **2,** 1–400.

Liu, C. C., and J. C. Li. 1930. *Peking Soc. Nat. Hist. Bull.* **4,** 67–94.

Lobo, L. 1961. *Zoologica* **46,** 103–104.

Loeffler, C. A. 1969. *J. Morphol.* **128,** 403–425.

Loveridge, J. P., and G. Crayé. 1979. *S. Afr. J. Sci.* **75,** 18–20.

Løvtrup, S. 1965. *Acta Zool. (Stockholm)* **46,** 119–165.

Løvtrup, S. 1983. *Biol. Rev.* **58,** 91–130.

Løvtrup, S., Landstrom, U., and H. Løvtrup-Rein. 1978. *Biol. Rev.* **53,** 1–42.

Luckenbill, L. M. 1965. *Dev. Biol.* **11,** 25–49.

Lutz, B., and G. Orton. 1946. *Bol. Mus. Nac. Rio de Janeiro Zool.* **70,** 1–20.

Lynn, W. G. 1942. *Carneg. Inst. Washington Contr. Embryol.* No. 190, 27–62.

Lynn, W. G., and B. Lutz. 1946. *Bol. Mus. Nac. Rio de Janeiro Zool.* **71,** 1–46.

Lynn, W. G., and A. M. Peadon. 1955. *Growth* **19,** 263–286.

Lynn, W. G., and H. E. Wachowski. 1951. *Q. Rev. Biol.* **26,** 123–168.

Maclean, N., and R. D. Jurd. 1971. *J. Cell. Sci.* **9,** 509–528.

Maclean, N., and S. C. Turner. 1976. *J. Embryol. Exp. Morph.* **35,** 261–266.

Mairy, M., and H. Denis. 1971. *Dev. Biol.* **24,** 143–165.

Mak, L. L. 1978. *Dev. Biol.* **65,** 435–446.

Manelli, H., and F. Margaritora. 1961. *Rend. Accad. Naz. Lincei Ser. IV,* **12,** 183–195.

Maniatis, G. M., and V. M. Ingram. 1971. *J. Cell. Biol.* **49,** 380–401.

Manning, M. J., and J. D. Horton. 1969. *J. Embryol. Exp. Morph.* **22,** 265–277.

Marshall, J. A., and K. E. Dixon. 1978. *J. Anat.* **126,** 133–144.

Martin, A. A. 1965. *Victorian Nat.* **82,** 139–149.

Martin, A. A. 1967a. *Aust. Nat. Hist.* **15,** 326–330.

Martin, A. A. 1967b. *Proc. Linn. Soc. NSW.* **92,** 107–116.

Martin, A. A., and M. L. Littlejohn. 1966. *Proc. Linn. Soc. NSW.* **91,** 47–57.

Martin, A. A. Littlejohn, M. L., and P. A. Rawlinson. 1966. *Victorian Nat.* **83,** 312–315.

Martin, A. A., Tyler, M. T., and M. Davies. 1980. *Copeia No. I,* 93–99.

de Martino, G. N., and A. L. Goldberg. 1978. *Proc. Natl. Acad. Sci.* **75,** 1369–1373.

Maruyama, T., Watt, K. W. K., and A. Riggs. 1980. *J. Biol. Chem.* **255,** 3285–3293.

Marx, L. 1935. *Ergeb. Biol.* **II,** 244–334.

Masoni, A., and F. Garcia-Romeu. 1979. *Cell. Tissue Res.* **197,** 23–38.

Matsuda, H., and T. Kajashima. 1978. *Acta Embryol. Exp.* **3,** 309–318.

McAvoy, J. M. W., and K. E. Dixon. 1977. *J. Exp. Zool.* **202,** 129–138.

McAvoy, J. M. W., and K. E. Dixon. 1978. *J. Anat.* **125,** 155–169.

McClanahan, L. L., Shoemaker, V. H., and R. Ruibal. 1976. *Copeia* No. 1, 179–185.

McCutcheon, F. H. 1936. *J. Cell. Comp. Physiol.* **8,** 63–81.

McDiarmid, R. W., and M. S. Foster. 1981. *SouthWest. Nat.* **26,** 353–364.

McGarry, M. P., and J. W. Vanable. 1969. *Dev. Biol.* **20,** 426–434.

Medda, A. K., and E. Freiden. 1970. *Endocrinology* **87**, 356–365.
Medda, A. K., and E. Frieden. 1972. *Gen. Comp. Endocrinol.* **19**, 212–217.
Michael, M. I., and M. A. Adhami. 1974. *J. Zool.* **174**, 215–223.
Meyer, W. 1974. Dissertation. Dr. der Naturwissenschaften. University of Hanover
Michael, M. I., and A. Y. Yacob. 1974. *J. Zool.* **174**, 407–417.
Michael, P. 1981. *Dev. Growth Differ.* **23**, 149–156.
Michaels, J. E., Albright, J. T., and D. I. Patt. 1971. *Am. J. Anat.* **132**, 301–318.
Michniewska-Predygier, Z., and A. Pigon. 1957. *Stud. Soc. Sci, Torun Sect. E (Zool), III,* **8**, 1–11.
Millard, N. 1945. *Trans. R. Soc. S. Afr.* **30**, 217–234.
Miller, D. C. 1940. *Proc. Indiana Acad. Sci.* **49**, 209–214.
Mira-Moser, F. 1972. *Z. Zellforsch. Mikrosk. Anat.* **125**, 88–107.
Misumi, Y., Kurata, S., and K. Yamana. 1980. *Dev. Growth Differ.* **22**, 773–780.
Model, P. G. 1978. *Am. Zool.* **18**, 253–265.
Model, P. G., Jarrett, L. S., and R. Bonazzoli. 1981. *J. Embryol. Exp. Morph.* **66**, 27–41.
Mohanty-Hejmadi, P., and S. K. Dutta. 1979. *J. Bombay Nat. Hist. Soc.* **76**, 291–296.
Mohun, T. J., Tilly, R., Mohun, R., and J. M. W. Slack. 1980. *Cell* **22**, 9–15.
Mondou, P. M., and J. C. Kaltenbach. 1979. *Gen. Comp. Endocrinol.* **39**, 343–349.
Monod, J. 1972. *Chance and Necessity.* Collins. London.
Mookerjee, S. 1953. *J. Embryol. Exp. Morph.* **I**, 410–416.
Mookerjee, S., Deuchar, E. M., and C. H. Waddington. 1953. *J. Embryol. Exp. Morph.* **I**, 399–409.
Moore, J. A. 1957. *Heredity and Development.* Oxford University Press. New York.
Moore, J. A. 1960. In *Symposium on New Approaches to Cell Biology.* (ed. P. M. B. Walker). 1–14. Cambridge University Press. London.
Moore, J. A. 1962. *J. Cell. Physiol.* **60**, Suppl. I, 19–34.
Morris, S. M., and R. D. Cole. 1978. *Dev. Biol.* **62**, 52–64.
Morris, S. M., and R. D. Cole. 1980. *Dev. Biol.* **74**, 379–386.
Moser, H. 1950. *Rev. Suisse Zool.* **57**, 1–144.
Moss, B., and V. M. Ingram. 1965. *Proc. Natl. Acad. Sci.* **54**, 967–974.
Moss, B., and V. M. Ingram. 1968a. *J. Mol. Biol.* **32**, 481–492.
Moss, B., and V. M. Ingram. 1968b. *J. Mol. Biol.* **32**, 493–504.
Moulton, J. M., Jurand, A., and H. Fox. 1968. *J. Embryol. Exp. Morph.* **19**, 415–431.
Muchmore, W. B. 1962. *Am. Zool.* **2**, 542–543.
Muchmore, W. B. 1965. *Am. Zool.* **5**, 721.
Munger, B. L. 1977. *J. Invest. Dermatol.* **69**, 27–40.
Munro, A. F. 1939. *Biochem. J.* **33**, 1957–1965.
Muntz, L. 1975. *J. Embryol. Exp. Morph.* **33**, 757–774.
Myauchi, H., Larochelle, F. T., Suzuki, M., Freeman, M., and E. Frieden. 1977. *Gen. Comp. Endocrinol.* **33**, 254–266.

Nafstad, P. H., and R. E. Baker. 1973. *Z. Zellforsch. Mikrosk. Anat.* **139**, 451–462.
Nagata, S. 1976. *J. Fac. Sci, Hokkaido Univ. Ser. Zool.* **20**, 263–271.
Nagata, S. 1977. *Cell. Tissue Res.* **179**, 87–96.
Nakagawa, H., and P. P. Cohen. 1967. *J. Biol. Chem.* **242**, 462–469.
Nakagawa, H., Kim, K. H., and P. P. Cohen. 1967. *J. Biol. Chem.* **242**, 635.
Nakamura, O. 1969. *Annls Embryol. Morphogen. Suppl. 1,* 261–262.
Nakao, T. 1974. *J. Anat.* **140**, 533–549.

Nakao, T. 1976. *Cell. Tissue Res.* **166**, 241–254.

Nanba, H. 1972. *Arch. Histol. Japon* **34**, 277–291.

Nanba, H. 1973. *Arch. Histol. Japon* **35**, 313–322.

Nelson, B. D., Joste, V., Wielburski, A., and U. Rosenqvist. 1980. *Biochem. Biophys. Acta* **608**, 422–426.

Neuenschwander, P. 1972. *Z. Zellforsch. Mikrosk. Anat.* **130**, 553–574.

Nielsen, H. I., and J. Bereiter-Hahn. 1982. *J. Zool.* **130**, 363–381.

Nieuwkoop, P. D. 1962. *Acta Biotheor.* **16**, 57–68.

Nieuwkoop, P. D. 1966. In *Lecture Course on Cell Differentiation and Morphogenesis.* Weningen, Netherlands, 120–143. Wiley. New York.

Nieuwkoop, P. D., and J. Faber. 1956. *Normal Table of Xenopus laevis (Daudin).* North Holland. Amsterdam.

Nieuwkoop, P. D., and S. J. van d. Grinten. 1961. *Embryologia* **6**, 51–66.

Niijima, M., and R. Hirakow. 1964. *Acta Anat. Japon* **39**, 4–10.

Nishio, Y., Shiokawa, K., and K. Yamana. 1978. *Devel. Growth Differ.* **20**, 35–40.

Niu, M. C. 1958a. *Anat. Rec.* **131**, 585.

Niu, M. C. 1958b. *Proc. Natl. Acad. Sci.* **44**, 1264–1274.

Noble, S. K. 1954. *The Biology of The Amphibia.* Dover Publ. New York.

Noguchi, T., and S. Kikuyama. 1979. *Endocrinol. Japon* **26**, 617–621.

Nomura, S., Shiba, Y., Muneoka, Y., and Y. Kanno. 1979. *Hiroshima J. Med. Sci.* **28**, 79–86.

Nonidez, J. F. 1932. *Am. J. Anat.* **49**, 479–505.

Nordlander, R. H., Singer, J. F., Beck, R., and M. Singer. 1981. *J. Comp. Neurol.* **199**, 535–552.

Novikoff, A. B. 1963. In *Lysosomes—Ciba Foundation Symposium* (eds. A. V. S. de Reuck and M. P. Cameron). 36–77. Little, Brown. Boston.

Novikoff, A. B., Essner, E., and N. Quintana. 1964. In *Symposium on Lysosomes Fed. Proc. Fed. Am. Soc. Exp. Biol.* **23**, 1010–1022.

Novikoff, A. B., and E. Holtzman. 1970.. *Cells and Organelles.* Holt Rinehart and Winston. New York.

Nussbaum, R. A. 1969. *Herpetologica* **25**, 277–278.

Nussbaum, R. A., and G. W. Clothier. 1973. *Northwest. Sci.* **47**, 218–227.

Nyholm, M., Saxen, L., Toivonen, S., and T. Vainio. 1962. *Exp. Cell. Res.* **28**, 209–212.

Oates, C. L. 1977. PhD Thesis. University of Newcastle-upon-Tyne. England.

Oguro, C. 1973. *Gen. Comp. Endocrinol.* **21**, 565–568.

Okada, Y. K., and M. Ichikawa. 1947. *Jap. J. Exp. Morph.* **3**, 1–6.

Oksch, A., and M. Ueck. 1976. In *Physiology of the Amphibia* (ed. B. Lofts). **3**, 313–417. Academic Press. London.

Oliphant, L. W. 1973. *Can. J. Zool.* **51**, 1007–1009.

van Oordt, P. G. W. J. 1974. In *Physiology of the Amphibia* (ed. B. Lofts). **2**, 53–106. Academic Press. London.

van Oordt, P. G. W. J., Goos, H. J. T., Peute, J., and M. Terlou. 1972. *Gen. Comp. Endocrinol. Suppl. 3,* 41–50.

Oota, Y. 1978. *Rep. Fac. Sci. Shizuoka Univ.* **(12)**, 75–86.

Oppenheimer, J. H. 1979. *Science* **203**, 971–979.

Oppenheimer, J. H., Koerner, D., Schwartz, H. L., and M. I. Surks. 1972. *J. Clin. Endocrinol. Metab.* **35**, 330–333.

Oppenheimer, J. H., Schwartz, H. L., Surks, M. I., Koerner, D., and W. H. Dillman. 1976. *Recent. Progr. Hormone Res.* **32**, 529–563.

Orton, G. L. 1949. *Ann. Carneg. Mus.* **31**, 257–277.
Orton, G. L. 1951. *Copeia* No. 1, 62–66.
Osborn, M., Geisler, N., Shaw, G., Sharp, G., and K. Weber. 1982. In *Cold Spring Harbor Symposium* **46**, 413–429.
Ovalle, W. K. 1979. *Cell. Tissue Res.* **204**, 233–241.
Oyama, J. 1929. *Copeia* 92–94.
Oyama, J. 1930. *Zool. Mag. (Tokyo)* **42**, 465–473.

Packer, W. C. 1966. *Copeia* No. 1, 92–97.
Page, S. 1965. *J. Cell. Biol.* **26**, 477–479.
Paik, W. K., and R. Baserga. 1971. *Exp. Cell. Res.* **64**, 190–194.
Paik, W. K., and P. P. Cohen. 1960. *J. Gen. Physiol.* **43**, 683–696.
Paik, W. K., Metzenberg, R. L., and P. P. Cohen. 1961. *J. Biol. Chem.* **236**, 536–541.
Parducz, A., Leslie, R. A., Cooper, E., Turner, C. J., and J. Diamond. 1977. *Neuroscience* **2**, 511–521.
Parry, C. R., and R. Cavill. 1978. *Trans. Rhod. Sci. Assoc.* **58**, 55–58.
Pasteels, J. 1940. *Biol. Rev.* **15**, 59–106.
Patterson, K. 1978. *Copeia* No. 4, 649–655.
Peacock, R. L., and R. A. Nussbaum. 1973. *J. Herpetol.* **7**, 215–224.
Pearson, D. B., and W. K. Paik. 1972. *Exp. Cell. Res.* **73**, 208–220.
Pehlemann, F. W. 1972. In *Pigmentation: Its Genesis and Biological Control* (ed. V. Riley). 295–305. Appleton-Century-Crofts. New York.
Pehlemann, F. W., Evennett, P. J., Fliedner, T., and B. Gottfriedsen. 1974. *Gen. Comp. Endocrinol.* **22**, 336.
Pehlemann, F. W., and L. Hemme. 1972. *Gen. Comp. Endocrinol.* **18**, 615.
Peixoto, O. L. 1981. *Rev. Bras. Biol.* **41**, 515–520.
Peixoto, O. L., Izecksohn, E., and C. A. G. da Cruz. 1981. *Rev. Bras. Biol.* **41**, 553–555.
Pellegrino, C., and C. Franzini. 1963. *J. Cell. Biol.* **17**, 327–349.
Peng, N. B., Wolosewick, J. J., and P. C. Cheng. 1981. *Dev. Biol.* **88**, 121–136.
Perriard, J. C. 1971. *Arch. Entwicklungsmech. Org.* **168**, 39–62.
Perry, M. M., and C. H. Waddington. 1966. *J. Cell. Sci.* **1**, 193–200.
Pesetsky, I. 1962. *Gen. Comp. Endocrinol.* **2**, 229–235.
Pesetsky, I. 1966. *Z. Zellforsch. Mikrosk. Anat.* **75**, 138–145.
Pesetsky, I., and J.J. Kollros. 1956. *Exp. Cell. Res.* **11**, 477–482.
Pesetsky, I., and P. G. Model. 1969. *Exp. Neurol.* **25**, 238–245.
Pesetsky, I. and P. G. Model. 1971. In *Hormones in Development* (eds. M. Hamburg and E. J. W. Barrington). 335–343. Appleton-Century-Crofts.New York.
Peter, K. 1931. *J. Linn. Soc. London (Zool.)* **37**, 515–523.
Peute, J. 1974. Doctoral Thesis, University of Utrecht. (ref. from M. H. I. Dodd and J. M. Dodd, 1976.)
Pflugfelder, O., and G. Schubert. 1965. *Z. Zellforsch. Mikrosk. Anat.* **67**, 96–112.
Piatka, B. A., and C. W. Gibley. 1967. *Tex. Rep. Biol. Med.* **25**, 83–89.
Piatt, J. 1969. *Dev. Biol.* **19**, 608–616.
Piatt, J. 1971. *Anat. Rec.* **169**, 41–44.
Picard, J. J. 1976. *J. Morphol.* **148**, 193–208.
de Piceis Polver, P., Barni, S., and R. Nano. 1981. *Monit. Zool. Ital.* **15**, 123–132.
Pilkington, J. B., and K. Simkiss. 1966. *J. Exp. Biol.* **45**, 329–341.
del Pino, E. M., and B. Escobar. 1981. *J. Morphol.* **167**, 277–295.

Pitt-Rivers, R. 1978. In *Hormonal Proteins and Peptides* (ed. C. H. Li). **6,**391–422. Academic Press. London, New York.

Platt, J. E., and M. A. Christopher. 1977. *Gen. Comp. Endocrinol.* **31,** 243–248.

Platt, J. E., Christopher, M. A., and C. A. Sullivan. 1978. *Gen. Comp. Endocrinol.* **35,** 402–408.

Polansky, J. R., and B. P. Toole. 1976. *Dev. Biol.* **53,** 30–35.

Pollack, E. D., and M. Richmond. 1981. *J. Morphol.* **169,** 253–257.

Pollack, E. D., Muhlach, W. L., and V. Liebig. 1981. *J. Comp. Neurol.* **200,** 393–405.

Pollister, A. W., and J. A. Moore. 1937. *Anat. Rec.* **68,** 489–496.

Poole, T. J., and M. S. Steinberg. 1981. *J. Embryol. Exp. Morph.* **63,** 1–16.

Pope, P. H. 1924. *Carneg. Mus.* **15,** 305–368.

Pouyet, J. C., and A. Beaumont. 1975. *C. R. Seances Soc. Biol. Paris* **169,** 846–850.

Pouyet, J. C., and J. Hourdry. 1977. *Biol. Cellulaire* **29,** 123–134.

Power, J. H. 1926. *Trans. R. Soc. S. Afr.* **13,** 107–117.

Prahlad, K. V., and L. E. Delanney. 1965. *J. Exp. Zool.* **160,** 137–146.

Prestige, M. C. 1965. *J. Embryol. Exp. Morph.* **13,** 63–72.

Prestige, M. C. 1967. *J. Embryol. Exp. Morph.* **18,** 359–387.

Prestige, M. C. 1970. In *Neurosciences, Second Study Programme* (ed. F. O. Schmidt). 73–82. The Rockefeller Univ. Press. New York.

Prestige, M. C. 1973. *Brain Res.* **59,** 400–404.

Prestige, M. C., and M. A. Wilson. 1974. *J.Embryol. Exp. Morph.* **32,** 819–833.

Price, H. M. 1969. In *Disorders of the Voluntary Muscles* (ed. J. M. Walton). 29–56. Churchill. London.

Pyburn, W. F. 1980. *Pap. Avulsos Zool. (Sao Paulo)* **33,** 231–238.

Race, J. 1961. *Gen. Comp. Endocrinol.* 1, 322–331.

Race, J., Robinson, C., and R. J. Terry. 1966. *J. Exp. Zool.* **162,** 181–192.

Rafelson, M. E., and S. B. Binkley. 1968. *Basic Biochemistry.* MacMillan. New York.

Raff, E. C., and R. A. Raff. 1978. *Am. Zool.* **18,** 237–251.

Raff, M. 1976. *Nature (London)* **259,** 265–266.

Ramaswami, L. S. 1944. *Proc. Zool. Soc. London* **114,** 117–138.

Ramaswami, L. S. 1954. *Anat. Anz.* **101,** 120–122.

Ramaswami, L. S. 1957. *Zool. Anz.* **161,** 271–280.

Ramaswami, L. S., and A. B. Lakshman. 1959. *Proc. Nat. Inst. Sci. India* **25B,** 68–79.

Rappaport, R. 1955. *J. Exp. Zool.* **128,** 481–487.

Rau, R. E. 1978. *Ann. S. Afr. Mus.* **76,** 247–264.

Razarihelisoa, M. 1974. *Bull. Acad. Malag.* **51,** 113–128.

Reeves, R. 1977. *Dev. Biol.* **60,** 163–179.

Regard, E. 1978. *Int.Rev. Cytol.* **52,** 81–118.

Regard,E., and J. Mauchamp. 1971. *J. Ultrastruct. Res.* **37,** 664–678.

Regard, E., and J. Mauchamp. 1973. *J. Microsc. (Paris)* **18,** 291–306.

Regard, E., Taurog, A., and T. Nakashima. 1978. *Endocrinology* **102,** 674–684.

Reichel, P. 1976. *Differentiation* 5, 75–83.

Rémy, C., and J. J. Bounhiol. 1971. *C. R. Seances Acad. Sci. Paris* **272,** 455–458.

Rémy, C., and P. Disclos. 1970. *C. R. Seances Soc. Biol. Paris* **164,** 1989–1993.

de Reuck, A. V. S., and J. Knight. 1966. *Histones. Their Role in the Transfer of Genetic Information.* Ciba Foundation Study Group. No. 24. Churchill. London.

Reynolds, W. A. 1963. *J. Exp. Zool.* **153,** 237–250.

Reynolds, W. A. 1966. *Gen. Comp. Endocrinol.* **6,** 453–465.

Richards, C. M. 1982. *Gen. Comp. Endocrinol.* **46,** 59–67.

Richter, J. D., Wasserman, W. J., and L. Dennis Smith. 1982. *Dev. Biol.* **89,** 159–167.

Riggs, A. 1951. *J. Gen. Physiol.* **35,** 23–40.

Riggs, A. 1960. *J. Gen. Physiol.* **43,** 737–752.

Riviere, H. B., and E. L. Cooper. 1973. *Proc. Soc. Exp. Biol. Med.* **143,** 320–322.

Roberts, A. 1972. *Z. Vergleich. Physiol.* **75,** 388–401.

Roberts, A., and A. R. Blight. 1975. *Proc. R. Soc. London Ser. B.* **192,** 111–127.

Roberts, A., and J. D. W. Clarke. 1982. *Phil. Trans. R. Soc. London Ser. B.* **296,** 195–212.

Roberts, A., and E. P. Hayes. 1977. *Proc. R. Soc. London Ser. B.* **196,** 415–429.

Robertson, D. R. 1970. *Endocrinology* **87,** 1041–1050.

Robertson, D. R. 1972. *Gen. Comp. Endocrinol. Suppl. 3,* 421–429.

Robertson, D. R., and G. E. Schwartz. 1964. *Trans. Am. Microsc. Soc.* **83,** 330–337.

Robinson, H. 1972. *J. Exp. Zool.* **180,** 127–140.

Robinson, H., Chaffee, S., and V. A. Galton. 1977. *Gen. Comp. Endocrinol.* **32,** 179–186.

Romer, A. S. 1946. *Vertebrate Palaeontology.* University of Chicago Press. Illinois.

Rosen, D. E., Forey, E. L., Gardiner, B. G., and C. Patterson. 1981. *Bull. Am. Mus. Nat. Hist.* **167,** 163–275.

Rosen, S., and N. J. Friedley. 1973. *Histochemie* **36,** 1–4.

Rosenberg, M. 1970. *Proc.Natl. Acad. Sci.* **67,** 32–36.

Rosenberg, M., Lewinson, D., and M. R. Warburg. 1982. *J. Morphol.* **174,** 275–281.

de Rosier, D. J., and L. G. Tilney, 1982. In *Cold Spring Harbor Symposium,* **46,** 525–540.

Rossi, A. 1958. *Monit. Zool. Ital.* **66,** 133–149.

Rothman, T. E., Fries, E., Dunphy, W. G., and L. J. Urbani. 1982. In *Cold Spring Harbor Symposium* **46,** 797–805.

Rotman, E. 1940. *Arch. Entwicklungsmech. Org.* **140,** 124–156.

Roux,W. 1888. *Virchows Arch.* **114,** 22–60.

Rugh, R. 1948. *Experimental Embryology.* Burgess. Minneapolis.

Russell, I. J. 1976. In *Frog Neurobiology* (eds. R. Llinás and W. Precht). 513–550. Springer Verlag. Heidelberg.

Ryffel, G., Hagenbuchle, O., and R. Weber. 1973. *Cell. Differ.* **2,** 191–198.

Ryffel, G., and R. Weber. 1973. *Exp. Cell. Res.* **77,** 79–88.

Sagata, N., Okuyama, K., and K. Yamana. 1981. *Dev. Growth Differ.* **23,** 23–32.

Saint-Aubain, M. L. 1981. *Z. Zool. Syst. EvolutionsForsch.* **19,** 175–194.

Saint-Aubain, M. L. 1982. *Zoomorphologie (Berlin)* **100,** 55–63.

Sakai, J., and S. Horiuchi. 1979. *Comp. Biochem. Physiol.* **62B,** 269–273.

Saleem, M., and B. G. Atkinson. 1978. *J. Biol. Chem.* **253,** 1378–1384.

Saleem, M., and B. G. Atkinson. 1979. *Gen. Comp. Endocrinol.* **38,** 441–450.

Saleem, M., and B. G. Atkinson. 1980. *Can. J. Biochem.* **58,** 461–468.

Salpeter, M. M., and M.Singer. 1959. *J.Biophys. Biochem. Cytol.* **6**, 35–40.

Salzmann, R., and R. Weber. 1963. *Experientia* **19**, 352.

Sampson, L. V. 1904. *Am. J. Anat.* **3**, 473–504.

Samuels, H. H., and J. S. Tsai. 1973. *Proc. Natl. Acad. Sci.* **70**, 3488–3492.

Sanders, E. J. 1973. *Protoplasma* **76**, 115–122.

Sanders, E. J., and S. E. Zalik. 1972. *Arch. Entwicklungsmech. Org.* **171**, 181–194.

Sasaki, F. 1974. *Acta Histochem. Cytochem.* **7**, 239–256.

Sasaki, F. 1979. *Acta Histochem. Cytochem.* **12**, 7–19.

Sasaki, F., and K. Watanabe. 1983. *Histochemistry* **78**, 11–20.

Sasayama, Y., and C. Oguro. 1974. *Annot. Zool. Japon* **47**, 232–238.

Sasayama, Y., and C. Oguro. 1976. *Comp. Biochem. Physiol.* **55A**, 35–37.

Sasayama, Y., Noda, H., and C. Oguro. 1976. *Dev. Growth Differ.* **18**, 467–471.

Sato, A. 1976. *Anat.Rec.* **186**, 565–583.

Sato, A., and I. Kawakami. 1976. *Annot. Zool. Japon* **49**, 131–141.

Savage, J. M. 1980a. *Bull. S. Calif. Acad. Sci.* **79**, 45–54.

Savage, J. M. 1980b. *Proc. Biol. Sci. Wash.* **93**, 1177–1183.

Saxen, L., and J. Kohonen. 1969. *Int. Rev. Exp. Path.* **8**, 57–128.

Saxen, L., Saxen, E., Toivonen, S., and K. Salimaki. 1957a. *Endocrinology* **61**, 35–44.

Saxen, L., Saxen, E., Toivonen, S., and K. Salimaki. 1957b. *Ann. Zool. Soc. Zool. Bot. Fenn. Vanamo* **18**, 1–44.

Saxen, L., and S. Toivonen. 1955. *J. Embryol. Exp. Morph.* **3**, 376–386.

Saxen, L., Lehtonen, E., Karkinen-Jäskelainen, M., Nordling,S., and J. Wartiovaara. 1976. *Nature (London)* **259**, 662–663.

Schaeffer, B. 1965. *Am. Zool.* **5**, 267–276.

Schiaffino, S., and V. Hanzlikova. 1972. *J.Ultrastruct. Res.* **39**, 1–14.

Schliwa, M., Blerkom, J. van., and K. B. Pryzwansky. 1982. In *Cold Spring Harbor Symposium* **46**, 51–67.

Schmidt, A. A. 1978. *Salamandra* **14**, 49–57.

Schreckenberg, M. G. 1956. *Growth* **20**, 295–313.

Schreckenberg, G. M., and A. G. Jacobson. 1975. *Dev. Biol.* **42**, 391–400.

Schreiber, G. 1937. *Rend. Accad. Naz. Lincei Ser. VI,* **25**, 342–348.

Schroeder, T. E. 1970. *J. Embryol. Exp. Morph.* **23**, 427–462.

Schultheis, H., and W. Hanke. 1978a. *Comp. Biochem. Physiol.* **61A**, 567–570.

Schultheis, H., and W. Hanke. 1978b. *Comp. Biochem. Physiol.* **61A**, 571–576.

Scott, S. A., Cooper, E., and J. Diamond. 1981. *Proc. R. Soc. London Ser. B.,* **211**, 455–470.

Sedra, S. N. 1950. *Proc. Zool. Soc. London* **120**, 405–449.

Sedra, S. N., and M. I. Michael. 1957. *Verh. K. Ned. Akad. Wet. Natuurk.* **51**, 1–80.

Sedra, S. N., and M. I. Michael. 1958. *J. Morphol.* **103**, 1–13.

Sedra, S. N., and M. I. Michael. 1959. *J. Morphol.* **104**, 359–376.

Sedra, S. N., and M. I. Michael. 1961. *Cesk. Morfol.* **9**, 333–351.

Seshimo, H., Ryuzaki, M., and K. Yoshizato. 1977. *Dev. Biol.* **59**, 96–100.

Shaffer, B. M. 1963. *J. Embryol. Exp. Morph.* **11**, 77–90.

Shellabarger, C. J., and J. R. Brown. 1959. *J.Endocrinol.* **18**, 98–101.

Shelton, P. M. J. 1970. *J. Embryol. Exp. Morph.* **24**, 511–524.

Shepherd, G. W., and R. Flickinger. 1979. *Biochim. Biophys. Acta* **563**, 413–421.

Shiba, Y., Sumomogi, H., Nomura, S., Muneoka, Y., and Y. Kanno. 1979. *J. Hiroshima Univ. Dent. Soc.* **11**, 236–242.

Shiba, Y., Sumomogi, H., Nomura, S., Muneoka, Y., and Y. Kanno. 1980. *Dev. Growth Differ.* **22**, 209–217.

Shih, R. J., O'Connor, C. M., Keem, K., and L. D. Smith. 1978. *Dev. Biol.* **66**, 172–182.

Shih, R. J., Smith, L. D., and K. Keem. 1980. *Dev. Biol.* **75**, 329–342.

Shiokawa, K., Misumi, Y., and K. Yamana. 1981a *Dev. Growth Differ.* **23**, 579–581.

Shiokawa, K., Misumi, Y., Yasuda, Y., Nishio, Y., Kurata, S., Sameshima, M., and K. Yamana. 1979. *Dev. Biol.* **68**, 503–514.

Shiokawa, K., Tashiro, K., Misumi, Y., and K. Yamana. 1981b. *Dev. Growth Differ.* **23**, 589–597.

Shiokawa, K., and K. Yamana. 1979. *Dev. Growth Differ.* **21**, 501–508.

Shivpal, and I. A. Niazi. 1979. *Univ. Stud. Zool. Univ. Rajasthan* **1**, 8–17.

Shukuya, R. 1966. In *Proteins, Nucleic Acid, Enzyme* **11**, 228–237.

Shumway, W. 1940. *Anat. Rec.* **78**, 139–147.

Siboulet, P. 1970. *Vie Milieu Ser. C. Biol. Terr.* **21**, 179–188.

Singh-Jande, S. 1966. *J. Ultrastruct. Res.* **15**, 496–509.

Singer, M., and M. Salpeter. 1961. *J. Exp. Zool.* **147**, 1–19.

Singer, S. J. 1975. In *Cell Membranes: Biochemistry, Cell Biology and Pathology* (eds. G. Weissmann and R. Claiborne). 35–44. H. P. Publ. New York.

Sit, K., and R. Kanagasuntheram. 1973. *Malay Nat. J.* **26**, 4–18.

Smidova, A., Hurychova, D., Romanovsky, A., and J. Macha. 1974. *Folia Biol. (Prague)* **20**, 177–181.

Smit, A. L. 1953. *Ann. Univ. Stellenbosch Sect. A.* **29**, 79–136.

Smith, G. B. 1912. *J. Morphol.* **23**, 61–156; 455–579.

Smith, K. B., and J. R. Tata. 1976. *Exp. Cell. Res.* **100**, 129–146.

Smith, M. A. 1924. *Rec. Ind. Mus.* **25**, 309–310.

Smith, P. E. 1916. *Anat. Rec.* **11**, 57–64.

Smith-Gill, S. J. 1979. *Dev. Growth Differ.* **21**, 291–301.

Smith-Gill, S. J., and V. Carver. 1981. In *Metamorphosis: A Problem in Developmental Biology* (eds. L. I. Gilbert and E. Frieden). 2nd. Ed. 491–544. Plenum. New York.

Smith-Gill, S. J., Garrett-Reilly, J., and E. M. Weber. 1979. *Dev. Growth Differ.* **21**, 281–290.

Snyder, B. W., and B. E. Frye. 1972. *J. Exp. Zool.* **179**, 299–314.

Snyder, R. C. 1956. *Copeia* No. 1, 41–50.

Sokol, O. M. 1962. *Copeia* No. 2, 272–284.

Sokol, O. M. 1977. *J. Morphol.* **154**, 357–426.

Sokol, O. M. 1981. *J.Morphol.* **169**, 161–184.

Sorimachi, K., and N. Ui. 1974. *Gen. Comp. Endocrinol.* **24**, 38–43.

Spearman, R. I. C. 1977. In *Comparative Biology of the Skin* (ed. R. I. C. Spearman). *Symp. Zool. Soc. Lond.* No. 39. 335–352. Academic Press. London.

Spemann, H. 1901. *Arch. Entwicklungsmech. Org.* **12**, 224–264.

Spemann, H. 1903. *Anat. Anz.* **23**, 457–464.

Spemann, H. 1921. *Arch. Entwicklungsmech. Org.* **48**, 533–570.

Spemann, H. 1928. *Z. Wiss. Zool.* **132**, 105–134.

Spemann, H., and H. Mangold. 1924. *Arch. Mikrosk. Anat. Entwicklungsmech.* **100**, 599–638.

Spiegel, E. S., and M. Spiegel. 1970. *Exp. Cell. Res.* **61**, 103–112.

Spornitz, U. M. 1978. *Anat. Embryol.* **154**, 1–25.

Starrett, P. 1960. *Misc. Publ. Mus. Zool. Univ. Michigan* No.10, 1–37.
Stebbins, R. C. 1949. *Copeia* 161–168.
Stefanelli, A. 1951. *Q. Rev. Biol.* **26,** 17–34.
Steinman, R. M. 1968. *Am. J. Anat.* **122,** 19–55.
Steinmetz, C. 1952. *Endocrinology* **51,** 154–156.
Steinmetz, C. 1954. *Physiol. Zool.* **27,** 28–40.
Stephens, L. B. 1965. *Am. Zool.* **5,** 222–223.
Stephens, L. B. 1968. *Trans. Am. Microsc.Soc.* **87,** 106–109.
Stephens, R. E. 1968. *J. Mol. Biol.* **33,** 517–519.
Stephenson, N. G. 1951. *J. Linn. Soc. London (Zool),* **42,** 18–28.
Sterling, R. E. 1977. *Bull. N.Y. Acad. Med.* **53,** 260–276.
Sterling, R. E. 1979. *N. Engl. J. Med.* **300,** 173–177.
Streb, M. 1967. *Z. Zellforsch. Mikrosk. Anat.* **82,** 407–433.
Strum, J. M., and D. Danon. 1974. *Anat. Rec.* **178,** 15–40.
Stuart, E. S., and M. S. Fischer. 1978. *Biochem. Biophys. Res. Commun.* **82,** 621–626.
Sugimoto, K., Ichikawa, Y., and I. Nakamura. 1981. *J. Exp. Zool.* **215,** 53–62.
Sullivan, B. 1974. In *Chemical Zoology* (eds. M. Florkin and B. J. Scheer). **9,** 77–122. Academic Press.New York.
Sumiya, M., and S. Horiuchi. 1980. *Zool. Mag. (Tokyo)* **89,** 176–182.
Sussman, P., and T. W. Betz. 1978. *Can. J. Zool.* **56,** 1540–1545.
Sutherland, E. W. 1972. *Science* **177,** 401–408.
Suzuki, S., and M. Suzuki. 1981. *Gen. Comp. Endocrinol.* **45,** 74–81.
Swanepoel, J. H. 1970. *Ann. Univ. Stellenbosch Sect. A.* **45,** 1–119.
Sweet, S. S. 1977. *Herpetologica* **33,** 364–375.
Szarski, H. 1962. *Q. Rev. Biol.* **37,** 189–241.
Szarski, H. 1968. In *Nobel Symposium* No. 4 (ed. T. Ørvig). 445–453. Almqvist and Wiksell. Stockholm.
Szarski, H., and J. Czopek. 1965. *Zool. Pol.* **15,** 51–64.
Szepsenwol, J. 1935a. *C. R. Seances Soc. Biol. Paris* **118,** 944–946.
Szepsenwol, J. 1935b. *C. R. Seances Ass. Anat.* **30,** 475–482.

Tachibana, T. 1978. *Anat. Rec.* **191,** 487–502.
Tachibana, T. 1979. *Arch. Histol. Japon* **42,** 129–140.
Tachibana, T., Sakakura, Y., and T. Nawa. 1980. *Acta Anat. Japon* **55,** 588–599.
Tahara, Y. 1959. *Jap. J. Exp. Morph.* **13,** 49–60.
Tahara, U., and M. Ichikawa. 1965. In *Embryology of Vertebrates* (ed. M. Kume) 178–195. Barfukan Publ. Company. Tokyo.
Tajima, I. 1977. *Zool. Mag. (Tokyo)* **86,** 48–53.
Takata, K., Yamamoto, K. Y., and R. Ozawa. 1981. *Arch. Devel. Biol.* **190,** 92–96.
Takaya, H. 1973. *Dev. Growth Differ.* **15,** 33–37.
Takisawa, A., Shimura, Y., and K. Kaneko. 1976. *Okajima Folia Anat. Japon* **53,** 253–277.
Tarkowski, A. K., and J. Wroblewska. 1969. *J. Embryol. Exp. Morph.* **18,** 155–180.
Tata, J. R. 1965. *Nature (London)* **207,** 378–381.
Tata, J. R. 1966. *Dev. Biol.* **13,** 77–94.
Tata, J. R. 1967. *Biochem. J.* **104,** 1–15.
Tata, J. R. 1968a. *Nature (London)* **219,** 331–337.
Tata, J. R. 1968b. *Dev. Biol.* **18,** 415–441.

Tata, J. R. 1969. *Gen. Comp. Endocrinol. Suppl.* **2**, 385–397.

Tata, J. R. 1970. *Nature (London)* **227**, 686–689.

Tata, J. R. 1971. In *Current Topics in Developmental Biology* (eds. A. A. Moscona and A. Monroy). **6**, 79–110. Academic Press, London.

Tata, J. R. 1972. *Symp. Soc. Exp. Biol.* **25**, 163–181.

Tata, J. R. 1975. *Nature (London)* **257**, 18–23.

Tata, J. R. 1978. In *Hormones and Cell Regulation* (eds. J. Dumont and J. Nunez). **2**, 37–53.

Taylor, A. C., and J. J. Kollros. 1946. *Anat. Rec.* **94**, 7–24.

Taylor, E. H. 1968. *The Caecilians of the World, A Taxonomic Review*. University of Kansas Press. Lawrence, Kansas.

Taylor, J. D., and J. T. Bagnara. 1972. *Am. Zool.* **12**, 43–62.

Ten Donkelaar, H. J., and R. de Boer-Van Huizen. 1982. *Anat. Embryol.* **163**, 461–474.

Terent'ev, P. V. 1950. *The Frog*. Sovietskaya, Nauka, Moscow.

Terent'ev, P. V., and S. A. Chernov. 1965. *Key to Amphibians and Reptiles*. Israel Program for Scientific Translations. S. Monson. Jerusalem.

Terry, G. S. 1918. *J. Exp. Zool.* **24**, 567–587.

Tesik, I. 1978. In XIX *Morphology Congress Symposium* (ed. E. Klika). 451–456. Charles University, Prague.

Theil, E. C. 1967. *Biochem. Biophys. Acta* **138**, 175–185.

Theil, E. C. 1970. *Comp. Biochem. Physiol.* **33**, 717–720.

Thomson, K. S. 1968. In *Nobel Symp. No. 4. Current Problems in Lower Vertebrate Phylogeny* (ed. T. Ørvig), 285–306.

Thomson, K. S. 1969. *Biol. Rev.* **44**, 91–154.

Thomson, K. S. 1981. *Paleobiology* **7**, 153–156.

Thors, F., de Kort, E. J. M., and R. Nieuwenhuys. 1982. *Anat. Embryol.* **164**, 427–442; 443–454.

Tiedemann, H. 1976. *J. Embryol. Exp. Morph.* **35**, 437–444.

Tiedemann, H., Born, J., and H. Tiedemann. 1972. *Arch. Entwicklungsmech. Org.* **171**, 160–169.

Tiedemann, H., Tiedemann, H., Born, J., and U. Kocher-Becker. 1969. *Arch. Entwicklungsmech. Org.* **163**, 316–324.

Tilney, L. G., and R. R. Cardell. 1970. *J. Cell Biol.* **47**, 408–422.

Ting, H. P. 1951. *Copeia No.* **1**, 82.

Ting, H. P. 1970. *J. Singapore Nat. Acad. Sci.* **2**, 38–46.

Tochinai, S. 1975. *J. Fac. Sci. Hokkaido Univ.* **19**, 803–811.

Toivonen, S. 1978. In XIX *Morphology Congress Symposium* (ed. E. Klika). 167–175. Charles University, Prague.

Toivonen, S., and L. Saxen. 1968. *Science* **159**, 539–540.

Toivonen, S., Tarin, D., and L. Saxen. 1976. *Differentiation* **5**, 49–55.

Toivonen, S., Tarin, D., Saxen, L., Tarin, P. J., and J. Wartiovaara. 1975. *Differentiation* **4**, 1–7.

Toivonen, S., and J. Wartiovaara. 1976. *Differentiation* **5**, 61–66.

Tomita, Y., Goto, Y., Okazaki, T., and R. Shukuya. 1979. *Biochem. Biophys. Acta* **562**, 504–514.

Toth, E., and M. Tabachnik. 1979. *Gen. Comp. Endocrinol.* **38**, 345–355.

Toth, E., and M. Tabachnik. 1980. *Gen. Comp. Endocrinol.* **42**, 57–62.

Townes, P. L., and J. Holtfreter. 1955. *J. Exp. Zool.* **128**, 53–150.

Trader, C. D., and E. Frieden. 1966. *J. Biol. Chem.* **241**, 357–366.

Travis, J. 1980. *Growth* **44**, 167–181.

Travis, J. 1981. *Brimleyana* **6**, 119–128.

Trevison, P. 1972. *Atti Accad. Naz. Lincei Rend.* **52**, 965–968.

Trevison, P. 1973. *Atti Accad. Naz. Lincei Rend.* **53**, 217–220.

Trowbridge, M. S. 1941. *Trans. Am. Microsc. Soc.* **60**, 508–526.

Trowbridge, M. S. 1942. *Trans.Am. Microsc. Soc.* **61**, 66–83.

Truman, D. E. S. 1974. *The Biochemistry of Differentiation*. Blackwell Scientific Publications. London.

Tschernoff, N. D. 1907. *Anat. Anz.* **30**, 593–612.

Turner, R. J. 1969. *J. Exp. Zool.* **170**, 467–480.

Turner, S. C. 1973. PhD Thesis. University of London.

Turner, S. C. 1978. In *XIX Morphology Congress Symposium* (ed. E. Klika). 243–246. Charles University, Prague.

Turpen, J. B., Turpen, C. J., and M. Flajnik. 1979. *Dev. Biol.* **69**, 466–479.

Turpen, J. B., and C. M. Knudson. 1982. *Dev. Biol.* **89**, 138–151.

Tusques, J. 1951. *C. R. Seances Soc. Biol. Paris* **145**, 1555–1556.

Tweedle, C. D. 1977. *Cell. Tissue Res.* **185**, 191–197.

Tweedle, C. D. 1978. *Neuroscience* **3**, 481–486.

Twitty, V. C., and D. Bodenstein. 1948. In *Experimental Embryology* (R. Rugh). Burgess. Minneapolis.

Tyler, M. J., and A. A. Martin. 1975. *Trans. R. Soc. S. Aust.* **99**, 93–100.

Tyler, M. J., Davies, M., and A. A. Martin. 1982. *Copeia* No. 2, 260–264.

Uchiyama, M., and P. K. T. Pang. 1981. *Gen. Comp. Endocrinol.* **44**, 428–435.

Ueck, M. 1967. *Z. Wiss. Zool.* **176**, 173–270.

Ultsch, G. R. 1973. *Herpetologica* **29**, 304–305.

Uray, N. J., and A. G. Gona. 1978. *J. Comp. Neurol.* **180**, 265–276.

Urch, U. A., and J. L. Hedrick. 1981. *J. Supramol. Struct. Cell. Biochem.* **15**, 111–118.

Usui, M., and M. Hamasaki. 1939. *Zool. Mag.* (Tokyo) **51**, 195–206.

Usuku, G., and J. Gross. 1965. *Dev. Biol.* **11**, 352–370.

Utsunomiya, Y., and T. Utsunomiya. 1977. *J. Fac. Fish. Anim. Husb. Hiroshima Univ.* **16**, 65–76.

Vainio, T., Saxen, L., and S. Toivonen. 1960. *Experientia* **16**, 27–29.

Vainio, T., Saxen, L., Toivonen, S., and J. Rapola. 1962. *Exp. Cell. Res.* **27**, 527–538.

Valett, R. D., and D. L. Jameson. 1961. *Copeia* No. 1, 103–109.

Vanable, J. W., and R. D. Mortensen. 1966. *Exp. Cell. Res.* **44**, 436–442.

Vargas-Lizardi, P., and K. M. Lyser. 1974. *Dev. Biol.* **38**, 220–228.

Varute, A. T., and N. K. More. 1971. *Comp. Biochem. Physiol.* **38**, 225–234.

Vietti, M., Perone, A. C., and A. Guardabassi. 1973. *Boll. Zool.* **40**, 401–404.

Vilirnkova, V., and J. Nedvidek. 1962. *Fol. Biol. (Prague)* **8**, 381–389.

Villani, L. 1980. *Arch. Ital. Anat.Embriol.* **85**, 149–157.

Viperina, S., and J. J. Just. 1975. *Copeia* No. 1, 103–109.

Vogt, W. 1925. *Arch. Entwicklungsmech. Org.* **106**, 542–610.

Vogt, W. 1929. *Arch. Entwicklungsmech. Org.* **120**, 385–706.

Voitkevitch, A. A. 1962. *Gen. Comp. Endocrinol. Suppl. 1,* 133–147.

Volpe, E. P., and J. L. Dobie. 1959. *Tulane Stud. Zool. Bot.* **7**, 145–152.

Volpe, E. P., and S. M. Harvey. 1958. *Copeia* No. 3, 197–207.

Volpe, E. P., Wilkins, M. A., and J. L. Dobie. 1961. *Copeia* No. 3, 340–349.

Voss, W. J. 1961. *Southwest. Nat.* **6,** 168–174.
Voûte, C. L., Thummel, J., and M. Brenner. 1975. *J. Steroid Biochem.* **6,** 1175–1179.

Waddington, C. H. 1947. *Organisers and Genes.* Cambridge University Press. London.
Waddington, C. H., and M. M. Perry. 1962. *Proc. R. Soc. London Ser. B.* **156,** 459–482.
Wahli, W., Abraham, I., and R. Weber. 1978. *Arch. Devel. Biol.* **185,** 235–248.
Wakahara, M. 1981. *J. Embryol. Exp. Morph.* **66,** 127–140.
Wake, M. H. 1967. *Bull. South Calif. Acad. Sci.* **66,** 109–116.
Wake, M. H. 1969. *Copeia* No. 1, 183–184.
Wake, M. H. 1976. *J. Morphol.* **148,** 33–46.
Wake, M. H. 1977. *J. Herpetol.* **11,** 379–386.
Wake, M. H. 1978a. *Pap. Avulsos Zool. (Sao Paulo)* **32,** 1–13.
Wake, M. H. 1978b. *J. Herpetol.* **12,** 121–133.
Wake, M. H. 1980. *Copeia* No. 2, 193–209.
Wald, G., 1981. In *Metamorphosis: A Problem in Developmental Biology* (eds. L. I. Gilbert and E. Frieden). 2nd. Ed. 1–39. Plenum. New York.
Wang, V. B., and E. Frieden. 1973. *Gen. Comp. Endocrinol.* **21,** 381–389.
Warburg, M. R., and D. Lewinson. 1977. *Cell. Tissue Res.* **181,** 369–393.
Wassersug, R. J., Frogner, K. J., and R. F. Inger. 1981. *J. Herpetol.* **15,** 41–52.
Watanabe, K. 1969. *Acta Anat. Japon* **44,** 80.
Watanabe, K., and F. Sasaki. 1974. *Cell. Tissue Res.* **155,** 321–336.
Watanabe, K., Sasaki. F., and M. A. Khan. 1978a. *Histochemistry* **55,** 293–306.
Watanabe, K., Khan, M. A., Sasaki, F., and H. Iseki. 1978b. *Histochemistry* **58,** 13–22.
Watanabe, Y. G. 1966. *J. Fac. Sci. Hokkaido Univ.* **16,** 85–89.
Watanabe, Y. G. 1971a. *Annot. Zool. Japon* **44,** 145–152.
Watanabe, Y. G. 1971b. *J. Fac. Sci. Hokkaido Univ.* **18,** 1–8.
Watson, D. M. S. 1926. *Phil. Trans. R. Soc. London Ser. B.* **214,** 189–257.
Watson, D. M. S. 1940. *Trans. R. Soc. Edinburgh* **60,** 195–231.
Watson, G. F., and A. A. Martin, 1973. *Trans. R. Soc. S. Aust.* **97,** 33–45.
Watson, G. F., and A. A. Martin. 1979. *Aust. Zool.* **20,** 259–268.
Watson, J. D. 1976. *Molecular Biology of The Gene.* 3rd Ed. W. Benjamin. California.
Watt, K. W. K., Maruyama, T., and A. Riggs. 1980. *J. Biol. Chem.* **255,** 3294–3301.
Webb, A. C., and C. J. Camp. 1979. *Exp. Cell. Res.* **119,** 414–418.
Webb, R. G. 1978. *Nat. Hist. Mus. Los Angeles City Contrib. Sci.* (300), 1–13.
Webb, R. G., and J. K. Korky. 1977. *Herpetologica* **33,** 73–82.
Weber, R. 1961. *Experientia* **17,** 365–366.
Weber, R. 1962. *Experientia* **18,** 84.
Weber, R. 1964. *J. Cell. Biol.* **22,** 481–487.
Weber, R. 1965. *Experientia* **21,** 665–666.
Weber, R. 1967. In *Biochemistry of Animal Development* (ed. R. Weber). **2,** 227–301. Academic Press. New York.
Weber, R. 1969a. In *Lysosomes in Biology and Pathology* (eds. J. T. Dingle and H. B. Fell). **2,** 437–461. North Holland. Amsterdam.
Weber, R. 1969b. *Gen. Comp. Endocrinol. Supp.* **2,** 408–416.

Weber, R. 1977. In *Mécanismes de la Rudimentation des Organes chez les Embry-ons de Vertébrés* (ed. A. A. Raynaud). Colloq. Int. CNRS No. 266, 137–146.

Weber, R., and E. J. Boell. 1955. *Exp. Cell. Res.* **9**, 559–567.

Weber, R., and E. J. Boell. 1962. *Dev. Biol.* **4**, 452–472.

Webster, H., and S. M. Billings. 1972. *J. Neuropathol. Exp. Neurol.* **31**, 102–112.

Weets, J., and J. J. Picard. 1979. *J. Exp. Zool.* **207**, 305–314.

Weiss, P., and W. Ferris. 1954. *Proc. Natl. Acad. Sci.* **40**, 528–540.

Weissmann, G., and R. Claiborne. 1975. In *Cell Membranes: Biochemistry, Cell Biology and Pathology*. H. P. Publ. Co. New York.

Weisz, P. B. 1945a. *Anat. Rec.* **93**, 161–169.

Weisz, P. B. 1945b. *J. Morphol.* **77**, 163–192; 193–217.

Westley, B., and J. Knowland. 1978. *Cell* **15**, 367–374.

White, B. A., and C. S. Nicoll. 1979. *Science* **204**, 851–853.

White, B. A., and C. S. Nicoll. 1981. In *Metamorphosis: A Problem in Developmental Biology* (eds. L. I. Gilbert and E. Frieden). 2nd Ed. 363–396. Plenum. New York.

Whitear, M. 1974. *J. Zool.* **172**, 503–529.

Whitear, M. 1975. *J. Zool.* **175**, 107–149.

Whitear, M. 1976. *Cell. Tissue Res.* **175**, 391–402.

Whitear, M. 1977. In *Comparative Anatomy of the Skin* (ed. R. I. C. Spearman). *Symp. Zool. Soc. Lond.* No. 39, 291–313. Academic Press. London.

Wilczynska, B. 1981. *Acta Biol. Cracov* **23**, 13–46.

Wilder, I. W. 1913. *Biol. Bull. (Woods Hole, Mass.),* **24**, 251–292; 293–343.

Wilder, I. W. 1924. *J. Exp. Zool.* **40**, 1–112.

Wilder, I. W. 1925. *Smith College 50'th. Aniv. Publ.* **6**, 1–161.

Wilkie, D. R. 1976. *Muscle. Inst. Biol. Stud. in Biology*. No. 11, Ed. Arnold. London.

Willis, A. G. 1947. *Nature (London)* **159**, 410.

Willis, A. G. 1948. *Nature (London)* **161**, 65.

Wilson, L., and R. L. Margolis. 1982. In *Cold Spring Harbor Symposium* **46**, 199–205.

Wingstrand, K. G. 1966. In *The Pituitary Gland* (eds. G. W. Harris and B. T. Donovan). **1**, 58–148. University of California. Berkeley.

Winkelmann, R. K. 1977. *J. Invest. Dermatol.* **69**, 41–46.

Winkelmann, R. K., and A. S. Breathnach. 1973. *J. Invest. Dermatol.* **60**, 2–15.

Wintrebert, M. P. 1905a. *C. R. Seances Soc. Biol. Paris* **59**, 576–578.

Wintrebert, M. P. 1905b. *C. R. Seances Soc. Biol. Paris* **59**, 690–692.

Witschi, E. 1953. In *International Congress on Zoology Copenhagen*. 260–262. Danish Science Press. Copenhagen.

Witschi, E. 1955. *J. Morphol.* **96**, 497–511.

Witschi, E. 1956. *Development of Vertebrates*. W. B. Saunders. Philadelphia.

Witschi, E. 1962. In *Growth, including Reproduction and Morphological Development* (eds. P. L. Altman and D. S. Dittmar). *Fed. Am. Soc. Exp. Biol.* 272–273. Washington.

de Witt, W. 1968. Appendix to B. Moss and V. M. Ingram. 1968b. *J. Mol. Biol.* **32**, 502–504.

Wittle, L. W., and J. N. Dent. 1979. *Gen. Comp. Endocrinol.* **37**, 428–439.

Wixom, R. L., Reddy, M. K., and P. P. Cohen. 1972. *J. Biol. Chem.* **247**, 3684–3692.

Woodland, H. R. 1974. *Dev. Biol.* **40**, 90–101.

Woody, A. Y., and J. T. Justis. 1974. *J. Exp. Zool.* **188**, 215–224.
Worthington, D. D., and D. B. Wake. 1971. *Am. Midl. Nat.* **82**, 349–365.
Wright, A. H. 1929. *Proc. US Natl. Mus.* **74**, (No. 11), 1–70.
Wright, M. L., 1977. *J. Exp. Zool.* **202**, 223–234.
Wright, M. L., Majerowski, M. A., Lukas, S. M., and P. A. Pike. 1979. *Gen. Comp. Endocrinol.* **39**, 53–62.
Wunderer, H. 1910. *Zool.Jahrb.Abt.Anat.Ontog.* **29**, 367–444.
Wunderlich, F., Berezney, R., and H. Kleenig. 1976. In *Biological Membranes* (eds. D. Chapman and F. H. Wallach). **3**, 241–333. Academic Press. New York, London.

Yamada, T. 1958. *Experientia* 14, 81–87.
Yamada, T. 1961. In *Advances in Morphogenesis* (eds. M. Abercrombie and J. Brachet). **1**, 1–53. Academic Press. London.
Yamada, T. 1962. *J. Comp. Physiol.* **60**, Suppl. I, 49–64.
Yamada, T. 1981. *Neth. J. Zool.* **31**, 78–98.
Yamada, T., and K. Takata. 1961. *Dev. Biol.* **3**, 411–423.
Yamaguchi, K., and I. Yasumasu. 1977a. *Dev. Growth Differ.* **19**, 149–159.
Yamaguchi, K., and I. Yasumasu. 1977b. *Dev. Growth Differ.* **19**, 161–169.
Yamaguchi, K., and I. Yasumasu. 1978. *Dev. Growth Differ.* **20**, 61–69.
Yamamoto, K., and S. Kikuyama. 1982. *Endocrinol. Jap.* **29**, 81–85.
Yamamoto, K., Kikuyama, S., and I. Yasumasu. 1979. *Dev. Growth Differ.* **21**, 255–261.
Yamamoto, K. Y., Ozawa, R., Takata, K., and J. Kitoh. 1981. *Arch. Devel. Biol.* **190**, 313–319.
Yoon, B. W., Keller, R. E., and G. M. Malacinski. 1980. *J. Embryol. Exp. Morph.* **59**, 223–247.
Yoon, B. W., and G. M. Malacinski. 1981a. *Dev. Biol.* **83**, 339–352.
Yoon, B. W., and G. M. Malacinski, 1981b. *J. Embryol. Exp. Morph.* **66**, 1–26.
Yoshizaki, N. 1973.. *J. Fac. Sci. Hokkaido Univ.* **18**, 469–480.
Yoshizaki, N. 1974. *J. Fac. Sci. Hokkaido Univ.* **19**, 309–314.
Yoshizaki, N. 1975. *Zool. Mag. (Tokyo)* **84**, 39–47.
Yoshizaki, N. 1976. *Dev. Growth Differ.* **18**, 133–144.
Yoshizaki, N. 1979. *Dev. Growth Differ.* **21**, 11–18.
Yoshizaki, N. 1981. *J. Embryol. Exp. Morph.* **61**, 249–258.
Yoshizaki, N., and C. Katagiri, 1975. *J. Exp. Zool.* **192**, 203–212.
Yoshizato, K., and E. Frieden. 1975. *Nature (London)* **254**, 705–707.
Yoshizato, K., Kistler, A., and E. Frieden. 1975a. *Endocrinology* **97**, 1030–1035.
Yoshizato, K., Kistler, A., and E. Frieden. 1975b. *J. Biol. Chem.* **250**, 8337–8343.
Yoshizato, K., and I. Yasumasu. 1970. *Dev. Growth Differ.* **11**, 305–317.

Zabroda, W. N., and E. P. Il'enko. 1981. *Vestn. Zool.* (**4**), 66–71.
Zug, G. R. 1978. *Smithson. Contrib. Zool.* **276**, 1–31.
Zweifel, R. G. 1964a. *Copeia* No.1, 201–208.
Zweifel, R. G. 1964b. *Copeia* No. 2, 300–308.

Appendix

Appendix

Section I. Less is known about the thyroid hormonal circulatory concentration in urodeles than in anurans during larval development. Rosenkilde et al. (1982) showed that in *Ambystoma mexicanum* the T_4 serum concentration rises from the beginning of hind-limb development, from 15 to 20 nM to peak during final hind-limb differentiation at 40 nM, and thence it falls after hind-limb completion to around adult levels at 3–6 nM.

    Section II. *Ascaphus truei* from Montana, USA, metamorphoses after 4 yr, first reproduces when 7–8 yr old, and has an extended life history of 15–20 yr, one of the longest known in anurans (Daugherty and Sheldon, 1982). In contrast, the Australian frog *Lymnodynastes tasmaniensis* spawns only 80–100 d after metamorphosis and is one of the quickest forms to reach sexual maturity (Horton, 1982). Information on the larval life history of *Onychodactylus japonicus* was published by Hyase and Yamane (1982).

    The larva and mouth parts of defined Gosner (1960) stages of *Atelopus flavescens* (Bufonidae) from French Guiana, were described and illustrated by Lescure (1981), and likewise of the Brazilian species *Sphaenorhynchus orophilus* (Hylidae) and *Atelopus pernambucensis* (Bufonidae) by da Cruz and Peixoto (1980, 1982), *Crossodactyles pintoi* and *Pleurodema diplolistris* (Leptodactylidae) by Peixoto (1981, 1982) and *Proceratophrys appendiculata* (Leptodactylidae) by Peixoto and da Cruz (1980).

    Section III. Hb chemistry during early developing stages of *Rana catesbeiana* has been investigated by Futamura et al. (1982). Okazaki et al. (1982) showed that the densities of larval and adult erythrocytes of *Rana catesbeiana* differ, and that during climax erythrocytes with tadpole Hb coexist with those containing adult Hb. Frog Hb is preferentially synthesized during transition at climax and is associated with the replacement of tadpole erythrocytes by those of adults. These authors appear to support the view of separate populations of larval and adult red cells, each population synthesizing either larval or adult hemoglobins.

Delfino et al. (1982) have described the development of the cutaneous glands in *Salamandra terdigitata* larvae.

Section IV. Protein synthesis from transcribed mRNA occurs during oogenesis of *Xenopus laevis*, and Capco and Jaeckle (1982) localized it in the cortical regions of the vitellogenic and postvitellogenic oocytes. The spatial arrangement disappears as the egg matures. DNA polymerase is localized in germinal vesicles of full grown oocytes and is released into the cytoplasm to bind to the ER when the germinal vesicle breaks down (Nagano et al., 1982).

Richmond and Pollack (1983) cultured tadpole spinal cord with blastemas of regenerating hind-limbs of *Rana pipiens* tadpoles, and showed that blastema mesenchyme enhanced nerve fiber growth and its survival.

Finally, Zeng (1982) showed nuclear ultrastructural changes in presumptive ectodermal cells of gastrulae of *Cynops orientalis,* after explants were cultured in mesodermal-inducing guinea pig bone marrow extract. Nuclear chromatin of induced cells was more dispersed, in contrast to the condensed chromatin masses of non-induced control tissue nuclei. These results may indicate nuclear transcription phenomena, elicited by inductive substances during the process of cellular differentiation.

References

Capco, D. G., and H. Jaeckle. 1982. *Dev. Biol.* **94,** 41–50.

da Cruz, C. A. G., and O. L. Peixoto. 1980. *Rev. Bras. Biol.* **40,** 383–386.

da Cruz, C. A. G., and O. L. Peixoto. 1982. *Rev. Bras. Biol.* **42,** 627–629.

Daugherty, C. D., and A. L. Sheldon. *Herpetologica* **38,** 461–468.

Delfino, G., Brizzi, R., and C. Calloni. 1982. *Z. Mikrosk.-Anat. Forsch. (Leipzig),* **96,** 948–971.

Futamura, M., Terashi, Y., Okazaki, T., and R. Shukuya, 1982. *Biochem. Biophys. Acta* **704,** 37–42.

Horton, P. 1982. *Herpetologica* **38,** 486–489.

Hyase, N., and S. Yamane. 1982. *Jap. J. Ecol.* **32,** 395–403.

Lescure, J. 1981. *Amphibia-Reptilia* **2,** 209–215.

Nagano, H., Okano, K., Ikegami, S., and C. Katagiri. 1982. *Biochem. Biophys. Res. Comm.* **106,** 683–690.

Okazaki, T., Ishihara, H., and R. Shukuya. 1982. *Comp. Biochem. Physiol.* **73,** 309–312.

Peixoto, O. L. 1981. *Rev. Bras. Biol.* **41,** 339–341.

Peixoto, O. L. 1982. *Rev. Bras. Biol.* **42,** 631–633.

Peixoto, O. L., and C. A. G. da Cruz. 1980. *Rev. Bras. Biol.* **40,** 491–493.

Richmond, M. J., and E. D. Pollack. 1983. *J. Exp. Zool.* **225,** 233–242.

Rosenkilde, P., Mogensen, E., Centervall, G., and O. S. Jørgensen, 1982. *Gen. Comp. Endocrinol.* **48,** 504–514.

Zeng, M. 1982. *Scient.Sinica* **25,** 725–729.

Index of Species

Index of Species*

*All species are fully referenced except for *Xenopus laevis, Rana pipiens, R. catesbeiana,* and *R. temporaria,* that are featured throughout the work. After their initial references, including that in Section II, they are not paged again in the Index.

Subject Index

Subject Index